D1035146

THE LOEB CLASSICAL LIBRARY

FOUNDED BY JAMES LOEB, LL.D.

LUCIUS JUNIUS MODERATUS COLUMELLA

ON AGRICULTURE

II

LUCIUS JUNIUS MODERATUS COLUMELLA

ON AGRICULTURE

WITH A RECENSION OF THE TEXT AND AN ENGLISH TRANSLATION BY

†E. S. FORSTER, M.B.E., M.A.(Oxon.), F.S.A.
EMERITUS PROFESSOR OF GREEK, SHEFFIELD UNIVERSITY
AND
EDWARD H. HEFFNER, A.M., Ph.D.
PROFESSOR OF LATIN, UNIVERSITY OF PENNSYLVANIA

IN THREE VOLUMES

II

RES RUSTICA V–IX

CAMBRIDGE, MASSACHUSETTS
HARVARD UNIVERSITY PRESS
LONDON
WILLIAM HEINEMANN LTD
MCMLIV

Printed in Great Britain

CONTENTS

PREFATORY NOTE

Owing to the death of Dr. Harrison Boyd Ash of the University of Pennsylvania shortly after the publication of the first volume (Books I–IV) of the *De Re Rustica* of Columella, the Editors entrusted me with the remainder of the work.

There has been no complete modern edition of the text since J. G. Schneider's (Leipzig 1794), but the principles laid down by Dr. Ash (Vol. I, p. xxi) appear to me to be entirely satisfactory. He describes them as follows: "The text and manuscript readings of the present edition, for Books I–II, VI–VII, X–XI and the *De Arboribus*, rest substantially on the work of Lundström. For Books III–V, VIII–IX and XII, the translator has attempted to construct a critical text in some approximation to that of Lundström by the collation of four major manuscripts with the text of Schneider." It was natural to conclude from these words that a text constructed by Dr. Ash would be available for the rest of the work, but no traces of the existence of such a text have been found in America. It has, therefore, been necessary to undertake the construction of a new text, and I have tried to conform as far as possible with Dr. Ash's system, using Lundström's edition for those books which he has edited and attempting a new text for Books V, VIII, IX and XII. For this purpose I have been fortunate, through the good offices of Professor L. A. Post, in obtaining from America photostats of the four most

important MSS. (see p. xvi of Vol. I), which fall into two classes, (*a*) the two 9th–10th century MSS. and (*b*) the two best of the 15th-century MSS. The photostats, which were used by Dr. Ash for his collation of Books III and IV, were purchased with a grant provided by the Faculty Research Fund of the University of Pennsylvania. The only point in which my text of these books differs from that of Dr. Ash is that I have not had an opportunity, which Dr. Ash had, of comparing my text with that of the MS. known as *Morganensis* 138, formerly *Hamiltonensis* 184 in the Pierpont Morgan Library in New York.

For some unexplained reason the text of Book V, especially Chapter VIII to the end, is in a worse condition than in any other part of the work, and there is the further complication that, from Chapter X to the end, the text, though slightly longer, is closely identical with that of *De Arboribus*, Chapter XVIII to the end. It seems certain that the *De Arboribus* is part of an earlier and shorter treatise which was afterwards superseded by the *De Re Rustica*. It is a question how far the text of these similar chapters in the *De Re Rustica* and the *De Arboribus* should be corrected from one another. There are numerous places in which the text of Book V is deficient or careless, and these can be corrected from the *De Arboribus*, but it also appears that the author made a good many verbal changes as well as inserting new matter. I have, therefore, refrained from making the two slightly different versions correspond exactly and have kept the MS. reading in both treatises where it makes sense—very often the same sense in slightly different

words—but the fact that there are these two versions has necessitated a larger *apparatus criticus* in these chapters of Book V than for any other part of the work.

I have to thank His Grace the Duke of Devonshire for lending me [M. C. Curtius], *L. Junius Moderatus Columella on Husbandry in Twelve Books and his Book concerning Trees* (London 1745) (a very rare work) from the Chatsworth Library, and Mademoiselle Hélène Rousseau for obtaining for me in Paris a copy of M. Nisard, *Les Agronomes Latins* (Paris, 1844), for which I had been searching for many months.

EDWARD S. FORSTER.

Upon the death of Professor Forster, the Editors of this Library entrusted to me the responsibility of completing the unfinished project. In the circumstances this assignment naturally extended to the making of a thorough examination of every aspect of the work. The photostats mentioned by my predecessor in the above lines were in due time returned to America and were fully utilized in the process of examination and study. In the checking between these manuscripts, as well as in the verification of references to important earlier editions of Columella, very substantial assistance was furnished to me by my wife, which I desire gratefully to acknowledge here. It is to be hoped that the process of restudying and reviewing has resulted in an improved product. It is always a serious thing to find yourself differing with another person on matters of a scholarly nature; to handle such materials when left

by the hand of one who is no longer able to speak in defence of his interpretation imposes many a delicate task. Naturally there are numerous passages in the text of Columella, and also in the English version, which I would have handled somewhat differently from the manner in which they were treated by my predecessor if I had been free to shape things *de novo*. However, this statement applies rather to materials involving the factors of taste and judgment than to those where the essential thought was an issue.

The reader might be reminded of the Bibliography prepared by the late Professor Ash and included in Vol. I of this Library. The works pertaining to Columella that are there cited were obviously made use of by Professor Forster, as they were also utilized by me.

EDWARD H. HEFFNER.

SIGLA

S = Cod. Sangermanensis Petropolitanus 207 (9th cent.).
A = Cod. Ambrosianus L 85 sup. (9th–10th cents.)
R = all or consensus of the 15th cent. MSS.
a = Cod. Laurentianus plut. 53. 32 (15th cent.).
c = Cod. Caesenas Malatestianus plut. 24. 2 (15th cent.).
ed. pr. = editio princeps (Jensoniana), Venice, 1472.
Ald. = the first Aldine edition, Venice, 1514.
Gesn. = J. M. Gesner, *Scriptores Rei Rusticae Veteres Latini*, Leipzig, 1735.
Schneider = J. G. Schneider, *Scriptores Rei Rusticae Veteres Latini*, Leipzig, 1794.
Lundström = V. Lundström, *L. Iun. Mod. Columella Lib. I–II, VI–VII, X–XI, de Arboribus*, Upsala-Göteborg, 1897–1940.

Note.—In Books VI and VII, where the *apparatus criticus* is based on Lundström's recension, his siglum *R* is used as representing the reading of all or the majority of the twenty-five 15th-century MSS. collated by him. In Books V and VIII a new collation has been made of only the two best 15th-century MSS., for which the sigla *a* and *c* are used.

LUCIUS JUNIUS MODERATUS COLUMELLA

ON AGRICULTURE

L. IUNI MODERATI COLUMELLAE

REI RUSTICAE

LIBER V

I. Prioribus libris, quos ad te de constituendis colendisque vineis, Silvine, scripseram, nonnulla defuisse dixisti, quae agrestium operum studiosi desiderarent; neque ego infitior aliqua me praeteriisse, quamvis inquirentem sedulo, quae nostri saeculi cultores quaeque veteres literarum monumentis prodiderunt: sed cum sim professus [1] rusticae rei praecepta, nisi fallor, asseveraveram, quae vastitas [2] eius scientiae contineret, non cuncta me dicturum, sed plurima. Nam illud in unius hominis
2 prudentiam cadere non poterat. Neque enim est ulla disciplina, non ars, quae singulari [3] consummata sit ingenio. Quapropter ut in magna silva boni venatoris est indagantem feras quamplurimas capere; nec cuiquam culpae fuit non omnes cepisse: ita nobis

[1] sim professus *c*: sum professus *a*: summo festus *SA*.
[2] vastitas *ac*: unitas *SA*.
[3] singulari *ac*: consulari *SA*.

LUCIUS JUNIUS MODERATUS COLUMELLA

ON AGRICULTURE

BOOK V

I. You have said, Silvinus, that in the earlier books, which I had written to you about establishing and cultivating vineyards, some things were omitted of which those who devote themselves to agriculture felt the want; and indeed I do not deny that, although I carefully studied what the agriculturists of our own age and also the ancients have handed down in written records, there are some topics which I have passed over. But when I undertook to teach the precepts of husbandry, if I mistake not, I did not assert that I would deal with all but only with very many of those subjects which the vast extent of that science embraces; for it could not fall within the scope of one man's knowledge, and there is no kind of learning and no art which has been completely mastered by a single intellect. Therefore, just as the task of a good sportsman, tracking his prey in a vast forest, is to catch as many wild beasts as he can nor has blame ever attached to anyone if he did not catch them all, so it is amply sufficient for us to have treated of the greatest part of

The measurement of land.

2

abunde est, tam diffusae materiae, quam suscepimus, maximam partem tradidisse. Quippe cum ea velut omissa desiderentur, quae non sunt propria nostrae professionis, ut proxime, cum de commetiendis agris rationem M. Trebellius noster requireret a me, vicinum atque adeo coniunctum esse censebat demonstranti, quemadmodum agrum pastinemus, praecipere etiam pastinatum quemadmodum metiri de-

3 beamus. Quod ego non agricolae sed mensoris officium esse dicebam: cum praesertim ne architecti quidem, quibus necesse est mensurarum nosse rationem, dignentur consummatorum aedificiorum, quae ipsi disposuerint, modum comprehendere, sed aliud existiment professioni suae convenire, aliud eorum, qui iam exstructa [1] metiuntur, et [2] imposito [3] calculo perfecti operis rationem computant. Quo magis veniam tribuendam esse nostrae disciplinae censeo,[4] si eatenus progreditur, ut dicat, qua quidque ratione faciendum, non quantum id sit quod effecerit.

4 Verum quoniam familiariter a nobis tu quoque, Silvine, praecepta mensurarum desideras, obsequar voluntati tuae, cum eo, ne dubites id opus geometrarum magis esse quam rusticorum, desque veniam, si quid in eo fuerit erratum, cuius scientiam mihi non vindico.

Sed ut ad rem redeam, modus omnis areae pedali mensura comprehenditur, qui [5] digitorum est XVI.

[1] exstructa *ac* : exstructam *S* : extrunctam *A*.
[2] *post* et *add. SA* iuncturae : *om. ac.*
[3] imposito *Aac* : posito *S*.
[4] censeo *add. edd.*
[5] qui *Aac* : quia *S*.

the extensive material with which we have undertaken to deal. For indeed subjects, which do not properly belong to our profession, are demanded as though they had been left out; for example, only recently, when my friend Marcus Trebellius required from me a method of measuring land he expressed the opinion that it was a kindred and indeed closely connected task for one who was showing how we ought to trench land to give instructions also how we ought to measure the land thus trenched. I replied 3 that this was the duty not of a farmer but of a surveyor, especially as even architects, who must necessarily be acquainted with the methods of measurement, do not deign to reckon the dimensions of buildings which they have themselves planned, but think that there is a function which befits their profession and another function which belongs to those who measure structures after they have been built and reckon up the cost of the finished work by applying a method of calculation. Therefore I hold that excuse should rather be made for our system of instruction if it only goes as far as to state by what method each of the operations of farming should be carried out and not the area over which it has been performed. But since, Silvinus, you also ask us in a 4 friendly spirit for instructions about measurements, I will comply with your wish, on condition that you harbour no doubt that this is really the business of geometricians rather than of countrymen, and make allowances for any errors that may be committed in a sphere where I do not claim to possess scientific knowledge.

But to return to my subject, the extent of every area is reckoned by measurement in *feet*, and a foot consists

Pes multiplicatus in passus et actus et climata et iugera et stadia centuriasque mox etiam in maiora spatia procedit. Passus pedes habet v. Actus minimus (ut ait M. Varro) latitudinis[1] pedes quattuor, longitudinis habet pedes cxx. Clima quoquo versus 5 pedum est lx. Actus quadratus undique finitur pedibus cxx. Hoc duplicatum facit iugerum, et ab eo, quod erat iunctum, nomen iugeri usurpavit: sed hunc actum provinciae Baeticae rustici acnuam[2] vocant: itemque triginta pedum latitudinem et 6 clxxx longitudinem porcam dicunt. At Galli candetum[3] appellant in areis urbanis spatium centum pedum, in agrestibus autem pedum cl.[4] Semiiugerum quoque arepennem vocant. Ergo (ut dixi) duo actus iugerum efficiunt longitudinem pedum ccxl, latitudinem pedum cxx. Quae utraeque summae in se multiplicatae quadratorum faciunt pedum milia viginti octo et octingentos. Stadium deinde habet passus cxxv, id est pedes dcxxv, quae mensura octies efficit mille passus, sic veniunt quinque milia 7 pedum.[5] Centuriam nunc dicimus (ut idem Varro ait) ducentorum iugerum modum. Olim autem ab centum iugeribus vocabatur centuria, sed mox duplicata nomen retinuit: sicuti tribus dictae primum a partibus populi tripartito divisi, quae tamen nunc 8 multiplicatae pristinum nomen possident. Haec non aliena, nec procul a ratiocinio, quod tradituri sumus, breviter praefari oportuit.

[1] latitudinis *ac* : latitudinem *A* : latitudine *S*.
[2] agnuam *SAac*.
[3] candetum *Aa* : candentum *c* : gandetum *S*.
[4] *post* cl *add.* quod aratores candetum nominant *SAac*.
[5] sic veniunt quinque millia pedum *ac*: sunt campum *SA*.

of 16 *fingers*. The multiplication of the foot produces successively the *pace*, the *actus*, the *clima*, the *iugerum*, the *stadium* and the *centuria*, and afterward still larger measurements. The *pace* contains five feet. The *smallest actus* (as Marcus Varro says) is four feet wide and 120 feet long. The *clima* is 60 feet each way. The *square actus* is bounded by 120 feet each way; when doubled it forms a *iugerum*, and it has derived the name of *iugerum* from the fact that it was formed by joining.[a] This *actus* the country folk of the province of Baetica call *acnua*; they also call a breadth of 30 feet and a length of 180 feet a *porca*. The Gauls give the name *candetum* to areas of a hundred feet in urban districts but to areas of 150 feet in rural districts; they also call a half-*iugerum* an *arepennis*. Two *actus*, as I have said, form a *iugerum* 240 feet long and 120 feet wide, which two numbers multiplied together make 28,800 square feet. Next a *stadium* contains 125 paces (that is to say 625 feet) which multiplied by eight makes 1000 paces, which amount to 5000 feet. We now call an area of 200 *iugera* a *centuria*, as Varro again states; but formerly the *centuria* was so called because it contained 100 *iugera*, but afterwards when it was doubled it retained the same name, just as the tribes were so called because the people were divided into three parts but now, though many times more numerous, still keep their old name. It was proper that we should begin by briefly mentioning these facts first, as being relevant to and closely connected with the system of calculation which we are going to set forth.

5

6

7

8

[a] *I.e. iugerum* is derived from the verb *iungere* "to join", because it consists of two square *actus* joined together.

Nunc veniamus ad propositum. Iugeri partes non omnes posuimus,[1] sed eas, quae cadunt in aestimationem facti operis. Nam minores persequi supervacuum fuit, pro quibus nulla merces dependitur.[2] Igitur (ut diximus) iugerum habet quadratorum pedum viginti octo milia et octingentos: qui pedes 9 efficiunt scripula CCLXXXVIII. Ut autem a minima parte, id[3] est, ab dimidio scripulo incipiam, pars quingentesima septuagesima sexta pedes efficit quinquaginta; id est iugeri dimidium scripulum. Pars ducentesima octogesima octava pedes centum; hoc est scripulum.[4] Pars CXLIV pedes CC, hoc est scripula duo. Pars septuagesima et secunda pedes CCCC, hoc est[5] sextula in qua sunt scripula quattuor. Pars quadragesima octava[6] pedes DC, hoc est sici-

[1] posuimus *edd.*: possumus *SAac.*
[2] dependitur *SAc*: deprehenditur *a.*
[3] ut autem—id *om. A*: autem *om. Sac.*
[4] *post* est scripulum *add.* pars septuagesima et secunda *SAac.*
[5] sextula *Aac*: sextulam *S.*
[6] quadragesima octava *om. a.*

The divisions of a iugerum *mentioned by Columella with the number of square feet, both Roman and English, in each division.*

Latin name of the divisions of the *iugerum.*	Number of *scripula* in each division.	Fractions of *iugerum.*	Roman square feet.	English square. feet.
Dimidium scripulum	$\frac{1}{2}$	$\frac{1}{576}$	50	48·35
Scripulum	1	$\frac{1}{288}$	100	96·70
Duo scripula	2	$\frac{1}{144}$	200	193·40
Sextula	4	$\frac{1}{72}$	400	386·80
Sicilicus	6	$\frac{1}{48}$	600	580·20

Let us now come to our real purpose. We have not put down all the parts of the *iugerum* but only those which enter into the estimation of work done. For it was needless to follow out the smaller fractions on which no business transaction depends. The *iugerum*, therefore, as we have said, contains 28,800 square feet, which number of feet is equivalent to 288 *scripula*. But to begin with the smallest fraction, the half-*scripulum*, the 576th part of a *iugerum*, contains 50 feet; it is the half-*scripulum* of the *iugerum*. The 288th part of the *iugerum* contains 100 feet; this is a *scripulum*. The 144th part contains 200 feet, that is two *scripula*. The 72nd part contains 400 feet and is a *sextula*, in which there are four *scripula*. The 48th part, containing 600 feet, is a *sicilicus*, in which there

9

The divisions of a iugerum (*continued*).

Latin name of the divisions of the *iugerum*.	Number of *scripula* in each division.	Fractions of *iugerum*.	Roman square feet.	English square feet.
Semuncia	12	$\frac{1}{24}$	1,200	1,160·40
Uncia	24	$\frac{1}{12}$	2,400	2,320·80
Sextans	48	$\frac{1}{6}$	4,800	4,641·60
Quadrans	72	$\frac{1}{4}$	7,200	6,962·40
Triens	96	$\frac{1}{3}$	9,600	9,283·20
Quincunx	120	$\frac{5}{12}$	12,000	11,604·0
Semis	144	$\frac{1}{2}$	14,400	13,924·80
Septunx	168	$\frac{7}{12}$	16,800	16,245·60
Bes	192	$\frac{2}{3}$	19,200	18,566·40
Dodrans	216	$\frac{3}{4}$	21,600	20,887·20
Dextans	240	$\frac{5}{6}$	24,000	23,208·0
Deunx	264	$\frac{11}{12}$	26,400	25,528·80
Iugerum	288	1	28,800	27,849·60

10 licus, in quo sunt scripula sex. Pars vigesima quarta pedes mille ducentos, hoc est semuncia, in qua scripula XII. Pars duodecima duo milia et quadringentos, hoc est uncia, in qua sunt scripula XXIV. Pars sexta pedes quattuor milia et octingentos, hoc est sextans, in quo sunt scripula XLVIII. Pars quarta pedes[1] septem milia et ducentos, hoc est quadrans, in 11 quo sunt scripula LXXII. Pars tertia pedes novem milia, et sexcentos, hoc est triens, in quo sunt scripula XCVI. Pars tertia et duodecima pedes duodecim milia hoc est quincunx, in quo sunt scripula CXX. Pars dimidia pedes quattuordecim milia et quadringentos, hoc est semis, in quo sunt scripula CXLIV. Pars dimidia et duodecima, pedes sexdecim milia et octingentos, hoc est septunx, in quo sunt scripula CLXVIII. Partes duae tertiae pedes decem novem milia et ducentos, hoc est bes, in quo sunt scripula CXCII. Partes tres quartae pedes unum et viginti milia et sexcentos, hoc est dodrans, in quo sunt 12 scripula CCXVI. Pars dimidia et tertia ped. viginti quattuor milia, hoc est dextans,[2] in quo sunt scripula CCXL.[3] Partes duae tertiae[4] et una quarta pedes viginti sex milia et quadringentos, hoc est deunx, in quo sunt scripula CCLXIV. Iugerum pedes viginti octo milia et octingentos, hoc est as,[5] in quo sunt 13 scripula CCLXXXVIII. Iugeri autem modus[6] si semper quadraret, et in agendis mensuris in longitudinem haberet pedes CCXL,[7] in[8] latitudinem pedes CXX, expeditissimum esset eius ratiocinium. Sed quoniam diversae formae agrorum veniunt in disputa-

[1] pedes *om. A.*
[2] destas *SA.*
[3] pars dimidia—CCXL *om. ac.*
[4] tertiae et II *SA* : duae tertiae et *ac.*

are six *scripula*. The 24th part, containing 1200 feet, 10 is a *semi-uncia*, in which there are 12 *scripula*. The 12th part, containing 2400 feet, is the *uncia*, in which there are 24 *scripula*. The 6th part, containing 4800 feet, is a *sextans*, in which there are 48 *scripula*. The 4th part, containing 7200 feet is a *quadrans*, in which there are 72 *scripula*. The 3rd part, containing 9600 11 feet, is a *triens*, in which there are 96 *scripula*. The 3rd part plus the 12th part, containing 12,000 feet, is the *quincunx*, in which there are 120 *scripula*. The half of a *iugerum*, containing 14,400 feet, is a *semis*, in which there are 144 *scripula*. A half plus a 12th part, containing 16,800 feet, is a *septunx*, in which there are 168 *scripula*. Two-thirds of a *iugerum*, containing 19,200 feet, is a *bes*, in which there are 192 *scripula*. Three-quarters, containing 21,600 feet, is a *dodrans*, in which there are 216 *scripula*. A half plus a third, 12 containing 24,000 feet, is a *dextans*, in which there are 240 *scripula*. Two-thirds plus a quarter, containing 26,400 feet, is a *deunx*, in which there are 264 *scripula*. A *iugerum*, containing 28,800 feet, is the *as*,[a] in which 13 there are 288 *scripula*. If the form of the *iugerum* were always rectangular and, when measurements were being taken, were always 240 feet long and 120 feet wide, the calculation would be very quickly done; but since pieces of land of different shapes come to be the subjects of dispute, we will give below

[a] The *as* is the unit which forms the standard in Roman measures, weights and coinage.

---

[5] as *SAc* : axis *a*.
[6] modus *ac* : modum *SA*.
[7] CCXL *a* : CXL *c* : CCXLVIII *SA*.
[8] in *add. edd.*

tionem, cuiusque generis species subiciemus, quibus quasi formulis utemur.

II. Omnis ager aut quadratus, aut longus, aut cuneatus, aut triquetrus, aut rotundus, aut etiam semicirculi vel arcus, nonnunquam etiam plurium angulorum formam exhibet.[1] Quadrati mensura facillima est. Nam cum sit undique pedum totidem, multiplicantur in se duo latera, et quae summa ex multiplicatione effecta est, eam dicemus esse quadratorum pedum. Tanquam est locus quoquo versus C pedum: ducimus centies centenos, fiunt decem 2 milia. Dicemus igitur eum locum habere decem milia pedum quadratorum, quae efficiunt iugeri trientem, et sextulam,[a] pro qua portione operis effecti numerationem facere oportebit.

3 At si longior fuerit, quam latior, ut exempli causa iugeri forma pedes habeat longitudinis CCXL, latitudinis pedes CXX, ita ut paulo ante dixi, latitudinis pedes cum longitudinis pedibus sic multiplicabis. Centies vicies duceni quadrageni fiunt viginti octo milia et octingenti. Dicemus iugerum agri tot

[1] exibet *ac*: exiget *SA*.

[a] *I.e.* 9600 + 400 Roman square feet = 10,000 square feet.

specimens of every kind of shape which we will use as patterns.

The shapes of land and their dimensions.

II. Every piece of land is square, or long, or wedge-shaped, or triangular, or round, or else presents the form of a semi-circle or of the arc of a circle, sometimes also of a polygon. The measuring of a square is very easy; for, since it has the same number of feet on every one of its sides, two sides are multiplied together and the product of this multiplication we shall say is the number of square feet. For example

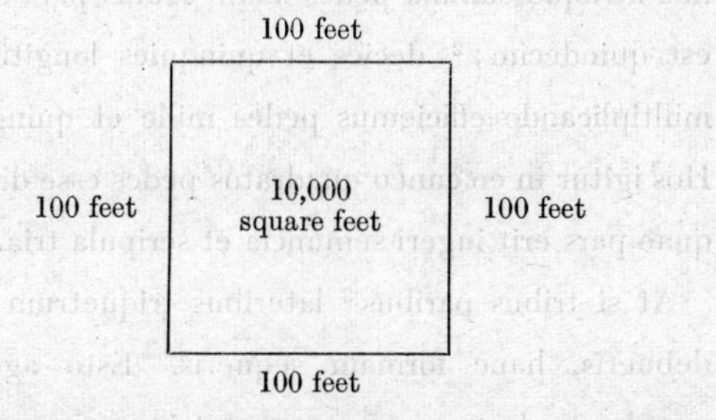

if an area were 100 feet each way, we multiply 100 by 100 and the result is 10,000. We shall, therefore, say 2 that the area contains 10,000 square feet, which make a *triens* ($\frac{1}{3}$) plus a *sextula* ($\frac{1}{72}$) of a *iugerum*,[a] and on the basis of this fraction we shall have to calculate the amount of work done.

If it is longer than it is broad (for example let the 3 form of the *iugerum* have 240 feet of length and 120 of breadth), as I said just now, you will multiply the feet of the breadth with the feet of the length in the following manner: 120 times 240 make 28,800, and we shall say that the *iugerum* of land contains this number

4 pedes quadratos habere. Similiterque omnis longitudinis pedes cum pedibus latitudinis multiplicabimus.[1]

Sin autem cuneatus ager fuerit, ut puta longus pedes centum, latus ex una parte pedes xx, et ex altera pedes x: tunc duas latitudines componemus, fiet utraque summa pedes xxx. Huius pars dimidia est quindecim;[2] decies et quinquies longitudinem multiplicando efficiemus pedes mille et quingentos. Hos igitur in eo cuneo quadratos pedes esse dicemus, quae pars erit iugeri semuncia et scripula tria.

5 At si tribus paribus[3] lateribus triquetrum metiri debueris, hanc formam sequeris. Esto ager triangulus pedum quoquo versus tricentorum. Hunc numerum in se multiplicato. Fiunt pedum nonaginta milia. Huius summae partem tertiam sumito,

[1] omnis longitudinis pedes cum pedibus latitudinis multiplicabimus *SAac*: fiet de omnibus agris, quorum longitudo maior sit latitudine *Schneider*.
[2] quindecim *edd.*: decus quinquies *SA*: *om. ac.*
[3] paribus *ac*: *om. SA.*

---

[a] *I.e.* 1200 + 300 Roman square feet = 1500 square feet.

of square feet. Similarly we shall always multiply the feet of the length with those of the width. 4

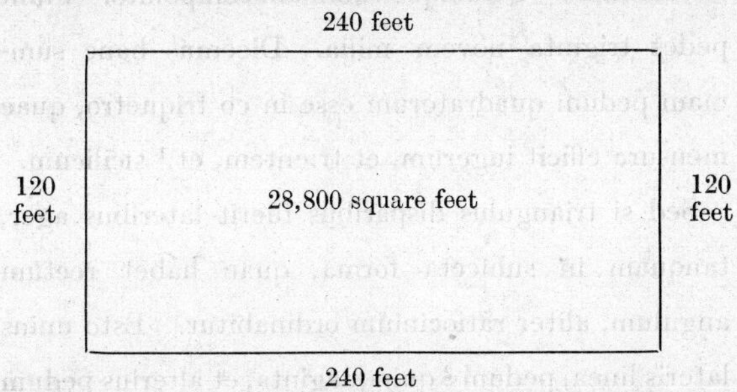

But if the field is wedge-shaped (for instance, suppose it to be 100 feet long and 20 feet broad on one side and 10 feet on the other side) we shall add the two breadths together, making a total of 30 feet. Half of this sum is 15, and by multiplying the longitude by 15 we shall obtain the result of 1500. We shall say then that this is the number of square feet in the wedge-shaped field which will be a *semuncia* plus three *scripula* ($\frac{3}{288}$ of a *iugerum*).[a]

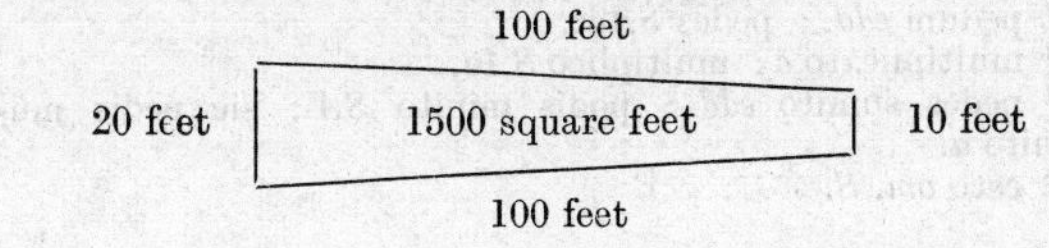

5 But if you have to measure a triangle with three equal sides, you will follow this formula. Suppose the field to be triangular, three hundred feet on every side. Multiply this number by itself and the result is 90,000 feet. Take a third part of this sum, that is

id est triginta milia. Item sumito decumam, id est novem milia. Utramque summam componito. Fiunt pedes triginta novem milia. Dicemus hanc summam pedum quadratorum esse in eo triquetro, quae mensura efficit iugerum, et trientem, et [1] sicilicum.

Sed si triangulus disparibus fuerit lateribus ager, tanquam in subiecta forma, quae habet rectum angulum, aliter ratiocinium ordinabitur. Esto unius lateris linea, pedum [2] quinquaginta, et alterius pedum centum. Has duas summas in se multiplicato; [3] quinquagies centeni fiunt quinque milia. Horum pars dimidia duo milia quingeni, quae pars iugeri unciam et scripulum efficit. Si rotundus ager erit, ut circuli speciem habeat, sic pedes sumito.[4] Esto [5] area

6

[1] et *ac*: *om. SA.*
[2] pedum *edd.*: pedes *SAac.*
[3] multiplicato *c*: multiplico *SAa.*
[4] pedes sumito *edd.*: podis minito *SA*: sic pedis muminito *a.*
[5] esto *om. S.*

---

[a] *I.e.* 28,800 + 9600 + 600 (Roman) square feet = 39,000 square feet.
[b] *I.e.* 2400 + 100 (Roman) square feet = 2500 square feet.

30,000. Likewise take a tenth part, that is 9,000. Add the two numbers together; the result is 39,000. We shall say that this is the total number of square feet in this triangle, which measure makes a *iugerum*, plus a *triens* ($\frac{1}{3}$), plus a *sicilicus* ($\frac{1}{48}$).[a]

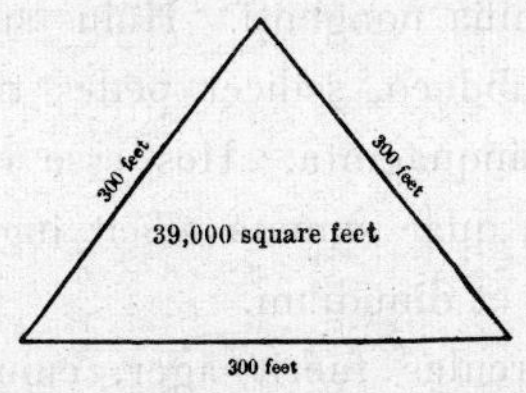

But if your field is triangular with unequal sides, as in the figure given below, which has a right angle, the calculation will be ordered differently. Let the line on one side of the right angle be 50 feet long and that on the other side 100 feet. Multiply these two numbers together; 50 times 100 makes 5000; half of this is 2500, which makes an *uncia* ($\frac{1}{12}$ of a *iugerum*) + a *scripulum* ($\frac{1}{288}$).[b] 6

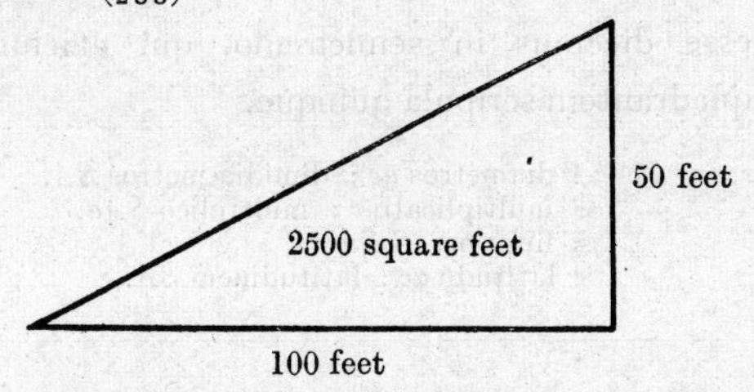

If the field is to be round, so as to have the appearance of a circle, reckon the number of feet as follows. Let there be a circular area of which the

7 rotunda, cuius diametros,[1] id est dimensio, habeat pedes LXX. Hoc in se multiplicato,[2] septuagies septuageni fiunt[3] quattuor milia et nongenti. Hanc summam undecies multiplicato, fiunt pedes quinquaginta tria milia nongenti. Huius summae quartam decimam subduco, scilicet pedes tria milia octingenti et quinquaginta. Hos esse quadratos in eo circulo dico, quae summa efficit iugeri sexcunciam, scripula duo et dimidium.

8 Si semicirculus fuerit ager, cuius basis habeat pedes CXL, curvaturae autem latitudo[4] pedes LXX: oportebit multiplicare latitudinem cum basi. Septuagies centeni quadrageni fiunt novem milia et octingenti. Haec undecies multiplicata fiunt centum septem milia et octingenti. Huius summae quarta decima est septem milia et septingenti. Hos pedes esse dicemus in semicirculo, qui efficiunt iugeri quadrantem scripula quinque.

[1] diametros *ac*: dimidia metros *SA*.
[2] multiplicato *c*: multiplico *SAa*.
[3] fiunt *om. SAac*.
[4] latitudo *ac*: latitudinem *SA*.

---

[a] *I.e.* a *sexcuncia* (36 *scripula*) and $2\frac{1}{2}$ *scripula* = $38\frac{1}{2}$ *scripula* = 3850 square feet.

[b] A *quadrans* (72 *scripula*) and 5 *scripula* = 77 *scripula* = 7700 square feet.

diameter (that is, the measurement across) is 70 feet. Multiply this number by itself: 70 times 70 makes 4900. Multiply this sum by 11 and the result is 53,900 feet. I subtract a fourteenth part of this

7

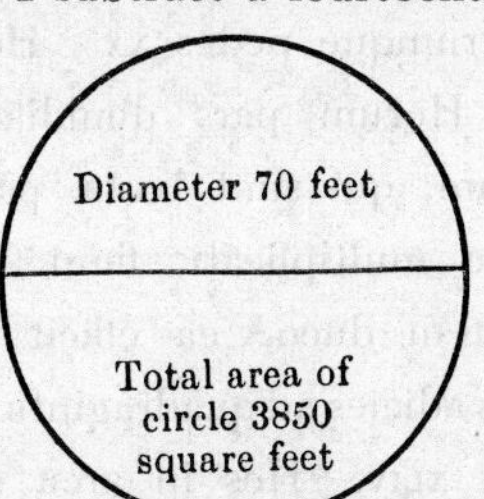

sum, namely 3850, and this I declare to be the number of square feet in the circle, which sum amounts to a *sexcuncia* of a *iugerum* and two *scripula* ($\frac{1}{144}$) and a half *scripulum* ($\frac{1}{576}$).[a]

8

If the piece of land is to be semi-circular and its base measures 140 feet and the depth of the circular portion is 70 feet, it will be necessary to multiply the

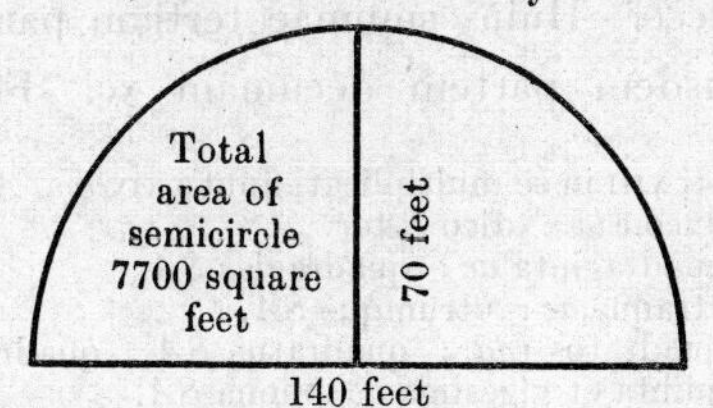

depth by the base. 70 times 140 makes 9800. This sum multiplied by 11 makes 107,800, and a fourteenth part of this is 7700. This we shall say is the number of square feet in the semi-circle, which makes a *quadrans* ($\frac{1}{4}$) of a *iugerum* and 5 *scripula* ($\frac{5}{288}$).[b]

9 Si autem minus quam semicirculus erit, arcum sic metiemur. Esto arcus, cuius basis habeat pedes XVI, latitudo autem pedes IV. Latitudinem cum basi pono. Fit utrumque pedes XX. Hoc duco quater. Fiunt LXXX. Horum pars dimidia est XL. Item sedecim pedum, qui sunt basis, pars dimidia VIII. Hi VIII in se multiplicati, fiunt LXIV.[1] Quartam decimam partem duco,[2] ea efficit pedes IV paulo amplius. Hos adicies ad quadraginta.[3] Fit utraque[4] summa pedes XLIV. Hos in arcu quadratos[5] esse dico, qui faciunt iugeri dimidium scripulum, quinta et vigesima[6] parte minus.

10 Si fuerit sex angulorum, in quadratos pedes sic redigitur. Esto hexagonum quoquo versus lineis pedum XXX. Latus unum in se multiplico. Tricies triceni fiunt DCCCC. Huius summae tertiam partem statuo CCC, eiusdem partem decumam XC. Fiunt CCCXC.

[1] Hi VIII in se multiplicati fiunt LXIV *om. A.*
[2] duco *edd.* : dico *SAac.*
[3] quadraginta *ac* : quadragies *SA.*
[4] utraque *ac* : utrumque *SA.*
[5] quadratos *edd.* : quadratus *SA* : quadrato *ac.*
[6] quinta et vigesima *ac* : nona *SA.*

---

[a] Actually 4·57.
[b] Half a *scripulum* is 50 square feet, $\frac{1}{25}$ of a *scripulum* is 4 square feet, whereas the actual total is 44 square feet.

But if the area is to be less than a semicircle, we shall measure the arc as follows: let there be an arc the base of which measures 16 feet and the depth 4 feet. I add the base to the depth, which together make 20 feet. This I multiply by 4, making 80, of which the half is 40. Again, the half of 16 feet, which 9

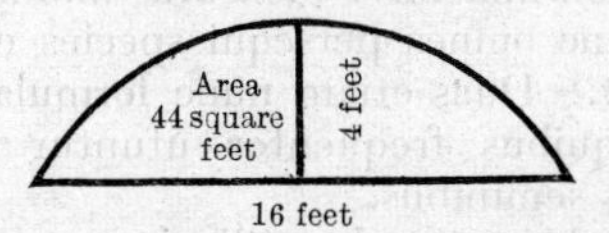

form the base, is 8. This I multiply by itself, making 64. I then take a fourteenth part of this, which make 4 feet and a little more.[a] This you will add to 40, and together they make a total of 44. This I declare to be the number of square feet in the arc, which is equivalent to half a *scripulum* ($\frac{1}{576}$ of a *iugerum*) less $\frac{1}{25}$ of a *scripulum*.[b]

10 If the area has six angles, it is reduced to square feet in the following manner. Let there be a

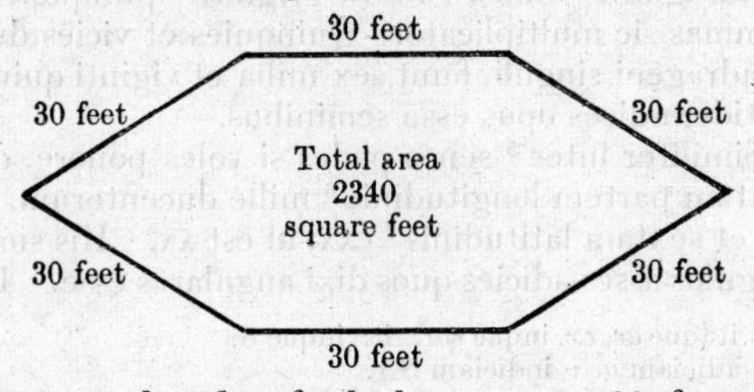

hexagon, each side of which measures 30 feet. I multiply one side by itself: 30 times 30 makes 900. Of this sum I take one-third, which is 300, a tenth part of which is 90: total 390. This must be multiplied

Hoc sexies ducendum est, quoniam sex latera sunt, quae consummata efficiunt duo milia trecenteni et quadraginta. Tot igitur pedes quadratos esse dicemus. Itaque [1] erit iugeri uncia dimidio scripulo et decima parte scripuli minus.

III. His igitur velut primordiis talis ratiocinii perceptis non difficiliter mensuras inibimus agrorum, quorum nunc omnes persequi species et longum et arduum est. Duas etiam nunc formulas praepositis adiciam,[2] quibus frequenter utuntur agricolae in disponendis seminibus.

Esto ager longus pedes mille ducentos, latus pedes CXX. In eo vites disponendae sunt ita, ut quini pedes inter ordines relinquantur. Quaero [3] quot [4] seminibus opus sit, cum quinum pedum spatia inter semina

2 desiderantur. Duco quintam partem longitudinis, fiunt CCXL; et quintam partem latitudinis, hoc est XXIV. His utrisque summis semper singulos asses adicito, qui efficiunt extremos ordines, quos vocant angulares. Fit ergo altera summa ducentorum quadraginta unius, altera viginti quinque. Has summas sic multiplicato. Quinquies et vicies duceni quadrageni singuli, fiunt sex milia et viginti quinque. Totidem dices opus esse seminibus.

3 Similiter inter [5] senos pedes si voles ponere, duces sextam partem longitudinis [6] mille ducentorum, fiunt CC, et sextam latitudinis [7] CXX, id est XX. His summis singulos asses adicies quos dixi angulares esse. Fiunt

[1] itaque *ac, ex.* inque *cor. A* : inque *S.*
[2] adiciam *ac* : indiciam *SA.*
[3] quaero *om. S.*
[4] quod *SAac.*
[5] inter *ac* : *om. SA.*
[6] sextam partem longitudinis *Aac* : longī *S.*
[7] sextam partem *post* latitudinis *add. S.*

by 6, because there are 6 sides: the product is 2340. We shall say, therefore, that this is the number of square feet. It will, then, be equivalent to an *uncia* ($\frac{1}{12}$ of a *iugerum*) less half a *scripulum* ($\frac{1}{596}$) plus $\frac{1}{10}$ of a *scripulum*.[a]

How many plants can a iugerum of land contain at intervals of three to ten feet apart?

III. Having grasped what may be called the first principles of this kind of calculation, we shall have no difficulty about entering upon the measurement of pieces of land, with the various kinds of which it is a long and arduous task to deal at this point. I will now also add, in addition to those which I have already set forth, two rules which husbandmen often employ in the setting out of plants.

Suppose that you have a piece of land 1200 feet long and 120 feet wide, in which vines have to be so arranged that five feet are left between the rows. How many plants, I ask, are necessary when spaces of five feet are required between the plants. 2 I take a fifth of the length, which makes 240, and a fifth of the breadth, which makes 24. To each of these numbers always add one unit, which forms the outermost row, and which they call the angular row; one number, therefore amounts to 241, the other to 25. Multiply these figures as follows: 25 times 241 makes 6025. 3 This, you will say, is the number of plants required.

Similarly, if you wish to set them six feet apart, you will take a sixth of the longitude (which is 1200), that is 200, and a sixth of the breadth (which is 120), that is 20. To each of these figures you will add what I called the angular units. The numbers are

[a] $\frac{1}{12}$ of a *iugerum* = 2400 square feet: half a *scripulum* = 50 square feet: $\frac{1}{10}$ of a *scripulum* = 10: therefore 2400 − 60 = 2340 square feet.

CCI, et XXI. Has summas inter se multiplicabis, vicies et semel ducentos et unum, atque ita efficies quattuor milia ducentos et viginti unum.[1] Totidem seminibus opus esse dices.

4 Similiter si inter septenos pedes ponere voles, septimam partem longitudinis et latitudinis duces, et adicies asses angulares, eodem modo eodemque[2] 5 ordine[3] consummabis numerum seminum. Denique quotcunque pedum spatia facienda censueris,[4] totam partem longitudinis et latitudinis duces, et praedictos asses adicies. Haec cum ita sint, sequitur ut iugerum agri, quod habet pedes CCXL longitudinis et latitudinis pedes[5] CXX, recipiat inter pedes ternos (hoc enim spatium minimum esse placet vitibus ponendis) per longitudinem semina LXXXI, per latitudinem inter quinos pedes semina XXV. Qui numeri inter se multiplicati fiunt seminum duo milia et viginti quinque.

6 Vel si[6] quoquo versus inter quaternos pedes vinea erit disposita, longitudinis ordo habebit semina LXI, latitudinis XXXI, qui numeri efficiunt in iugero vites mille octingentas et nonaginta unam. Vel si in longitudinem per quaternos pedes, in latitudinem per quinos pedes fuerit disposita, ordo longitudinis 7 habebit semina LXI, latitudinis XXV. Quod si[7] inter quinos pedes consitio fuerit, per longitudinem ordinis habebit semina XLIX, et rursus per latitudinem semina XXV. Qui numeri duo inter se multiplicati efficiunt mille ducentum et viginti quinque. At si per senos

[1] XXI *ac* : XI *SA*.
[2] -que *Aac* : quae *S*.
[3] ordine *ac* : ordines *SA*.
[4] censueris *Aac* : censeris *S*.
[5] CCXL longitudinis et latitudinis pedes *ac* : *om. SA*.
[6] vel si XXV *a* : versi XXV *SA* : *om. c*.

201 and 21. These sums you will multiply together, 21 times 201, and you will get 4221. This, you will say, is the number of plants required.

Similarly, if you wish to set them seven feet apart, 4 you will take a seventh of the length and of the breadth and you will add the angular units, and by the same method and the same arrangement you will make up the number of the plants. In a word, how- 5 ever many feet you have decided for the distance between the plants, you will take the total length and the total breadth and add the units mentioned above. This being so, it follows that the *iugerum* of land, which is 240 feet long and 120 feet broad, if the distance between the plants is three feet (and this we consider to be the smallest distance which should be left when planting vines), will accommodate 81 plants in its length, and in its breadth, with a distance of five feet between them, it will hold 25 plants. These numbers when multiplied together make 2025.

If the vineyard is arranged with intervals of four 6 feet each way, the row which runs lengthways will contain 61 plants, and the row which runs breadthways 31 plants; this gives 1891 vines to a *iugerum*. If the vineyard be laid out so that there are intervals of four feet lengthways and five feet breadthways, the row which runs lengthways will have 61 plants and that which runs breadthways 25 plants. If the 7 planting is carried out with intervals of five feet, the row will contain 49 plants lengthways and 25 breadthways; the two numbers multiplied together make 1225. If, however, you have decided to lay out the same area

[7] si *ac* : *om. SA.*

pedes eundem vitibus locum placuerit ordinare, nihil dubium est quin longitudini dandae sint XLI vites, 8 latitudini autem viginti una. Quae inter se multiplicatae efficiunt numerum DCCCLXI. Sin autem inter septenos pedes vinea fuerit constituenda, ordo per longitudinem recipiet capita triginta quinque, per latitudinem XVIII. Qui numeri inter se multiplicati efficiunt DCXXX. Totidem dicemus semina praeparanda. At si inter octonos pedes vinea conseretur, ordo per longitudinem recipiet semina XXXI, per latitudinem autem XVI. Qui numeri inter se multi- 9 plicati efficiunt CCCCXCVI. At si inter novenos pedes, ordo in longitudinem recipiet semina viginti septem, et in latitudinem quattuordecim. Qui numeri inter se multiplicati faciunt CCCLXXVIII. Vel inter denos pedes, ordo longitudinis recipiet semina XXV, latitudinis XIII. Hi numeri inter se multiplicati faciunt CCCXXV. Et ne in infinitum procedat disputatio nostra, eadem portione, ut cuique placuerint[1] laxiora spatia, semina faciemus. Ac de mensuris agrorum numerisque seminum dixisse abunde sit. Nunc ad ordinem redeo.

IV. Vinearum provincialium plura genera esse comperi. Sed ex iis, quas ipse incognovi,[2] maxime probantur velut arbusculae brevi crure sine adminiculo per se stantes: deinde quae pedaminibus adnixae singulis iugis imponuntur: eas rustici canteriatas appellant. Mox quae defixis arundinibus circum-

[1] placuerit *SAac*.
[2] incognovi *SA* : cognovi *ac*.

---

[a] See note on Book IV. 12. 2.

with the vines at intervals of six feet, there is no doubt that 41 vines must be assigned to the length and 21 to the breadth; these numbers multiplied together make a total of 861. But if the vineyard 8 has to be arranged with intervals of seven feet, a row will accommodate 35 heads lengthways and 18 breadthways; these numbers multiplied together make 630, and this, we shall say, is the number of plants which must be got ready. But if the vineyard is to be planted with intervals of eight feet, a row will accommodate 31 plants lengthways and 16 breadthways; these numbers multiplied together make 496. If the interval is to be nine feet, a row 9 will hold 27 plants lengthways and 14 breadthways; these numbers multiplied together make 378. With intervals of ten feet, a row will hold 25 plants lengthways and 13 breadthways; these numbers multiplied together make 325. So that our discussion may not be infinitely prolonged, we shall carry out our planting by using the same proportion to suit the wider spacing which any one of us prefers. Let what we have said about the measurement of land and the number of plants suffice. I now return to my proposed order of subjects.

Of the culture of provincial vineyards.

IV. I have found that there are several kinds of vines in the provinces; but of those of which I have personal knowledge those resembling small trees and standing by themselves on a short stock without any suppport are the most highly approved. Next come those which are supported by props and placed each on a single frame; these the peasants call "horsed"[a] vines. Next come those which are fastened round canes fixed in the ground and are bent into curves and circles, their firm-wood branches

vinctae [1] per statumina calamorum materiis ligatis in orbiculos gyrosque flectuntur [2]: eas non nulli 2 characatas vocant. Ultima est conditio stratarum vitium, quae ab enata stirpe confestim velut proiectae per humum porriguntur.

Omnium autem sationis fere eadem est conditio. Nam vel scrobe vel sulco semina deponuntur, quoniam pastinationis expertes sunt exterarum gentium agricolae: quae tamen ipsa paene supervacua est his locis quibus solum putre et per se resolutum est: namque hoc imitamur arando, ut ait Vergilius, id est 3 etiam pastinando. Itaque Campania, cum vicinum ex nobis capere possit exemplum, non utitur hac molitione terrae quia facilitas eius soli minorem operam desiderat. Sicubi autem densior ager provincialis rustici [3] maiorem poscit impensam, quod nos pastinando efficimus, ille sulco facto consequitur, ut laxius subacto solo deponat semina.

V. Sed ut singula earum quae proposui vinearum genera persequar, praedictum ordinem repetam. Vitis quae sine adminiculo suis viribus consistit, solutiore terra, scrobe; densiore, sulco ponenda est. Sed et scrobes et sulci plurimum prosunt, si in locis temperatis,[4] in quibus aestas non est praefervida, ante annum fiant, quam vineta conserantur.[5] Soli

[1] circumvinctae *SAac*: circummunitae *edd.*
[2] flectuntur *c*: flectentur *SAa.*
[3] provincialem rusticum *SAac*: provincialis rustici *edd.*
[4] temperatis *a*: tempatis *c*: terrae spatio *S*: prae spatio *A.*
[5] ante—conserantur *om. SA.*

---

[a] *Georg.* II. 184.
[b] *pastinare* is to dig with a *pastinum*, a fork.

being tied by means of props formed of reeds. These some people call "staked" vines. The type which comes last in esteem is the vine which lies flat on the ground and which, being as it were projected from the stock as soon as it grows out of the earth, stretches all over the ground.

2

The conditions under which all these vines are planted are almost identical. The plants are placed either in a plant-hole or in a furrow, since the farmers of foreign races are unacquainted with trenching, which indeed is almost superfluous in places where the soil crumbles and has fallen to pieces of its own accord, for, as Vergil says;[a]

'Tis this that with the plough we imitate,

that is to say in fact by trenching.[b] Thus the Campanians, though they might take a neighbouring example from us, do not employ this method of working the ground, because the ease with which their soil can be cultivated calls for less labour; but wherever a dense soil calls for a greater expenditure on the part of the provincial peasant, what we effect by soil-preparation he achieves by making a furrow in order that he may set his plants in soil which has already been worked into a looser condition.

3

Methods of cultivating vines.

V. But that I may deal particularly with each kind of the vine of which I have proposed to speak, I will resume the order already mentioned. The vine which stands by virtue of its own strength without any prop must in rather loose soil be placed in a planting-hole, in denser soil in a furrow, but both planting-holes and furrows are very beneficial, if, in temperate regions where the summer is not excessively hot, they are made a year before the vine-

tamen ante bonitas exploranda est.[1] Nam si ieiuno atque exili agro semina deponentur, sub ipsum 2 tempus sationis scrobis aut sulcus faciendus est. Si ante annum fiant, quam vinea conseratur, scrobis[2] in longitudinem altitudinemque defossus tripedaneus abunde est, latitudine tamen bipedanea: vel si quaterna pedum spatia inter ordines relicturi sumus, commodius habetur eandem quoquo versus dare mensuram scrobibus, non amplius tamen quam in tres pedes altitudinis depressis. Ceterum quattuor angulis semina applicabuntur subiecta minuta terra, et ita scrobes adobruentur.

3 Sed de spatiis ordinum eatenus praecipiendum habemus, ut intelligant agricolae, sive aratro vineas culturi sint, laxiora interordinia relinquenda, sive bidentibus, angustiora: sed neque spatiosiora quam decem pedum, neque contractiora quam quattuor. Multi tamen ordines ita disponunt, ut per rectam lineam binos pedes aut ut plurimum, ternos inter semina relinquant, transversa rursus laxiora spatia faciant,[3] per quae vel fossor vel arator incedat.

4 Sationis autem cura non alia debet esse, quam quae tradita est a me tertio volumine. Unum tamen huic consitioni Mago Carthaginiensis adicit,[4] ut semina ita deponantur, ne protinus totus scrobis terra compleatur, sed dimidia fere pars eius sequente biennio paulatim adaequetur. Sic enim putat vitem cogi

[1] si ante annum fiant *post* exploranda est *add. SA*.
[2] scrobis *SA*: scrobibus *ac*.
[3] faciant *ac*: fiat *SA*.
[4] adicit *Aac*: adigit *S*.

---

[a] Two-pronged instruments.
[b] Chapters 14–16.

yards are planted. Inquiry, however, must first be made into the excellence of the soil; for if the plants are going to be set in hungry and poor land, planting-holes or furrows must be made just before the time of planting. If they are made a year before the vine- 2 yard is planted, it is quite enough for the planting-hole to be dug three feet in length and depth, but two feet in width; or, if we are going to leave four feet between the rows, it is generally reckoned more convenient to give the planting-hole the same measurement in every dimension without, however, sinking them to a greater depth than three feet. Each plant, then, will be applied to the four corners after fine soil has been put into the bottom of the planting-holes, which will then be filled in.

3 As to the spaces between the rows we have this much advice to offer, that farmers should understand that, if they intend to cultivate their vineyards with the plough, wider intervals must be left, but they can be narrower if hoes [a] are used; but they should never be wider than ten feet or narrower than four. Many people, however, arrange the rows so as to leave two or at most three feet in a straight line between the plants, while on the other hand they make the transverse spaces wider, so that the digger or ploughman may pass freely.

4 The precautions taken in planting ought not to differ from those which I directed in my Third Book.[b] Mago, the Carthaginian, however, makes one addition to this system of planting, namely, that the plants should be put into the ground in such a way that the whole plant-hole is not immediately filled with soil but about half of it is gradually levelled up in the two following years; for he thinks that in this

deorsum agere radices. Hoc ego siccis locis fieri utiliter non negaverim; sed ubi aut uliginosa regio est, aut caeli status imbrifer, minime faciundum censeo. Nam consistens in semiplenis[1] scrobibus nimius[2] humor, antequam convalescant, semina 5 necat. Quare utilius[3] existimo, repleri quidem scrobes stirpe deposita, sed cum semina comprehenderint, statim[4] post aequinoctium autumnale debere diligenter atque alte ablaqueari, et recisis radiculis, si quas in summo solo citaverint, post paucos dies obrui.[5] Sic enim utrumque incommodum vitabitur, ut nec radices in superiorem partem evocentur, neque immodicis pluviis parum valida vexentur 6 semina. Ubi vero iam corroborata fuerint,[6] nihil dubium est, quin caelestibus aquis plurimum iuventur. Itaque locis, quibus clementia hiemis permittit, adapertas vites relinquere et tota hieme ablaqueatas habere eas conveniet.

De qualitate autem seminum inter auctores non convenit. Alii malleolo[7] protinus conseri vineam melius existimant, alii viviradice: de qua re quid 7 sentiam, iam superioribus professus sum. Et nunc tamen hoc adicio, esse quosdam agros, in quibus non aeque bene translata semina quam immota respon-

[1] semiplenis *ac* : semipleni *SA*.
[2] nimius *ac* : seminis *A* : *om. S*.
[3] quare utilius *ac* : quarum et ilius *SA*.
[4] comprehenderint statim *ac* : *om. SA*.
[5] obrui *SA* : adobrui *ac*.
[6] corroborata fuerint *Pontedera* : corroboraverint *ac* : comprobaverunt *SA*.
[7] malleolo *ac* : malleoli *SA*.

[a] *ablaqueare* is to dig round a plant so as to make a shallow furrow to hold water.
[b] Book III. 14. 2.

way the vine is forced to drive its roots downwards. I shall not deny that this can be done with advantage in dry places; but where either the district is marshy or the climate rainy, I am of opinion that it should certainly not be done, for excessive moisture standing in the half-filled plant-holes kills the plants before they can gain strength. Therefore I think that it 5 is more expedient that the plant-holes should be filled up again after the vine-stock has been put into them, but, when the plants have taken root, immediately after the autumn equinox, the soil round them ought to be carefully dug up [a] to a good depth and, after the rootlets which they may have put forth on the surface of the ground have been cut away, the earth ought to be filled in again after a few days. In this way two inconveniences will be avoided; firstly, the roots are not drawn to the upper part of the soil, and, secondly, the plants will not be troubled by excessive rains while they are still weak. When, 6 however, they have become quite strong, there is no doubt that they are greatly benefited by the rains from heaven; and so, in places where the mildness of the winter allows it, it will be expedient to leave the vines uncovered and to keep the soil round them loose the whole winter.

As regards the sort of vine-plants, the authorities are not agreed amongst themselves. Some think that it is better to plant a vineyard with mallet-shoots from the first, others think that it should be planted with quick-sets; I have already stated my opinion in the earlier part of this work.[b] However, I now add this further point, that there are 7 some lands where vines which have been transplanted do not answer as well as those which have not been

deant: sed istud rarissime accidere. Notandum item diligenter explorandum esse,[1]

quid quaeque ferat regio, quid quaeque recuset.

Depositam ergo stirpem, id est, malleolum vel vivi-
8 radicem,[2] formare sic convenit, ut vitis sine pedamine consistat. Hoc autem protinus effici non potest. Nam nisi adminiculum tenerae viti[3] atque infirmae[4] contribueris, prorepens pampinus terrae se applicabit.[5] Itaque posito semini arundo adnectitur, quae velut infantiam eius tueatur atque educet, producatque in tantam staturam, quantam permittit agricola. Ea porro non debet esse sublimis: nam usque in sesqui-
9 pedem coercenda est. Cum deinde robur accipit, et iam sine adiumento consistere valet, aut capitis aut bracchiorum incrementis adolescit. Nam duae species huius quoque culturae sunt. Alii capitatas vineas, alii bracchiatas magis probant. Quibus cordi est in bracchia vitem componere, convenit a summa parte, qua decisa novella vitis est, quicquid iuxta cicatricem citaverit, conservari, et in quattuor bracchia pedalis mensurae dividere, ita ut omnem partem caeli[6]
10 singula aspiciant. Sed haec bracchia non statim primo anno tam procera submittuntur, ne oneretur exilitas vitis; sed compluribus putationibus in praedictam mensuram educuntur. Deinde ex bracchiis quasi quaedam cornua prominentia relinqui oportet,

[1] *post* explorandum esse *add.* versibus aliter *SA*.
[2] viviradicem *ac*: viridem *SA*.
[3] vite *add. edd.*
[4] infirmae *ac*: infirmum *SA*.
[5] applicabit *ac*: adplicavit *SA*.
[6] celi *ac*: cesi *SA*.

[a] Verg. *Georg.* I. 53.
[b] Not, however, pointing straight upwards.

moved, but that this happens very rarely. It must also be noted that we ought to try diligently to discover:

What every clime may yield and what refuse.[a]

When, therefore, the plant has been put into the ground, whether it be a mallet-shoot or a quick-set, it is proper to adjust it in such a way that the vine may stand up without any prop. This, however, cannot be achieved immediately. For unless you have provided the vine with a support when it is tender and weak, the young shoots will creep along and keep close to the ground. So, when the plant is set in the earth, a reed is attached to it, so that it may, as it were, watch over its infancy and train it and raise it to such stature as the husbandman allows it to reach. This, moreover, ought not to be high, for it must be checked when it reaches a foot and a half. Afterwards, when it gains strength and can already stand without any help, it comes to maturity by the growth of its head or its branches. For here too there are two methods of cultivation, some people preferring vines which grow to a head, others those which grow out in arms. Those who delight in shaping a vine into arms should preserve whatever it puts forth near the scar where the young vine has had its top removed, and divide it into four arms a foot long in such a way that each of them looks towards a different region of the sky.[b] But these arms are not allowed to reach this height immediately in the first year, lest the vine be too heavily laden while it is still weak, but they must only reach the length which I have indicated after numerous prunings. Next there must be left projecting from these arms

8

9

10

atque ita totam vitem omni parte in orbem diffundi.
11 Putationis autem ratio eadem [1] est, quae in iugatis vitibus: uno tamen differt, quod pro materiis longioribus pollices quaternum aut quinum gemmarum relinquuntur: pro custodibus autem bigemmes reseces fiunt. In ea deinde vinea quam capitatam diximus, iuxta ipsam matrem usque ad corpus sarmentum [2] detrahitur, una aut altera tantummodo gemma
12 relicta, quae ipsi trunco adhaeret. Hoc autem riguis aut pinguissimis locis fieri tuto potest, cum vires terrae et fructum et materias valent praebere. Maxime autem aratris excolunt,[3] qui sic formatas vineas habent, eamque rationem sequuntur detrahendi vitibus bracchia, quod ipsa capita sine ulla extantia neque aratro neque bubus obnoxia sunt. Nam in bracchiatis plerumque fit, ut aut crure aut cornibus boum ramuli vitium defringantur, saepe etiam stiva, dum sedulus arator vomere perstringere ordinem, et quam proximam partem vitium excolere studet.

13 Atque haec quidem cultura vel bracchiatis vel capitatis antequam gemment, adhibetur. Cum deinde germinaverint, fossor insequitur, ac bidentibus eas partes subigit, quas bubulcus non potuit pertingere. Mox ubi materias vitis exigit, insequitur pampinator et supervacuos deterget fructuososque palmites sub-

[1] eadem *ac* : nam eodem *SA*.
[2] sarmentum *ac* : lumbrae *A* : inm̄tē (?) *S*.
[3] excolunt *ac* : cxcoli *S* : excolis *A*.

[a] *Cf.* Book IV. 21. 3. [b] *I.e.* sideways.

what may be called horns, and thus the whole vine must be spread in a circular form on all sides. The 11 method of pruning is the same as for vines which are trained on frames, though it differs in one respect, namely, that instead of longer firm-wood branches stumps with four or five "eyes" are left, and instead of "keepers"[a] short-cut branches with two "eyes" are formed. Then in the vine which we described as growing to a head, the shoot is pulled off close to the mother-vine right up to the stock, one or two "eyes" only being left which adhere to the trunk itself. This can be done with safety in well- 12 watered and very rich districts when the strength of the earth can supply both fruit and firm-wood. Those who have vineyards formed in this way cultivate them mainly with ploughs and follow this method of pulling off the arms from the vines, because the heads themselves, having nothing projecting[b] from them, are not liable to damage from the plough or from the oxen. For in vines which grow out into arms it generally happens that the small branches are broken off by the legs or horns of the oxen, and often too by the handle of the plough while the careful ploughman is striving to graze the edge of the row with the ploughshare and to cultivate the ground as near as possible to the vines.

Such then is the cultivation applied to vines 13 whether they grow to arms or to a head, before they bud. When they have budded, a digger follows the ploughman and breaks with a hoe the parts which the ploughman could not reach. Then, when the vine puts forth its firm-wood branches, the vine-trimmer follows and clears away the superfluous shoots and allows those which are fruitful to grow;

mittit, qui cum induruerunt, velut in coronam religantur. Hoc duabus ex causis fit: una, ne libero excursu in luxuriam prorepant,[1] omniaque alimenta pampini absumant; altera, ut religata vitis rursus aditum bubulco fossorique in excolenda[2] se praebeat.

14 Pampinandi autem modus is erit, ut opacis locis humidisque et frigidis aestate vitis nudetur, foliaque palmitibus detrahantur,[3] ut maturitatem fructus capere possit, et ne situ putrescat: locis autem siccis calidisque et apricis e contrario palmitibus uvae contegantur; et si parum pampinosa vitis est, advectis frondibus et interdum stramentis fructus muniatur.

15 M. quidem Columella patruus meus, vir illustribus disciplinis eruditus, ac diligentissimus agricola Baeticae provinciae, sub ortu Caniculae palmeis tegetibus vineas adumbrabat, quoniam plerumque dicti sideris tempore quaedam partes eius regionis sic infestantur Euro, quem incolae Vulturnum appellant, ut nisi teguminibus vites opacentur, velut halitu flammeo fructus uratur.

Atque haec capitatae bracchiataeque vitis cultura est. Nam illa, quae uni iugo superponitur, aut quae materiis[4] submissis arundinum statuminibus per orbem connectitur, fere eandem curam exigit, quam

16 iugata. Non nullos tamen in vineis characatis

[1] prorepant *SAac* : properent *edd.*
[2] excolendam *SAac.*
[3] detrahantur *c* : detrahuntur *SAa.*
[4] materiis *ac* : materies *SA.*

and when these have hardened they are tied up into a kind of crown. This is done for two reasons: firstly, lest, if they are allowed to run free, the shoots should creep forward and become over-luxuriant, and use up all the shoot's nourishment, and, secondly, in order that the vine, being tied back, may give the ploughman and the digger free access again for carrying on the cultivation of it.

The following will be the method of trimming. 14 In places which are shady and damp and cold, the vine should be stripped in summer and the leaves plucked from the shoots, so that the fruit may reach maturity and not become mouldy and rot away. In dry, warm and sunny places, on the contrary, the clusters of grapes should be covered by its shoots, and, if the vine is not sufficiently covered with foliage, the fruit should be protected with leaves brought from elsewhere and sometimes with straw. Indeed, my 15 paternal uncle, Marcus Columella, a man learned in the noble sciences and a most industrious farmer of the province of Baetica, used to shelter his vines about the rising of the Dogstar with palm-mats, because usually during the period of the said constellation some parts of that district are so troubled by the East wind, which the inhabitants call Vulturnus, that, unless the vines are shaded with coverings, the fruit is scorched as it were with a fiery breath.

Such is the method of cultivating both the vine which grows into a head and that which grows into arms. The vine which is placed on a single rail, or that of which the firm-wood is allowed to grow and which is tied in a circular form to props of reeds, requires almost the same treatment as that trained on a frame. I have, however, noticed that some 16

animadverti, et maxime elvenaci [1] generis, prolixos palmites quasi propagines summo solo adobruere, deinde rursus ad [2] arundines erigere, et in fructum submittere, quos nostri agricolae mergos, Galli candosoccos [3] vocant, eosque adobruunt simplici ex causa, quod existiment, plus alimenti terram [4] praebere fructuariis flagellis. Itaque post vindemiam velut inutilia sarmenta decidunt, et a stirpe submovent. Nos autem praecipimus easdem virgas, cum a matre fuerint praecisae, sicubi demortuis vitibus ordines vacent, aut si novellam quis vineam instituere velit, pro viviradice ponere. Quoniam quidem partes sarmentorum, quae fuerant obrutae, satis multas habent radices, quae depositae scrobibus confestim comprehendant.

17 Superest reliqua illa cultura prostratae vineae, quae nisi violentissimo caeli statu suscipi non debet. Nam et difficilem laborem colonis exhibet, nec unquam generosi saporis vinum praebet. Atque ubi regionis conditio solam eam culturam recipit,[5] bipedaneis scrobibus malleolus deponitur. Qui cum egerminavit, ad unam materiam revocatur : eaque primo anno compescitur [6] in duas gemmas : sequente deinde, cum palmites profudit, unus [7] submittitur, ceteri decutiuntur. At ille qui submissus est, cum

[1] eluaēnaci *a* : eluenaci *c* : et luennaci *SA*.
[2] ad *ac* : *om. SA*.
[3] andooccos *SA* : candos (*corr.* ocros) *c* : candos occos *a*.
[4] plus alimenti terram *ac* : eius alimenta terri *SA*.
[5] recipit *ac* : recepit *SA*.
[6] compescitur *edd.* : conspicitur *SAac*.
[7] *post* unus *add.* cum *SA*.

people when dealing with "staked" vines, especially those of the Helvenacan [a] kind, bury the sprawling shoots, as though they were layers, under the surface of the soil, and then again erect them on reeds and let them grow for fruit-bearing. These our husbandmen call *mergi* ("divers"), while the Gauls call them *candosocci* ("layers"), and they bury them for the simple reason that they think that the earth provides more nourishment for the fruit-bearing whips; and so after the vintage they cut them off as useless shoots and remove them from the stem. Our advice, however, is that these same rods, when they have been cut away from the mother-vine, should be planted as quick-sets in any vacant spaces in the rows where vines have died or in a new vineyard which anyone wants to establish; for indeed the parts of the shoots which had been buried have enough roots to take hold immediately if they are put into plant-holes.

17 There still remains the cultivation of the vine which grows on the ground; but this should not be undertaken except where the climate is very boisterous; for it presents a difficult task for the husbandmen and it never produces wine of a generous flavour. Where local conditions admit of this form of cultivation only, a hammer-shoot is put into plant-holes two feet deep. When it has budded, it is reduced to one firm-wood branch; this in the first year is confined to two "eyes." Then in the following year, when it has put forth a profusion of shoots, one is allowed to grow and the rest are struck off. The shoot which has been allowed to grow, when it has produced fruit, is pruned back to

[a] See Book III. 2. 25: Pliny, *N.H.* XIV. §§ 32–33; it produced a wine of a pale yellow colour.

fructum edidit, in eam longitudinem deputatur, uti iacens non excedat interordinii spatium. 18 Nec magna est putationis differentia cubantis et stantis rectae vineae: nisi quod iacenti viti breviores materiae submitti debent, reseces quoque angustius in modum furunculorum relinqui. Sed [1] post putationem, quam utique autumno in eiusmodi vinea fieri oportet, vitis tota deflectitur [2] in alterum interordinium: atque ita pars ea quae fuerat occupata vel foditur vel aratur, et cum exculta [3] est, eandem vitem recipit, ut altera quoque pars excoli possit. 19 De pampinatione talis vineae parum inter auctores convenit. Alii negant [4] esse nudandam vitem, quo melius contra iniuriam ventorum ferarumque [5] fructum abscondat: aliis placet parcius pampinari, ut et vitis non in totum supervacuis frondibus oneretur, et tamen fructum vestire aut protegere possit: quae ratio mihi quoque commodior videtur.

VI. Sed iam de vineis satis diximus. Nunc de arboribus praecipiendum est. Qui volet frequens et dispositum arbustum paribus spatiis fructuosumque habere, operam dabit, ne emortuis arboribus rarescat ac primam quamque senio aut tempestate afflictam submoveat, et in vicem novellam sobolem substituat. Id autem facile consequi poterit, si ulmorum semi-

[1] relinqui sed *edd.*: relictis et *SA*: relictis sed *a*: relinqui relictis *c*.
[2] deflectatur *ac*: deplectitur *SA*.
[3] exculta *ac*: excuncta *SA*.
[4] negant *ac*: negent *SA*. [5] -que *ac*: quo *SA*.

[a] *I.e.* for supporting vines.

such a distance that, as it lies on the ground, it does not reach beyond the space between the rows. Nor 18 is there a great difference between the pruning of a recumbent vine and of one which stands upright, except that the firm-wood branches in the vine which lies on the ground should be allowed to grow to a shorter length and the stumps ought to be left narrower so as to resemble knobs. But after the pruning, which in this kind of vine ought naturally to be carried out in the autumn, the whole vine is bent aside into one of the two spaces between the rows; and the part which was previously occupied is either dug up or ploughed, and when it has been thoroughly cultivated, it receives the same vine back again, so that the other space may also be cultivated. About 19 the trimming of this kind of vineyard, there is little agreement between the authorities. Some say that the vine ought not to be stripped, that it may the better conceal the fruit from injury by the wind and by wild beasts; others hold that it should be trimmed only sparingly, so that the vine may not be wholly burdened with superfluous leaves and yet may be able to cover or conceal the fruit. The latter method seems to me too to be the more expedient.

Plantations for the support of vines.

VI. We have now said enough about vines; we must now give directions about trees.[a] He who wishes to have a thick and profitable plantation for supporting vines with the trees set at equal distances from one another will take care that it does not grow sparse because the trees have died and will be careful to remove any tree as soon as it is afflicted with old age or damaged by storm and substitute a young growth in its place. This he will easily be able to achieve if he has a nursery for elms ready prepared.

narium paratum habuerit: quod[1] quomodo et qualis generis faciendum sit, non pigebit deinceps praecipere.

2 Ulmorum duo esse genera convenit, Gallicum et vernaculum: illud Atinia, hoc nostras dicitur. Atiniam ulmum Tremellius Scrofa non ferre sameram, quod est semen eius arboris, falso est opinatus. Nam rariorem sine[2] dubio creat, et idcirco plerisque et sterilis videtur, seminibus inter frondem, quam prima germinatione edit, latentibus. Itaque nemo iam 3 serit ex samera, sed ex sobolibus. Est autem ulmus longe laetior et procerior, quam nostra,[3] frondemque iucundiorem bubus praebet:[4] qua[5] cum assidue pecus paveris, et postea generis alterius frondem dare 4 institueris, fastidium bubus afferes.[6] Itaque si fieri poterit, totum agrum genere uno Atiniae ulmi conseremus: si minus, dabimus operam, ut in ordinibus disponendis pari numero vernaculas et Atinias alternemus. Ita semper mixta fronde utemur, et quasi hoc condimento illectae pecudes fortius iusta[7] cibariorum conficient.

Sed vitem maxime populus videtur alere, deinde 5 ulmus, post etiam fraxinus. Populus, quia raram, neque idoneam frondem pecori praebet, a plerisque repudiata est. Fraxinus, quia capris et ovibus gratissima est, nec inutilis bubus, locis asperis et

[1] quod *ac*: quo *SA*. [2] sine *ac*: semini *SA*.
[3] nostras *SAc*: nostra *a*. [4] praebet *ac*: *om.* *SA*.
[5] qua *a*: -que *SA*: quia *c*. [6] afferes *SAa*: adfert *c*.
[7] iusta *c*: iuxta *SAa*.

---

[a] From the town of Atina in Cispadane Gaul.
[b] A contemporary of Varro and one of the speakers in Varro's agricultural treatise.

In what manner and of what kind of trees it must be formed, I shall have no objection to stating forthwith.

2 It is generally agreed that there are two kinds of elms, the Gallic and the native; the former is called the Atinian,[a] the latter our own Italian. Tremellius Scrofa[b] was wrong when he expressed the opinion that the Atinian elm does not bear *samera*, which is the seed of that tree; it certainly produces it but rather thinly and for that reason most people think that it is actually barren, since the seeds are hidden among the foliage which it produces at its first budding. 3 That is why no one now grows it from seed but by means of shoots. This elm is much more luxuriant and taller than ours and produces foliage which is more acceptable to oxen; when you have fed cattle on it constantly and then begin to give them foliage of the other kind, you will cause them to feel a loathing for the latter. 4 Therefore, if possible, we shall plant a whole field with the Atinian kind of elm only, or, failing that, we shall take care, in arranging the rows, to plant native and Atinian elms to the same number alternately. In this way we shall always have a mixture of foliage for use and the cattle, attracted by this kind of seasoning for their food, will finish off with greater heartiness the full ration allotted to them.

But the poplar seems to sustain the vine best of all trees, then the elm, and after it the ash. 5 The poplar tree, because it provides foliage which is scanty and unsuitable for cattle, has been rejected by most people; the ash, because it is most acceptable to goats and sheep and of some use for oxen, is rightly planted in rough and mountainous places in which the

montosis, quibus minus laetatur ulmus, recte seritur. Ulmus, quod et vitem commodissime patitur, et iucundissimum pabulum bubus affert, variisque generibus soli provenit, a plerisque praefertur. Itaque si[1] arbustum novum instituere cordi est, seminaria ulmorum vel fraxinorum parentur ea ratione, quam deinceps subscripsimus. Nam populi melius cacuminibus in arbusto protinus deponuntur.

6 Igitur pingui solo et modice humido bipalio terram pastinabimus, ac diligenter occatam et resolutam humum verno tempore in areas componemus. Sameram deinde, quae iam rubicundi coloris erit, et compluribus diebus insolata iacuerit, ut aliquem tamen succum et lentorem habeat, iniciemus areis, et eas totas seminibus spisse contegemus, atque ita cribro putrem terram duos alte digitos incernemus, et modice rigabimus, stramentisque areas cooperiemus, ne prodeuntia cacumina seminum ab avibus

7 praerodantur. Ubi deinde prorepserint[2] plantae, stramenta colligemus, et manibus herbas carpemus: idque leviter et curiose faciendum est, ne adhuc tenerae brevesque radiculae ulmorum convellantur. Atque ipsas quidem areas ita anguste compositas habebimus, ut qui runcaturi sunt, medias partes earum facile manu contingant: nam si latiores fuerint, ipsa

8 semina[3] proculcata noxam capient. Aestate deinde prius quam sol oriatur, aut ad vesperum, seminaria

[1] si *S*: cui *ac*: *om. A.*
[2] prorepserit *SAac.*
[3] ipsa semina *Schneider*: ipsiseminibus *S*: ipsi seminibus *Aac.*

elm is less flourishing. The elm is preferred by most people, because it both accommodates itself very well to the vine and provides food most acceptable to oxen and flourishes in various kinds of soil. So if it is desired to establish a new plantation, nurseries of elms or ash-trees should be prepared on the system which we have described hereafter; for poplars are better put straight into the plantation in the form of tree-tops planted in the ground. We will, therefore, 6 prepare the ground with a double mattock where the earth is rich and moderately moist, and in the spring-time, after the soil has been carefully harrowed and broken up, we shall mark it out into beds. We shall then cast upon the beds the elm-seed which will now be of a ruddy colour and has been exposed to the sun for several days, but still retaining some juice and stickiness, and we shall thickly cover the beds all over with the seed and scatter crumbling earth over them with a sieve to the depth of two inches and give them a moderate watering and cover the beds with straw, so that the heads of the plants, when they come up, may not be pecked off by birds. Then, when the plants have crept forth, we shall 7 collect the straw and pull up the weeds by hand—a process which must be carried out gently and carefully, so that the still tender and short little roots of the elms may not be pulled up with the weeds. We shall have the beds themselves planned so as to be so narrow that those who are going to weed them can easily reach to the middle of them with their hands; for, if they are broader, the seedlings themselves will be trodden upon and receive damage. Then in the summer, before the sun rises 8 or towards evening, the nursery-beds ought to be

conspergi saepius[1] quam rigari debent:[2] et cum ternum pedum plantae fuerint, in aliud seminarium transferri, ac ne radices altius agant (quae res postmodum in eximendo magnum laborem affert, cum plantas in aliud seminarium transferemus) oportebit non maximos scrobiculos sesquipede inter se distantes fodere: deinde radices in nodum, si breves, vel in orbem coronae similem, si longiores erunt, inflecti, et oblitas fimo bubulo scrobiculis deponi, ac diligenter

9 circumcalcari. Possunt etiam collectae cum stirpibus plantae eadem ratione disponi: quod in Atinia ulmo fieri necesse est, quae non seritur e samera. Sed haec ulmus autumni tempore melius quam vere disponitur; paulatimque ramuli eius manu detorquentur, quoniam primo biennio ferri reformidat ictum. Tertio demum anno acuta falce abraditur,[3] atque ubi translationi iam idonea est, ex eo tempore autumni, quo terra imbribus permaduerit, usque in vernum tempus, antequam radix ulmi in eximendo

10 delibretur, recte seritur. Inde[4] in resoluta terra ternum pedum quoquo versus faciendi scrobes. At in densa, sulci eiusdem altitudinis et latitudinis, qui arbores recipiant,[5] praeparandi. Sed deinde in solo roscido et nebuloso conserendae sunt ulmi, ut earum rami ad orientem et occidentem dirigantur, quo plus

[1] saepius *edd.*: seminis *ac*: seminus *SA*.
[2] debet *SAac*.
[3] abraditur *a*: ablanditur *SAc*.
[4] inde *scripsi*: in se *SAac*.
[5] respiciat *Aac*: respiat *S*.

---

[a] *I.e* those which are planted in the form of cuttings as opposed to seedlings.

sprinkled from time to time rather than soaked, and when the plants have growth three feet high, they should be transferred to another nursery-bed, and that they may not strike their roots too deep (for this afterwards involves much labour in lifting them when we are going to transfer them to another nursery-bed), we shall have to dig not very large plant-holes a foot and a half apart. Next the roots, if they are short, will have to be bent as it were into a knot, or, if they are too long, into a circle resembling a crown and, after being smeared with ox-dung, they must be lowered into small plant-holes and carefully trodden down all round. The plants, too, which are 9 gathered on their stocks [a] can be set out in the same manner, and this is essential in the case of the Atinian elm which is not raised from seed. It is better to set this kind of elm in the autumn rather than in the spring, and its small branches are twisted little by little by hand, since in its first two years it dreads the blow of an iron implement. Finally, in its third year it is scraped with a sharp pruning-hook, and when it is fit for transplantation (that is, from the season of autumn, when the ground has been thoroughly soaked with rain, until the spring, before the root of the elm is likely to lose its bark while being removed from the soil), then is the proper time for planting it. Next plant-holes measuring three 10 feet each way must be made if the soil is loose, but, if it is dense, furrows of the same depth and width must be prepared to receive the trees. But also in a soil which is exposed to dew and mist the elms must be planted in such a way that their branches may be directed towards the east and west, in order that the middle of the trees, to which the

solis mediae arbores, quibus vitis applicata et religata innititur, accipiant.

11 Quod si etiam frumentis consulemus, uberi solo inter quadraginta pedes, exili, ubi nihil seritur, inter viginti, arbores disponantur. Cum deinde adolescere incipient, falce formandae, et tabulata instituenda sunt.[1] Hoc enim nomine usurpant agricolae ramos truncosque prominentes, eosque vel propius [2] ferro compescunt, vel longius promittunt, ut vites laxius diffundantur: hoc in solo pingui melius, illud in

12 gracili. Tabulata inter se ne minus [3] ternis pedibus absint, atque ita formentur, ne superior ramus in eadem linea sit, qua inferior. Nam demissum ex eo palmitem germinantem inferior atteret, et fructum decutiet.

Sed quamcunque arborem severis, eam biennio proximo putare non oportet. Post deinde si ulmus exiguum incrementum recipit,[4] verno tempore, antequam librum demittat, decacuminanda est iuxta ramulum, qui videbitur esse nitidissimus, ita tamen, uti supra eum trunco [5] stirpem dodrantalem [6] relinquas, ad quam ductus [7] et applicatus ramus alligetur, et correptus [8] cacumen arbori praebeat.

13 Deinde stirpem post annum praecidi et allevari oportet. Quod si nullum ramulum arbor idoneum habuerit, sat erit novem pedes a terra relinqui, et superiorem partem detruncari, ut novae virgae, quas

[1] instituenda sunt *Aa*: instituenda *c*: instituendis *S*.
[2] propius *c*: proprius *Aa*: prius *S*.
[3] se si minus *S*: seminibus *A*: se minus *ac*.
[4] recipit *om. SAac*. [5] truncum *SAac*.
[6] dodrantem *S*: drodant partem *A*: dodrantanitem *a*: dodrantanidem *c*.
[7] ductus *edd.*: dubitus *SAa*: dubius *c*.
[8] correptus *SAac*: correctus *edd.*

vine is applied and fastened, may receive more sunlight.

But if we have in view the sowing of cereals also, the trees should be placed, if the soil is rich, at intervals of forty feet from one another, but if it is thin and nothing is planted in it, at intervals of twenty feet. Then when they begin to grow tall, they must be shaped with the pruning-hook and successive "stages" must be arranged; for the husbandmen call prominent branches and trunks by this name and either cut them closer with the knife or let them grow longer, that the vines may spread more loosely, the latter process being better on rich soil, the former on thin soil. The "stages" should be not less than three feet apart from one another and so shaped that an upper branch may not be in the same line as a lower; for the lower branch will rub against the budding shoot let down from the upper branch and shake off the fruit. 11 12

But whatever tree you plant, you should not prune it during the next two years. Then afterwards, if the elm receives only a little growth, in the spring, before it sheds its bark, its top must be lopped off near the small branch which appears to be the most healthy, but in such a way as to leave above it on the trunk a stump nine inches long, towards which the branch can be trained and then applied and fastened, that, when it has been thus caught, it may provide a top for the tree. Then after a year the stump must be cut away and the place smoothed off. If, however, the tree has no suitable small branch, it will be enough if nine feet from the ground it is left standing and the upper part lopped off, in order that the new rods which it will have put forth may be safe 13

emiserit,[1] ab iniuria pecoris tutae sint. Sed si fieri poterit, uno ictu arborem praecidi; si minus, serra desecari, et plagam falce allevari oportebit, eamque plagam luto paleato contegi, ne sole aut pluviis 14 infestetur. Post annum aut biennium, cum enati ramuli recte convaluerint, supervacuos deputari, idoneos in ordinem submitti conveniet. Quae ulmus a positione bene provenerit,[2] eius summae virgae falce debent enodari. At si robusti ramuli erunt, ita ferro amputentur, ut exiguam stirpem prominentem trunco relinquas. Cum deinde arbor convaluerit, quicquid falce contingi poterit, exputandum est, allevandumque eatenus, ne plaga corpori matris applicetur. Ulmum autem novellam formare sic 15 conveniet. Loco pingui octo pedes a terra sine ramo relinquendi, vel in arvo gracili septem pedes: supra quod spatium deinde per circuitum in tres partes arbor dividenda est, ac tribus lateribus singuli ramuli 16 submittendi primo tabulato assignentur. Mox de ternis pedibus superpositis alii rami submittendi sunt, ita ne iisdem lineis, quibus in inferiore positi sint. Eademque[3] ratione usque in cacumen ordinanda erit arbor. Atque in frondatione cavendum, ne aut prolixiores pollices fiant, qui ex amputatis virgis

[1] emiserant *Sa*: emiserint *A*: emiserat *c*.
[2] provenerit *c*: provenerint *SAa*.
[3] positis in easdemque *S*: postas in eadem quae *A*: positis in eademque *ac*.

from injury by cattle. If possible, the tree should be cut through with a single blow; if not, it will have to be sawn through and the wound smoothed off with a pruning-hook and covered with mud mixed with straw, so that it may not be damaged by the sun or the rain. After a year or two, when the 14 little branches which have come forth have duly gained strength, it will be fitting that those which are superfluous should be pruned away and those which are suitable should be allowed to grow freely and take their place in the row. If an elm has made good progress since it was planted, its topmost rods should be freed from knots with a pruning-hook; but if the small branches are vigorous, they should be cut off with a knife in such a way that you leave a little stump projecting from the trunk. Then when the tree has gained strength, whatever can be reached with a pruning-hook should be cut away and smoothed off, without, however, any wound being inflicted on the body of the mother-tree. It will be proper to shape the young elm in the following manner. Where 15 the soil is rich, eight feet should be left from the ground, without any branches, or seven feet in poor soil; then above this the tree must be divided into three parts throughout its circumference, and small branches, one on each of the three sides, should be allowed to grow and be allotted to the first "stage." 16 Then, three feet above, other branches must be allowed to grow in such a manner that their position is not in the same line as in the stage underneath; and the tree will have to be arranged on the same principle right up to the top. In stripping the tree care must be taken that the knobs which are left where the rods have been cut away do not project too much, and that

relinquuntur, aut rursus ita alleventur, ut ipse truncus laedatur, aut delibretur ; nam parum gaudet ulmus [1] in corpus nuda. Vitandumque ne de duabus plagis una fiat, cum talem cicatricem non facile cortex

17 comprehendat. Arboris autem perpetua cultura est, non solum diligenter eam [2] disponere, sed etiam truncum circumfodere, et quicquid frondis enatum fuerit, alternis annis aut ferro amputare aut astringere, ne nimia umbra viti noceat. Cum deinde arbor vetustatem [3] fuerit adepta, propter terram [4] vulnerabitur ita, ut excavetur usque in medullam, deturque exitus humori, quem ex superiore parte conceperit. Vitem quoque, antequam ex toto arbor praevalescat, conserere convenit.

18 At si teneram ulmum maritaveris, onus iam non sufferet : si vetustae [5] vitem applicueris, coniugem necabit. Ita suppares esse aetate et viribus arbores vitesque convenit. Sed arboris maritandae causa scrobis viviradici fieri debet latus pedum duorum, altus levi terra totidem pedum (gravi, dupondio [6] et dodrante) longus pedum sex aut minimum quinque. Absit autem hic ab arbore ne minus sesquipedali spatio. Nam si radicibus ulmi iunxeris, male vitis comprehendet, et cum tenuerit, incremento arboris

19 opprimetur. Hunc scrobem, si res permittit, autumno facito, ut pluviis et gelicidiis maceretur. Circa vernum deinde aequinoctium binae vites, quo celerius

[1] *post* ulmus *add.* quae *SAac.*

[2] eam *S* : eadem *Aa* : *om. c.*

[3] vetustate *SAac.*

[4] terram *S* : ramum *Aac.*

[5] vetustae *Schneider* : vetustate *S* : vetustatem *A* : vetustam *a* : *om. c.*

[6] dupondio *a* : dupundio *c* : dupundiu *SA.*

they are not, on the other hand, so much smoothed away that the trunk itself is damaged or stripped of its bark; for an elm takes little pleasure in being bared to the quick. Also we must avoid making one wound out of two, for the bark does not easily grow over a scar of this kind. The elm requires constant 17 attention, not only in training it carefully but also in digging round the trunk and in alternate years cutting off with a knife or tying back any foliage which has grown from it, so that excessive shade may not harm the vine. Then when the tree has reached a good age, a wound will be made in it near the ground in such a way that a hole is made reaching to the pith and a passage thus given to the moisture, which it has formed in its upper portion. It is well also to plant the vine before the tree has reached its full strength.

But if you wed a tender young elm to a vine, it will 18 now not support the weight; if you couple a vine with an old elm, it will kill its mate. The trees and the vines, therefore, ought to be nearly equal in age and strength. In order to wed the tree and the vine, a trench ought to be made for the quick-set two feet wide and the same number of feet deep, if the soil is light (but if it is heavy, two feet and three-quarters deep) and six or at least five feet long. The trench, however, should not be less than a foot and a half from the tree; for if you put the vine close to the roots of the elm, it will not strike root properly and, when it has taken hold, it will be smothered by the growth of the tree. If circumstances allow, make 19 the trench in the autumn, that it may be softened by the rains and frosts; then, about the time of the spring equinox, in order more quickly to clothe the

ulmum vestiant, pedem inter se distantes scrobibus deponendae: cavendumque ne aut septentrionalibus ventis aut rorulentae sed siccae serantur.

20 Hanc observationem non solum in vitium positione, sed in ulmorum ceterarumque arborum praecipio: et uti cum de seminario eximuntur, rubrica notetur una pars, quae nos admoneat, ne aliter arbores constituamus, quam quemadmodum in seminario steterint. Plurimum enim refert, ut eam partem caeli spectent, cui ab tenero consueverunt.[1] Melius autem locis apricis, ubi caeli status neque praegelidus neque nimium pluvius est, autumni tempore et 21 arbores et vites post aequinoctium deponuntur. Sed eae ita conserendae sunt, ut summam terram,[2] quae aratro subacta sit, semipedem alte substernamus, radicesque omnes explicemus, et depositas stercorata, ut ego existimo, si minus, certe subacta operiamus, et circumcalcemus ipsum seminis codicem. Vites in ultimo scrobe deponi oportet, materiasque earum per scrobem porrigi, deinde ad arborem erigi; atque ab 22 iniuria pecoris caveis emuniri. Locis autem praefervidis semina septentrionali parte arbori applicanda sunt: locis frigidis a meridie, temperato[3] statu caeli aut ab oriente aut ab occidente, ne toto die solem vel umbram patiantur.

Proxima deinde putatione melius existimat Celsus ferro abstineri, ipsosque caules in modum coronae con-

[1] consueverunt *c*: consuerunt *SAa*.
[2] summam terram *Aac*: somnum a terra *S*.
[3] temperato *a*: tempato *c*: temperatu *SA*.

[a] See note on Vol. I. p. 35.

elm, two vines a foot apart should be put into the trench, and care should be taken that they are not planted when the north winds are blowing, nor when the vines are wet with dew, but when they are dry.

20 This rule I lay down not only when vines are being planted but also elms and the other trees; also, that, when they are removed from the nursery-bed, one side should be marked with ruddle to warn us not to plant trees in any position other than that in which they stood in the nursery-bed; for it is very important that they should face that quarter of the sky to which they have been accustomed from their early days. In sunny positions, however, when the climate is neither very cold nor too rainy, both trees and vines are better planted in the autumn after the equinox. 21 They should be planted on the principle of putting beneath them to a depth of half a foot top-soil which has been broken by the plough and uncoiling all the roots and covering the plants when they are set with dunged soil, which I consider the best course, or, if not, at least with broken soil, and treading round the actual stem of the plant. The vines should be set at the edge of the trench and their firm-wood branches stretched along the trench and then erected into the tree and protected by railings from damage by cattle. 22 In very hot localities the plants should be attached to the tree on the north side, in cold places to the south side, in a temperate climate either on the east or on the west side, so that they may not have to endure the sun or the shade all day.

Celsus [a] is of opinion that at the next pruning-season it is better to refrain from using the knife and that the shoots themselves should be twisted and wrapped round the tree in the shape of a crown, so

tortos arbori circumdari, ut flexura materias[1] profundat, quarum validissimam sequente anno caput vitis faciamus. 23 Me autem longus docuit usus, multo utilius esse primo quoque tempore falcem vitibus admovere, nec supervacuis[2] sarmentis pati silvescere. Sed eam quoque, quae primo submittetur, materiam ferro coercendam censeo usque in alteram vel tertiam gemmam, quo robustiores palmites agat, qui cum primum tabulatum apprehenderint, proxima putatione disponentur, omnibusque annis alioquin[3] in superius tabulatum excitabuntur,[4] relicta semper una materia, quae applicata trunco cacumen arboris spectet.

24 Iamque viti constitutae certa lex[5] ab agricolis imponitur: plerique ima tabulata materiis frequentant, uberiorem fructum et magis facilem cultum sequentes. At qui bonitati vini student, in summas arbores vitem promovent: ut quaeque materia se dabit,[6] ita in celsissimum quemque ramum extendunt, sic, ut summa vitis summam arborem sequatur, id est, ut duo palmites extremi trunco arboris applicentur, qui cacumen eius spectent, et prout quisque ramus convaluit, vitem accipiat.[7] 25 Plenioribus ramis plures palmites alius ab alio separati imponantur, gracilioribus pauciores; vitisque novella tribus toris ad arborem religetur, uno, qui est in crure

[1] materie *Sac*: materiae *A*.
[2] supervacuus *SAa*: supervacuis *c*.
[3] alioquin *SAac*: aliquis *edd.*
[4] excitabuntur *scripsi*: excitabitur *edd.*
[5] lex *om. S.*
[6] sedebit *S*: sed vetata *A*; sed evecta *a*; sed evetita *c*.
[7] accipit *SAac*.

that this bending-back may cause a profusion of firm-wood branches, the strongest of which we may make the head of the vine in the following year. But long experience has taught me that it is much 23 more expedient to apply the pruning-hook to the vines on the first possible opportunity and not allow them to become bushy with superfluous shoots. I also hold that the firm-wood branch which is to be allowed to grow at first, should be cut back with the knife as far as the second or third bud, so that it may put forth more vigorous shoots, which, when they have taken hold of the first " story " of the tree, will be trained in different directions at the next pruning, and furthermore will every year be raised to the story above, one firm-wood branch being always left which, applied to the trunk, will face towards the top of the tree.

Once the vine is set in its place a fixed rule is 24 applied to it by husbandmen. Most of them crowd the lower " stories " with firm-wood branches, their object being a more abundant yield of fruit and easier cultivation. But those whose chief object is high quality in the wine, encourage the vine to mount to the top of the trees, and, as each firm-wood shoot offers itself, they stretch it out to the highest possible branch in such a way that the top of the vine keeps pace with the top of the tree, that is, that the two furthest vine-shoots are applied to the trunk of the tree so that they face its top and, as each branch gathers strength, it takes up the burden of the vine. On the stouter branches more shoots should be placed, 25 separate from one another, but fewer on the slenderer branches, and the young vine should be attached to the tree with three bindings, one on the stem of the

arboris a terra quattuor pedibus distans;[1] altero, qui summa parte vitem capit; tertio, qui mediam vitem complectitur. Torum imum imponi non oportet, quoniam vires vitis. adimit. Interdum tamen necessarius habetur, cum aut arbor sine ramis truncata est, aut vitis praevalens in luxuriam evagatur.

26 Cetera putationis ratio talis est, ut veteres palmites, quibus proximi anni fructus pependit, omnes recidantur: novi, circumcisis undique capreolis et nepotibus, qui ex his nati sunt, amputatis, submittantur[2] et si laeta vitis est, ultimi potius palmites per cacumina ramorum praecipitentur; si gracilis, trunco proximi, si mediocris, medii; quoniam ultimus palmes plurimum fructum affert, proximus minimum vitemque exhaurit atque attenuat.

27 Maxime autem prodest vitibus, omnibus annis resolvi. Nam et commodius enodantur, et refrigerantur, cum alio loco alligatae sunt, minusque laeduntur, ac melius convalescunt. Atque ipsos palmites ita tabulatis superponi convenit, ut a tertia gemma vel quarta religati dependeant, eosque non 28 constringi, ne sarmentum vimine praecidatur. Quod si ita longe tabulatum est, ut[3] materia parum commode in id perduci possit, palmitem ipsum viti alligatum supra tertiam gemmam religabimus. Hoc ideo fieri praecipimus, quia quae pars palmitis prae-

[1] distant *SAac*.
[2] submittantur *Gesner*: committantur *SAac*.
[3] in *SAac*.

---

[a] *Cf.* Cato, *R.R.* 32, who warns his readers against this practice.

tree four feet from the ground, a second holding the vine at its top, and a third clasping it in the middle. A binding should not be placed at the bottom, since it takes away the strength of the vine; however, it is sometimes considered necessary when the tree has had its branches lopped off or when the vine, growing too strong, runs riot.

26 The other points to be observed in pruning are that the old shoots, upon which the fruit of the previous year has hung, should be all cut away, but the new ones should be allowed to grow after their tendrils have been cut back all round and the side-shoots which have grown from them have been lopped off—if the vine is in a flourishing state, the furthest shoots should be let down [a] through the top of the branches, if the vine is slender, the shoots nearest to the stock, and if it is of middling size, those in the middle. For the furthest shoot produces the most fruit, the nearest the least and exhausts and enfeebles the vine.

27 It is of great benefit to vines to unbind them every year; for they can then be more conveniently freed from knots and they are refreshed by being bound in another place and they are less damaged and recover strength better. Also it is expedient that the shoots themselves should be so placed upon the "stories" of the tree that they hang down, being attached at the third or fourth bud, and that they should not be bound too tightly, lest the vine-twig be cut by the osier. 28 But if the "story" is so far away that the firm-wood branch cannot conveniently be made to reach it, we shall bind the shoot itself to the vine, attaching it above the third bud. We give instructions that this should be done because it is the part of the shoot that is bent over which is clothed with

cipitata est, ea[1] fructu induitur: at quae vinculo adnexa[2] sursum tendit, ea materias sequenti anno 29 praebet. Sed ipsorum palmitum duo genera sunt: alterum, quod ex duro provenit, quod quia primo anno plerumque frondem sine fructu affert, pampinarium vocant; alterum, quod ex anniculo palmite procreatur: quod quia protinus creat, fructuarium appellant. Cuius ut semper habeamus copiam in vinea,[3] palmitum partes ad tres gemmas religandae sunt, ut quicquid intra vinculum est materias 30 exigat. Cum deinde annis et robore vitis convaluit, traduces in proximam quamque arborem mittendae, easque post biennium amputare atque alias teneriores transmittere convenit. Nam vetustate vitem fatigant. Nonnunquam etiam cum arborem totam vitis comprehendere nequit, ex usu fuit partem aliquam eius deflexam terrae immergere, et rursus ad eandem arborem duas vel tres propagines excitare, quo pluribus vitibus circumventa celerius vestiatur.

31 Viti novellae pampinarium immitti non oportet, nisi necessario loco natus est, ut viduum ramum maritet. Veteribus vitibus loco[4] nati palmites pampinarii utiles sunt, et plerique ad tertiam gemmam resecti optime submittuntur. Nam insequenti anno 32 materias fundunt. Quisquis autem pampinus loco natus in exputando vel alligando[5] fractus est, modo

[1] ex *SAac*. [2] abnexa *SAac*.
[3] inae *SA* : in ea *ac*.
[4] loco *a* : loca *SAc*.
[5] alligando *ac* : alligandi *SA*.

fruit, and it is the part which, being tied with a band, grows upwards that provides the firm-wood branches for the following year. There are two kinds of the fruit-bearing shoots themselves, one that comes out of the hard-wood of the vine, which, because in the first year it usually puts forth leaves but no fruit, is called a tendril-bearing shoot, and another which is produced from a one-year-old shoot and, because it bears fruit immediately, is called a fructuary shoot. In order that we may have plenty of shoots of this kind in our vineyard, the portions of the shoots up to three buds must be tied back, so that whatever is below the band may produce firm-wood. Then, afterwards, when the vine has increased in years and strength, the cross-branches must be conveyed to all the nearest trees and after two years must be cut away and others which are younger must be trained across; for when they grow old they wear out the vine. Sometimes too, when the vine cannot occupy the whole tree, it has been found useful to bend part of it down and sink it into the earth and raise two or three layers again into the same tree, so that it may be surrounded by several vines and so be more quickly covered. 29 30

A tendril-bearing shoot ought not to be allowed to grow on a young vine, unless it has grown in a place where it is required, so that it may be wedded to a branch which lacks a vine-shoot. Tendril-bearing shoots which grow in the right place on old vines are useful and are generally cut back to the third bud and allowed to grow with very good results; for in the following year they produce firm-wood in abundance. But if any tendril growing in the right place is broken in the process of pruning or tying, provided that it 31 32

ut aliquam gemmam habuerit, ex toto tolli non oportet, quoniam proximo anno vel validiorem 33 materiam ex una creabit.[1] Praecipites palmites dicuntur, qui de hornotinis [2] virgis enati in duro alligantur. Hi plurimum fructus afferunt, sed plurimum matri nocent. Itaque nisi extremis ramis, aut si vitis arboris cacumen superaverit, praecipitari palmitem 34 non oportet. Quod si tamen id genus colis propter fructum submittere quis velit, palmitem intorqueat. Deinde ita alliget et praecipitet. Nam et post eum locum quem intorseris, laetam materiam citabit, et praecipitata minus virium [3] in se trahet, quamvis fructu exuberet. Praecipitem vero plus anno pati non oportet.

35 Alterum [4] genus palmitis, quod de novello nascitur et in tenero alligatum dependet, materiam vocamus; ea et fructum et nova flagella bene procreat. Et iam si ex uno capite duae virgae submittantur, tamen utraque [5] materia dicitur; [6] nam pampinarius quam vim habeat, supra docui. Focaneus est, qui inter duo bracchia velut in furca de medio nascitur. Eum colem deterrimum esse comperi, quod neque fructum ferat, et utraque bracchia, inter quae natus est, attenuet. Itaque tollendus est.

36 Plerique vitem validam et luxuriosam falso crediderunt feraciorem fieri, si multis palmitibus submissis

[1] creabit *a*: creavit *SAc*.
[2] annotinis *codd.*
[3] vinum *SAac*.
[4] est *Sc*: et *Aa*: *om. Pontedera.*
[5] utramque *SAac*.
[6] deciditur *SA*: decitur *a*: decidunt *c*.

---

[a] See Book IV. 24. 10.

has some bud left, it should not be entirely removed, since in the following year it will produce an even stronger firm-wood branch from a single bud. Shoots [33] are called "precipitated" which, sprung from rods one year old, are tied to the hard wood. These bear fruit very freely but do much damage to the mother-vine; and so a shoot ought not to be "precipitated" except from the ends of the branches or if the vine has surmounted the top of the tree. If, however, any- [34] one wishes to let this kind of stem grow freely for the sake of the fruit, let him twist the shoot, and then tie it in that position and bend it over; for it will put forth flourishing firm-wood behind the point at which you have twisted it, and also, when it is bent over, it will attract less strength to itself, even though it bears an abundance of fruit. A shoot which has been bent over ought not to be allowed to continue so for more than one year.

Another kind of shoot which grows from a young [35] vine and hangs down tied to the tender part of the vine, we call firm-wood; it produces a good crop both of fruit and of new sprouts, and if two rods are allowed to grow from one head, both, nevertheless, are called firm-wood; for I have pointed out above what strength the leaf-bearing shoot possesses. The "throat-shoot" [a] is that which grows out of the middle between two branches, as it were in a fork. This I have found to be the worst kind of shoot, because it does not bear fruit and it weakens both of the branches between which it has grown. It must, therefore, be removed.

Most people have believed that a strong, luxuriant [36] vine becomes more fertile, if it is loaded with many shoots which are allowed to grow, but they are

oneretur. Nam ex pluribus virgis plures pampinos creat, et cum se multa fronde cooperit, peius defloret, nebulasque et rores [1] diutius continet, omnemque uvam perdit. Validam ergo vitem in ramos diducere censeo, et traducibus dispergere atque disrarare,[2] certosque vinearios coles praecipitare, et si minus [3] luxuriabitur, solutas materias relinquere; ea ratio vitem feraciorem faciet.[4]

37 Sed ut densum arbustum commendabile [5] fructu et decore est, sic ubi vetustate rarescit, pariter inutile et invenustum est. Quod ne fiat, diligentis patrisfamilias est, primam quamque arborem senio defectam tollere, et in eius locum novellam restituere,[6] nec eam viviradice frequentare,[7] ea etsi [8] sit facultas, sed,[9] quod est longe melius, ex proximo propagare. Cuius utriusque ratio consimilis est ei [10] quam tradidimus. Atque haec de Italico arbusto satis praecepimus.

VII. Est et alterum genus arbusti Gallici, quod vocatur rumpotinum. Id desiderat arborem humilem nec frondosam. Cui rei maxime videtur esse

[1] et rores *ac* : errores *SA*.
[2] dirrare *SA* : diradare *a* : durare *c*.
[3] nimis *SAac*.
[4] faceret *SAac*.
[5] commendabili *SAac*.
[6] *post* restituere *add.* vitem *SAac* : *om. Pontedera.*
[7] nec tam viviradice frequentare *Gesner* : que aut enectam viviradici frequenter *S* : quae aut nectam viviradici frequenter *A* : queat ut nectam viviradici frequenter *ac*.
[8] ut si *SAac*.
[9] sed *om. SAac*.
[10] ei *om. SAac*.

---

[a] The text here is quite doubtful.

wrong; for it produces more leaf-bearing shoots from its more numerous rods and, when it has covered itself with abundant foliage, it flowers less well and holds the fog and dew too long and loses all its clusters of grapes. I am, therefore, in favour of distributing a strong vine over the boughs of the supporting tree and spreading it in the form of cross-branches and thinning it out and bending over a certain number of grape-bearing shoots, and, if it is not luxuriant enough, leaving the firm-wood loose. This method will make the vine more productive.

37 Just as a dense plantation is commendable from the point of view of the fruit and for its fine appearance, so when it becomes thin through lapse of time it is equally unprofitable and ugly to look upon. To prevent this, it is the duty of a careful owner of property to remove every tree as soon as it becomes enfeebled by age and to plant a young tree in its place and not to crowd it round with quick-sets [a]—although there may be facilities for doing so—but, what is far better, to set layers from near at hand. In both cases the method is very similar to that which we have already set forth. We have now given enough instruction about Italian plantations.

Gaulish plantations of trees for supporting vines.

VII. There is another kind of plantation found in Gaul, which is called that of dwarf trees.[b] It requires a low and not very leafy tree, and the guelder-rose tree [c] seems

[b] This is derived from *rumpus* (Varro, *R.R.* I. 8. 4) meaning a "vine-branch" or "runner"—apparently the same as *tradux*—and *teneo*.

[c] *Viburnum opulus* is called the *cranberry-tree* or *high cranberry*, also *white dogwood*, *marsh-* or *water-elder*, or *gaiter-tree*.

idonea opulus[1]: ea est arbor corno[2] similis. Quin etiam cornus et carpinus et ornus non nunquam et salix a plerisque in hoc ipsum disponitur. Sed salix nisi aquosis locis, ubi aliae arbores difficiliter comprehendunt, ponenda non est, quia vini saporem infestat. Potest etiam ulmus sic disponi, ut adhuc[3] tenera decacuminetur, ne altitudinem quindecim 2 pedum excedat. Nam fere ita constitutum rumpotinetum animadverti, ut ad octo pedes locis siccis et clivosis, ad duodecim locis planis et uliginosis tabulata disponantur. Plerumque autem ea arbor in tres ramos dividitur, quibus singulis ab utraque parte complura bracchia submittuntur, tum omnes pene virgae, ne umbrent, eo tempore quo vitis putatur, abraduntur.

3 Arboribus rumpotinis, si[4] frumentum non inseritur, in utramque partem viginti pedum spatia interveniunt; at si segetibus indulgetur, in altera parte quadraginta pedes, in altera viginti relinquuntur. Cetera simili ratione atque in arbusto Italico administrantur, ut vites longis scrobibus deponantur, ut eadem diligentia curentur, atque in ramos diducantur,[5] ut novi[6] traduces omnibus annis inter se ex arboribus proximis committantur[7] et veteres deci-4 dantur. Si tradux traducem[8] non contingit, media virga inter eas deligetur. Cum deinde fructus pondere urgebit, subiectis adminiculis sustineatur. Hoc autem genus arbusti ceteraeque omnes arbores

[1] opulus *edd*: populus *SAac*.
[2] acerne *S*: cernae *A*: cerne *ac*.
[3] huc *SAac*. [4] Si *om. SAac*.
[5] deducatur *SAac*. [6] non novi *Aac*: non vita *S*.
[7] committantur *SAa*: commutantur *c*.
[8] ducem *Aac*: dulcem *S*.

[a] For dwarf planting, not for wet.

to be the most suitable for this purpose, a tree which closely resembles the cornel-tree. Indeed the cornel-tree, the horn-beam and sometimes the mountain-ash and the willow are planted by most people to this very end; but willows should not be planted except in watery places, where other trees take root with difficulty, because it spoils the flavour of the wine. The elm also can be adapted to this purpose [a] by having its top cut off while it is still young, so that it does not exceed the height of fifteen feet; for I have noticed that the plantation of dwarf trees is usually so ordered that the "stories" 2 are arranged at the height of eight feet in dry, sloping places, and twelve feet on flat, marshy ground. But usually this tree is divided up into three branches, upon each of which several arms are allowed to grow on both sides; then almost all the rods are pared off at the time when the vines are pruned, so that they may not cause a shade.

If no cereal is sown amongst the dwarf trees, spaces 3 of twenty feet are left on either side; but if one indulges in crops, forty feet are left on one side and twenty on the other. In all other respects operations are carried out on the same principle as in an Italian plantation, namely, that the vines are planted in long holes, that they may be looked after with the same care, and trained along the boughs of the trees, and the young cross-branches joined together every year from the nearest trees and the old ones cut off. If one cross-branch does not reach to another, it 4 should be connected by a rod running between them. When later the fruit bows the vine down with its weight, it should be supported by props put underneath it. This kind of plantation, just like all

quanto altius arantur et circumfodiuntur, maiore fructu exuberant; quod an expediat patrifamilias facere, reditus docet.

VIII. Omnis tamen arboris cultus simplicior quam vinearum est, longeque ex omnibus stirpibus minorem impensam desiderat olea, quae prima omnium 2 arborum est. Nam quamvis non continuis annis, sed fere altero quoque [1] fructum afferat, eximia tamen eius ratio est, quod levi cultu sustinetur, et cum se non induit, vix ullam impensam poscit. Sed et si quam recipit, subinde fructus multiplicat: neglecta compluribus annis non ut vinea deficit, eoque ipso tempore aliquid etiam interim patrifamilias praestat, et cum adhibita cultura est, uno anno emendatur. 3 Quare etiam nos in hoc genere arboris diligenter praecipere censuimus.

Olearum, sicut vitium, plura genera esse arbitror, sed in meam notitiam decem omnino pervenerunt: Posia,[2] Licinia,[3] Sergia,[4] Nevia,[5] Culminia,[6] Orchis,[7]

[1] quodque *SAac.* [2] posita *SAac.*
[3] licia *SAac.* [4] Sergia *SAac.*
[5] nevira *S* : nevi *Aac.*
[6] culminia *SA* : culmina *ac.*
[7] orces *SAa* : orches *c.*

[a] The MSS. readings of the names which follow have to be emended from the lists of olive-trees given by other authors, particularly Palladius (III. 18), who is obviously copying Columella. Whereas Columella says that he is going to give the names of *ten* kinds, *nine* only are named. To complete the number Schneider inserts *Algiana* as the second name, but he gives no indication of the source from which he derived this name. The meaningless *culi*, which in the MSS. precedes the last name, is possibly a corruption of the missing tenth name.

kinds of other trees, produces a greater abundance of fruit the deeper the ground is ploughed and dug round it; whether it pays the owner of the property to make it is shown by the profit which it returns.

The various kinds of olive-trees.

VIII. The cultivation of any kind of tree is simpler than that of the vine, and the olive-tree, the queen of all trees, requires the least expenditure of all. 2 For, although it does not bear fruit year after year but generally in alternate years, it is held in very high esteem because it is maintained by very light cultivation and, when it is not covered with fruit, it calls for scarcely any expenditure; also, if anything is expended upon it, it promptly multiplies its crop of fruit. If it is neglected for several years, it does not deteriorate like the vine, but even during this period it nevertheless yields something to the owner of the property and, when cultivation is again applied to it, it recovers in a single year. 3 We have, therefore, besides others thought it well to give careful instructions about this kind of tree.

I fancy that there are as many kinds of olive-trees as of vines, but ten in all have come under my notice:[a] the Posia,[b] the Licinian, the Sergian, the Nevian,[c] the Culminian,[d] the Orchis,[e] the Royal, the Shuttle,[f]

[b] *Posia*, or as it is sometimes spelt *Pausia* is called by Vergil (*Georg.* II. 86) *amara pausia bacca*: the derivation of the word is unknown.

[c] The Licinian, Sergian and Nevian olive-trees were called after the names of those who introduced them into Italy.

[d] The origin of this name is unknown: it is mentioned by Varro, *RR.*. I. 21. 1 and Pliny, *N.H.* XV. § 13.

[e] The Greek ὄρχις = Latin *testiculus*, and indicates the shape of the fruit.

[f] Also called *maiorina* from its great size (Pliny, *N.H.* XV. § 15). Gesner (*Index, s.v. RADIOLUS* and Vol. II. p. 1223) identifies *Cercitis* with *Radius* (below). Both words mean "shuttle".

4 Regia,[1] Cercitis, Murtea.[2] Ex quibus bacca iucundissima est Posiae,[3] speciosissima Regiae,[4] sed utraque potius escae, quam oleo est idonea. Posiae[5] tamen oleum saporis egregii, dum viride est, intra annum corrumpitur.[6] Orchis[7] quoque et Radius melius ad escam quam in liquorem stringitur. Oleum optimum Licinia dat, plurimum Sergia: omnisque olea maior fere ad escam, minor oleo est
5 aptior. Nulla ex his generibus, aut praefervidum, aut gelidum statum caeli patitur. Itaque aestuosis locis septentrionali colle, frigidis meridiano gaudet. Sed neque depressa loca neque ardua, magisque modicos clivos amat, quales in Italia Sabinorum vel tota provincia Baetica videmus. Hanc arborem plerique existimant ultra milliarium[8] centesimum[9] a mari aut non vivere aut non esse feracem.
6 Sed in quibusdam locis recte valet. Optime vapores sustinet Posia,[10] frigus Sergia.

Aptissimum genus terrae est oleis, cui glarea subest, si superposita creta sabulo admixta est. Non minus probabile est solum, ubi pinguis sabulo est. Sed et densior terra, si uvida et laeta est, commode recipit hanc arborem. Creta ex toto repudianda est, magis etiam scaturiginosa, et in qua semper uligo consistit. Inimicus est etiam ager sabulo macer, et

[1] regiona *SAac*.

[2] Cercitis mystea *edd.*: scrisis culi murtea *SA*: scrisis culimurtea *a*: scrisis culmurtea *c*.

[3] posiae *S*: positae *Aac*.

[4] regies *SAc*: reges *a*.

[5] posita *SAac*.

[6] intra annum corrumpitur *Codex Goesianus*: inani rumpitur *S*: imam rumpitur *A*: ima rumpitur *a*: una in rumpitur *c*.

[7] orceis *SA*: orces *ac*.

[8] milliarium *om. SAac*.

the Myrtle. Of these the berry of the Posia is the most agreeable, that of the Royal the showiest, and both are more suitable for eating than for oil. The oil from the Posia has an excellent flavour as long as it is green, but it goes bad within a year. The Orchis also and the Shuttle-olive are better gathered for eating than for their oil. The Licinian produces the best oil, the Sergian the most abundant, and, generally speaking, all the bigger olives are more suitable for eating, the smaller for oil. No olive-trees of these kinds can stand a very warm or a very cold climate; and so in very hot regions the olive-tree rejoices in the north side of a hill, in cool districts in the south side; but it does not like either low-lying or lofty situations but prefers moderate slopes such as we see in the Sabine territory in Italy and all over the province of Baetica.[a] Most people think that this tree either cannot live or is not productive more than a hundred miles from the sea, but in some places it thrives well. The Posia stands the heat best, the Sergian the cold.

4

5

6

The most suitable kind of ground for olive-trees is that which has gravel underneath, if chalk mixed with coarse sand forms the top-soil. Not less highly esteemed is ground where there is rich sand, but denser soil also is well adapted to receive this tree, if it is moist and fertile. Chalk must be wholly rejected, and even more land which abounds in springs and where ooze is always standing. Land which is lean because of sand is unfriendly to the olive-tree; so is

[a] Columella's native province in S.W. Spain.

[9] centesimum *S* : censimum *A* : sexagesimum *a* : LX *c*.
[10] postea *SA* : posita *ac*.

7 nuda glarea. Nam etsi non emoritur in eiusmodi solo, nunquam tamen convalescit. Potest tamen in agro frumentario seri, vel ubi arbutus, aut ilex steterant. Nam quercus etiam excisa radices noxias oliveto relinquit, quarum virus enecat oleam. Haec in universum de toto genere huius arboris habui dicere. Nunc per partes culturam eius exsequar.

IX. Seminarium [1] oliveto praeparetur caelo libero, terreno [2] modice valido et succoso, neque denso neque soluto solo, potius tamen resoluto; id genus fere terrae nigrae est. Quam cum in tres pedes pastinaveris, et alta fossa circumdederis, ne aditus [3] pecori detur,[4]
2 fermentari sinito; tum ramos [5] novellos proceros et nitidos, quos comprehensos manus possit circumvenire, hoc est manubrii crassitudine, feracissimis [6] arboribus adimito, et quam recentissimas [7] taleas recidito, ita ne corticem aut ullam aliam partem, quam qua [8] serra praeciderit, laedas. Hoc autem facile contingit, si prius varam feceris, et eam partem, supra quam ramum secaturus es, faeno aut stramentis texeris, ut molliter sine noxa corticis taleae super-
3 positae secentur. Taleae [9] deinde sesquipedales serra [10] praecidantur, atque earum plagae utraque parte falce leventur, et rubrica [11] notentur, ut sic quemadmodum in arbore steterat ramus, ita pars recte et cacumine caelum spectans deponatur. Nam

[1] seminario *SAac*.
[2] modo *SAac*.
[3] aditus *a* : traditus *SAc*.
[4] ater *SAac* : *post* ater *add.* inferetur *SAa* : infrecturi *c*.
[5] ramos *ac* : ramus *SA*.
[6] feracissimos *SAac*.
[7] recentissimos *c* : recentissimo *SAa*.
[8] -que *SAac*.
[9] tali *SAac*.
[10] terra *SAac*.
[11] rubrica *c* : lubrica *SAa*.

bare gravel: for, although it does not die in this kind of soil, yet it never acquires strength. It can, however, be planted on corn-land or where the strawberry-tree or holm-oak have stood; for the ordinary oak, even if it has been cut down, leaves behind roots harmful to the olive-grove, the poison from which kills the olive. So much for general remarks on this type of tree as a whole; I will now describe its cultivation in detail. 7

Nurseries for olive-trees.

IX. A nursery for your olive-grove should be prepared under the open sky on land which is moderately strong and juicy with soil which is neither dense nor loose but rather broken up. This kind of soil generally consists of black earth. When you have trenched it to the depth of three feet and surrounded it with a deep ditch, so that the cattle may have no access to it, allow the ground to loosen up. Then take from the most fruitful trees tall and flourishing young branches, such as the hand can grasp when it takes hold of them—that is to say of the thickness of a handle—and cut off from these the freshest slips in such a way as not to injure the bark or any other part except where the saw has made its cut. This is quite easy if you have first made a forked support and protect with hay or straw the part above which you are going to cut the branch, so that the slips which are placed in the fork may be severed gently without any damage to their bark. The slips then should be cut to the length of a foot and a half with the saw, and their wounds at each end smoothed with a pruning-knife and marked with ruddle, in order that the portion of the branch may be properly placed in the position which the branch had occupied on the tree, and with its top towards the sky; for, if it is 2 3

si inversa mergatur, difficulter comprehendet, et cum validius [1] convaluerit, sterilis in perpetuum erit.[2] Sed oportebit talearum [3] capita et imas partes mixto fimo cum cinere oblinire,[4] et ita totas eas immergere,[5] 4 ut putris terra digitis quattuor alte superveniat. Sed binis indicibus ex utraque parte muniantur: hi sunt de qualibet arbore [6] brevi [7] spatio iuxta eas positi, et inter se vinculo connexi, ne facile singuli deiciantur. Hoc facere utile est propter fossorum ignorantiam, ut cum bidentibus aut sarculis seminarium colere institueris, depositae [8] taleae [9] non laedantur.

5 Quidam melius existimant oculis excolere, et chorda [10] simili ratione disponere: sed utrumque debet post vernum aequinoctium seri, et quam frequentissime seminarium primo anno sarriri; postero et sequentibus, cum iam radiculae seminum convaluerint, rastris excoli. Sed biennio a putatione abstineri, tertio anno singulis [11] seminibus binos ramulos relinqui, et frequenter sarriri seminarium 6 convenit. Quarto anno ex duobus [12] ramis infirmior amputandus est. Sic excultae quinquennio arbusculae habiles [13] translationi sunt. Plantae autem

[1] validis *SAac*.
[2] esse *SAac*.
[3] palorum *SAac*.
[4] oblinire *ac*: oblinere *SA*.
[5] immergere *scripsi*: inmediri *SAc*: inmederi *a*: immergerei *Schneider*.
[6] arbore *a*: arbores *SAc*.
[7] brevi *ac*: breve *SA*.
[8] deposita *SAac*.
[9] et alere *Sac*: et alaerit *A*.
[10] et chorda *scripsi*: corde *SAac*.
[11] singuli *SAac*.
[12] duabus *SAa*: duobus *c*.
[13] stabiles *SAac*.

sunk into the ground in an inverted position, it will take root with difficulty and, when it has gained more strength, it will be barren for ever. You will have to smear the tops and lower ends of the slips with a mixture of dung and ashes and plunge them completely underground in such a way that there may be four inches of loose earth above them. But the slips should be provided with two marking-pegs, one on each side; these are of any kind of wood and are placed a little distance away from the slips and are tied together with a band, so that they may not easily be knocked over separately. It is expedient to do this because of the unobservance of the diggers, so that, when you start tilling your nursery with mattocks or hoes, the slips which you have planted may not be injured. 4

Some people think it better to cultivate olive-trees by means of buds and to arrange them by means of a cord on a similar principle;[a] but in either case the planting ought to take place after the spring equinox, and during the first year the nursery ought to be hoed over as often as possible. In the following and subsequent years, when the rootlets of the plants have gained strength, they should be cultivated with rakes; but for the first two years it is best to abstain from pruning, and in the third year two little branches should be left on each plant, and the nursery should be frequently hoed. In the fourth year the weaker of the two branches should be cut away. Thus cultivated the small trees are fit for transplantation in five years. In dry soil and 5 6

[a] The text here is apparently corrupt beyond emendation: the above is a translation of the reading of the MSS. with one slight change.

in oliveto disponuntur optime siccis minimeque uliginosis agris per autumnum, laetis et humidis 7 verno tempore, paulo antequam germinent. Atque ipsis scrobes quarternum pedum praeparantur anno ante, vel si tempus non largitur, priusquam deponantur arbores,[1] stramentis [2] atque virgis iniectis [3] incendantur scrobes, ut eos ignis putres faciat, quos sol et pruina [4] facere debuerat. Spatium intermedium esse debet [5] pingui et frumentario solo sexagenum pedum in alteram partem, atque in alteram quadragenum: macro nec idoneo segetibus, quinum [6] et vicenum [7] pedum. Sed in Favonium dirigi ordines convenit, ut aestivo perflatu refrigerentur.

8 Ipsae autem arbusculae hoc modo possunt transferri: antequam explantes arbusculam solo,[8] rubrica notato partem eius, quae meridiem spectat, ut eodem modo, quo in seminario erat,[9] deponatur. Deinde [10] arbusculae spatium pedale in circuitu relinquatur, atque ita cum suo caespite planta eruatur. Qui caespes in eximendo ne resolvatur,[11] modicos surculos [12] virgarum inter se connexos facere oportet, eosque pilae,[13] quae eximitur,[14] applicare, et viminibus ita

[1] deponantur arbores *a*: deponatur arbore *SAc*.

[2] stramentio *Pontedera*: sistam rectis *c*: sustam rectius *A*: si tam rectis *a*.

[3] atque virgis insectis *S*: *om Aac*.

[4] pruina *a*: ruina *SAc*.

[5] spatium intermedium esse debet pingui *Pontedera*: spatia ut vitis me deberit pingui *S*: spatium ut as me deberit pingui *A*: spatium minime debebit pingiu *a*: spatia vitis erit pungui *c*.

[6] quidam *SAa*: -em *c*.

[7] vicenum *Ursinus*: vicesimum *SA*: vigesimum *ac*.

[8] plantam pars arbuscula sole *SAac*.

[9] et id *SAac*.

where there is very little moisture the plants are best put out in the olive-grove during the autumn, but, where the soil is rich and damp, in the spring just before they come into bud. Four-foot plant-holes 7 are prepared for them a year earlier, or, if there is not an abundance of time before the trees are planted, let straw and twigs be thrown in and the plant-holes burnt, so that the fire may make them friable, as the sun and frost ought to have done. On ground which is rich and fit for growing corn the space between the rows ought to be sixty feet in one direction and forty in the other: if the soil is poor and not suitable for crops, twenty-five feet. But it is proper that the rows should be aligned towards the west, that they may be cooled by the summer-breeze blowing through them.

The small trees themselves may be transplanted 8 in the following manner. Before you pull up a little tree from the soil,[a] mark on it with ruddle the side of it which faces south, so that it may be planted in the same manner as in the nursery. Next let a space of one foot be left round the little tree in a circle and then let the plant be pulled up with its own turf, and that this turf may not be broken up in the process of removal, you must weave together moderate-sized twigs taken from rods and apply them to the lump of earth which is being removed and so bind it with

[a] The text here is quite uncertain, but the sense is obvious.

---

[10] *post* deinde *add.* ut *SAa* : aut *c.*
[11] solvatur *SAac.*
[12] modico surculos *SAac.*
[13] pilae quae *scripsi* : pila qua *S* : pila quae *Aac.*
[14] eximitur *c* : eximuntur *SAa.*

innectere, ut constricta terra[1] velut inclusa teneatur.
9 Tum subruta parte ima leviter pilam[2] commovere, et suppositis virgis alligare, atque plantam transferre. Quae antequam deponatur,[3] oportebit solum scrobis confodere[4] bidentibus: deinde terram aratro subactam, si tamen pinguior erit summa humus, immittere,[5] et ita seminibus substernere, et si[6] consistet[7] in scrobibus aqua, ea omnis haurienda est, antequam[8] demittantur arbores. Deinde ingerendi minuti lapides vel glarea mixta pingui solo, depositisque seminibus latera scrobis circumcidenda, et aliquid
10 stercoris interponendum. Quod si cum sua terra plantam[9] transferre[10] non convenit, tum optimum est omni fronde privare truncum, atque levatis plagis caenoque[11] et cinere oblitis,[12] in[13] scrobem vel sulcum deponere. Truncus autem aptior translationi est,[14] qui bracchii crassitudinem habet. Poterit etiam longe maioris incrementi et robustioris transferri. Quem ita convenit poni, ut, si non periculum a pecore habeat, exiguus admodum supra scrobem emineat: laetius enim frondet. Si tamen incursus pecoris aliter vitari non poterit, celsior[15] truncus constituetur, ut sit

[1] constrictae terrae *SAac* [2] pilam *SAac.*
[3] deponantur *SAac.*
[4] confodere *scripsi*: copia fodere *SAac.*
[5] mittere *SAac.*
[6] si *a*: sic *SAc.*
[7] constet *SAac.*
[8] nusquam *SAa*: nunquam *c.*
[9] plantam *SAac.*
[10] transferre *om. SAac.*
[11] caenoque *S*: cinoque *A*: acinoque *ac.*
[12] obrutis *SAac.* [13] in *om. SAac.*
[14] truncus autem aptior translationi est *Gesner*: truncos gratus autem maturis *SAac.*
[15] depressior *codex Goesianus.*

osiers that the soil, being pressed together, may be held as it were enclosed. Then having dug up the 9 lowest part, you must gently move the lump of earth and bind it to the rods put under it and transfer the plant. Before it is placed in the ground, you will have to dig up the soil in the plant-hole with hoes; then you should put in soil which has been broken up with the plough, provided that the top-soil shall be rather rich, and strew it with seeds underneath;[a] and, if there is any water standing in the plant-holes, it should all be drained away before the trees are put in. Next minute stones or gravel mixed with rich soil must be thrown in and, after seeds have been put in, the sides of the plant-hole must be pared away all round and some manure put in among them. If, however, it is not convenient to remove the plant 10 with its own earth, it is best to strip the stem of all its leaves and, after smoothing its wounds and daubing them with mud and ashes, place it in the plant-hole or furrow. A stem is quite ready for moving[b] which is as thick as a man's arm; one of much greater and stronger growth can also be transplanted, but it must be so placed if it is not in any danger from cattle, that only a little of it projects above the plant-hole; it then produces more luxuriant foliage. If, however, the attacks of cattle cannot be avoided in any other way, the stem will be planted so as to project further from the ground, so that it may

[a] Schneider, by a quotation from Palladius III. 18, who is there copying Columella, shows that it was customary to strew barley-seeds in the bottom of the hole in which a tree was about to be planted in order to cause fermentation; compare also (Aristotle) *Problems*, XX. 8, where it is said that barley-husks were sprinkled in the holes in which celery was to be planted.

[b] The reading here is uncertain.

11 innoxius ab iniuria pecorum. Atque etiam rigandae sunt plantae, cum siccitates incesserunt, nec nisi post biennium ferro tangendae.[1] Ac primo surculari debent, ita ut simplex stilus altitudinem maximi bovis[2] excedat; deinde arando ne[3] coxam bos, aliamve partem corporis offendat, optimum est etiam constitutas plantas circummunire[4] caveis.

Deinde constitutum iam et maturum olivetum in duas partes dividere, quae alternis annis fructu induantur. Neque enim olea continuo biennio uberat.[5] 12 Cum subiectus ager consitus non est, arbor[6] coliculum agit: cum seminibus repletur, fructum affert; ita sic divisum olivetum omnibus annis aequalem reditum adfert.[7] Sed id minime bis anno arari debet: et bidentibus alte circumfodiri. Nam post solstitium cum terra aestibus hiat, curandum est, ne 13 per rimas sol ad radices arborum[8] penetret. Post aequinoctium autumnale ita sunt arbores ablaqueandae, ut a superiore parte, si olea in clivo[9] sit, incilia[10] excitentur, quae ad codicem deducant aquam. Omnis deinde soboles, quae ex imo stirpe nata est, quotannis extirpanda est, ac tertio quoque fimo pabulandae sunt oleae. Atque eadem ratione stercorabitur olivetum, quam in secundo libro pro-

[1] tangeri de *SA* : tangi debeat *a* : tangi de hac *c*.
[2] bovis *Gesner* : scrobis *SAac*.
[3] ne *post* arando *om. SAac*.
[4] circumvenire *SAac*.
[5] eberat *S* : deberat *Aa* : debeat *c*.
[6] arbori *SAac*.
[7] ita sic—adfert *om. Sa*.
[8] arborem *SA*: arborum *ac*.
[9] clivoso *SAa*.
[10] incilia *S* : incilicia *Aac*.

be free from such injury by cattle. The plants must also be watered, when droughts occur, and they must not be touched with the knife unless two years have passed; and, firstly, they ought to be trimmed so that there is only a single stem which exceeds the height of the tallest ox; and, secondly, lest in ploughing an ox should hit it with his haunch or any other part of his body, it is best to protect the plants with fences, even plants that are established. 11

When the olive grove is established and has reached maturity, you must divide it into two parts, so that they may be clothed with fruit in alternate years; for the olive-tree does not produce an abundance two years in succession. When the ground underneath has not been sown with a crop, the tree is putting forth its shoots; when the ground is full of sown crop, the tree is bearing fruit; the olive-grove, therefore, being thus divided, gives an equal return every year. But it ought to be ploughed at least twice a year and dug deep all round the trees with hoes; for after the solstice, when the ground gapes open from the heat, care must be taken that the sun does not penetrate to the roots of the trees through the cracks. After the autumn equinox the trees ought to be trenched all round, so that, if the olive-grove is on a slope, ditches may be formed from the higher ground to convey water to the trunks of the trees. Next every shoot which springs from the lowest part of the stem must be removed each year, and every third year the olive-trees must be fed with dung. The olive-grove will be manured by the same method as that which I suggested in the second book,[a] if, 12 13

[a] Book II. 15. 1–3.

14 posui, si[1] tamen segetibus prospicietur. At si ipsis tantummodo arboribus, satisfacient[2] singulis stercoris caprini sex librae. vel[3] stercoris sicci modii singuli, vel amurcae insulsae congius[4] sufficient. Stercus autumno debet inici, ut permirtum hieme radices oleae calefaciat.[5] Amurca minus[6] valentibus infundenda est. Nam[7] per hiemem, si vermes atque alia suberunt animalia, hoc medicamento necantur.

15 Plerumque etiam locis siccis et humidis arbores musco infestantur. Quem nisi ferramento eam[8] raseris,[9] nec fructum nec laetam frondem[10] olea inducet. Quin etiam compluribus interpositis annis olivetum putandum est: nam veteris proverbii meminisse convenit, eum qui aret olivetum, rogare[11] fructum; qui stercoret, exorare; qui caedat, cogere. Quod tamen satis erit octavo anno fecisse, ne fructuarii rami subinde amputentur.

16 Solent etiam quamvis laetae arbores fructum non afferre. Eas terebrari gallica terebra convenit, atque ita in foramen[12] viridem taleam[13] oleastri arcte immitti.[14] Sic velut inita arbor fecundo semine

[1] si *om. SAac.*

[2] satisfaciant *edd.*: satis servari *S*: satis servaveri *A*: satis servaveris *a*: satis servaverimus *c.*

[3] vel *add. Schneider.*

[4] insulsae congius *Schneider*: in singulis condivis *SAc*: in singulis congiis *a.*

[5] calefacit *SAac.*

[6] minus *ex Palladio add. Schneider.*

[7] *Post* nam *add.* eius *SAac.*

[8] eam *scripsi*: ea *SAac.*

[9] reseris *SA*: resecaveris *ac.*

[10] laeta fronde *SAac.*

[11] rigare *SAac.*

that is, provision is going to be made for a crop of corn.[a] If you are providing only for the olive-trees 14 themselves, six pounds of goat's dung or a single *modius* of dry dung or a *congius* of unsalted lees of oil will suffice. The dung ought to be put in during the autumn, so that, being thoroughly mixed in, it may warm the roots of the olive in the winter. The lees of oil should be poured upon those trees which are not thriving very well; for during the winter, if worms and other creatures have got into them, they are killed by this treatment. Generally too in dry as well 15 as in moist places the trees are infested with moss, and unless you scrape it off with an iron instrument, the olive-tree will not put forth fruit or an abundance of leaves. Moreover, the olive-grove must be pruned at intervals of several years; for it is well to remember the old proverb "He who ploughs the olive-grove, asks it for fruit; he who manures it, begs for fruit; he who lops it, forces it to yield fruit." However, it will suffice to have pruned it every eighth year, so that the fruit-bearing branches may not be from time to time cut off.

It happens also frequently that, though the trees 16 are thriving well, they fail to bear fruit. It is a good plan to bore them with a Gallic auger and to put tightly into the hole a green slip taken from a wild olive-tree; the result is that the tree, being as it were impregnated with fruitful offspring, becomes more pro-

[a] *I.e.* if corn is being sown between the olive-trees.

[12] formem *SAac*.
[13] viridam talem *SA*: viridem talem *ac*.
[14] partem dimitti *Aac*: parte dimitti *S*.

fertilior extat. Sed[1] haec ablaqueatione adiuvanda est amurcaque insulsa cum suilla vel nostra urina[2] vetere, cuius utriusque modus servatur.[3] Nam maximae arbori, ut tantundem aquae misceatur,[4] urna abunde erit. Solent etiam vitio soli[5] fructum

17 oleae negare.[6] Cui rei sic medebimur. Altis gyris ablaqueabimus eas, deinde calcis pro magnitudine arboris plus minusve circumdabimus: sed minima arbor modium postulat. Hoc remedio si nihil fuerit effectum, ad praesidium insitionis confugiendum erit. Quemadmodum autem olea inserenda sit, postmodo dicemus. Non nunquam etiam in olea unus ramus ceteris aliquanto est laetior. Quem nisi recideris, tota arbor contristabitur.

Ac de olivetis hactenus dixisse satis est. Superest ratio[7] pomiferarum arborum, cui rei deinceps praecepta dabimus.

X. Modum pomarii, priusquam semina seras,[8] circummunire[9] maceriis vel saepe vel fossa praecipio,[10] nec solum pecori, sed et homini transitum negare, quoniam si saepius cacumina manu detracta aut a pecoribus praerosa sunt, in perpetuum semina in-

2 crementum capere nequeunt. Generatim autem dis-

[1] *Post* sed *add.* si *SAac.*
[2] nostra urina : natura *SAac.*
[3] servaturum *SAac.*
[4] misceatur *ac* : misatur *SA.*
[5] soli *a* : sol *SAc.*
[6] negare *S*: necare *Aac.*
[7] ratio *om. SAac.*

ductive. But it must also be assisted by being dug round and by unsalted lees of oil mixed with pigs' urine or stale human urine, a fixed quantity of each being observed; for a very large tree an *urn* will be fully enough, if the same quantity of water is mixed with it. Olive-trees also often refuse to bear fruit because of the badness of the soil. This we shall **17** remedy in the following manner. We shall dig deep trenches in circles round them and then put more or less lime round them according to the size of the tree, though the smallest tree requires a *modius*. If there is no result from this remedy, we shall have to have recourse to the assistance of grafting. How an olive-tree should be ingrafted we will describe hereafter. Sometimes also one branch of an olive-tree flourishes somewhat more than the rest and, unless you cut it back, the whole tree will languish.

This must suffice for our description of olive-groves. It remains to deal with the treatment of fruit-bearing trees, on which subject we will give instructions forthwith.

Pomiferous trees.

X. [a] Before you set the plants I advise you to protect the bounds of your orchard with walls or a fence or a ditch and to deny a passage not only to cattle but also to man, for if their tops are frequently pulled off by the hand of man or gnawed away by cattle, the plants are forever unable to reach their full growth. It is expedient to arrange the trees accord- **2**

[a] The rest of this book is slightly longer but almost identical with *de Arboribus*, Ch. 18 to the end.

---

[8] semina seras *a*: semiseras *SAc*.
[9] circumvenire *SAac*.
[10] praecipio *S*: praecipi *Ac*: praecipiti *a*.

ponere arbores utile est, maxime ne[1] etiam imbecilla a valentiore prematur, quia nec viribus nec magnitudine par est, imparique spatio temporis adolescit. Terra, quae vitibus apta est, etiam arboribus est utilis. Ante annum, quam seminare voles, scrobem fodies.[2] Ita sole pluviisve[3] macerabitur, et quod[4]
3 positum est cito comprehendet.[5] At si eodem anno et scrobem facere[6] et arbores serere properabis,[7] minime autem duos menses scrobes[8] fodito, postea stramentis incensis calefacito; quos si latiores patentioresque feceris,[9] laetiores uberioresque fructus
4 percipies. Sed scrobis clibano similis sit, imus summo[10] patentior, ut laxius radices vagentur ac minus frigoris hieme[11] minusque aestate vaporis per angustum os penetret,[12] etiam clivosis locis terra, quae in eum congesta est, a pluviis non abluatur.

5 Arbores raris intervallis serito, ut, cum creverint, spatium habeant, quo ramos extendant. Nam si spisse posueris, nec infra serere quid poteris, nec ipsae fructuosae erunt, nisi intervulseris: itaque inter ordines quadragenos pedes minimumque tri-
6 cenos relinquere convenit. Semina lege crassa non minus quam manubrium bidentis, recta, levia,

[1] ne *a* : *om. SAc.*
[2] fodies *SA* : fodi *c* : fodere *a* : fodito *edd.*
[3] pluviave *a* : pluviasve *c* : pluviasne *SA*.
[4] qua *c* : qua pedes *SAa*.
[5] comprehendet *S* : compedit *A* : competet *ac*.
[6] facere *om. SAac.*
[7] properabis *Brouckhusius* : proibis *S* : prohibis *A* : prohibes *a* : prohibe *c*.
[8] autem duos menses scrobes *add. edd. ex libro de Arboribus* 20, 1 : *om. SAac.*
[9] feceris *ac* : seris *SA*.
[10] imus summo *ex libro de Arboribus l.c. edd.* : humus summum *S* : humus summus *a* : umus summum *Ac*.

ing to their kinds, chiefly in order to prevent the weak from being overwhelmed by the stronger, because the former is not equal to the latter either in strength or in size and reaches maturity in a different period of time. Ground which is suitable for vines is also advantageous for trees. You will dig the plant-hole in which you wish to put a plant a year beforehand, for then it will be softened by the sun or the rain, and that which has been put into it will take root quickly. But if you are in a hurry to make the plant-hole and to set the plants in the same year, dig the plant-holes at least two months beforehand and afterwards warm the holes by burning straw in them. The broader and wider you make them, the more luxuriant and abundant will be the fruit which you will gather. Let your plant-hole be like an oven, wider at the bottom than at the top, so that the roots may spread more loosely, and less cold in winter and less heat in summer may penetrate through the narrow mouth, and also that on sloping ground the earth which is heaped up in it may not be washed away by rains. 3 4

Plant the trees at wide intervals, so that, when they have grown, they may have room to spread their branches. For if you set them thickly, neither will you be able to plant anything underneath them, nor will they be themselves fruitful unless you thin them out; and so it is well to leave forty or at least thirty feet between the rows. Choose plants at least as thick as the handle of a hoe and straight, smooth, 5 6

[11] hieme *om. SAac.*
[12] openetrum *SAc* : penetrum *a.*

procera, sine ulceribus, integro libro. Ea bene et celeriter comprehendent. Si ex arboribus [1] ramos [2] sumes,[3] de iis quae quotannis bonos et uberes fructus afferunt, eligito ab humeris [4] qui sunt contra solem orientem. Si cum radice plantam posueris, incrementum maius futurum quam ceteris senties. Arbor insita fructuosior est quam quae insita non est, id est, [5] quam quae ramis [6] aut plantis ponetur.[7]

7 Sed antequam arbusculas [8] transferas,[9] nota ventos quibus [10] antea fuerant constitutae, postea [11] manus adhibeto [12] ut de clivo et sicco in [13] humidum agrum transferas. Trifurcam [14] maxime ponito. Ea extet [15] minime tribus pedibus. Si eodem scrobe duas aut tres arbusculas voles [16] constituere,[17] curato ne inter [18]

8 se contingant,[19] nam ita vermibus interibunt.[20] Cum semina depones,[21] dextra sinistraque usque [22] in imum scrobem fasciculos [23] sarmentorum bracchii crassitudinis demittito, ita ut supra terram paulum extent,

[1] arboribus *Schneider ex libro de Arboribus* 20, 1 : veteribus *Aac* : veterius *S*.

[2] ramis *SAc* : rami *a*.

[3] sumes *SA* : summes *a* : sumus *c*.

[4] ab humeris *Schneider ex libro de Arboribus, l.c.* : ab illa *SAac*.

[5] orientem—id est *add. Schneider ex libro de Arboribus, l.c.* : *om. SAac*.

[6] ramis *a* : rimis *SAc*.

[7] ocnentur *S* : conentur *Aac*.

[8] arbuscula *SAac*.

[9] transferres *SAa* : transferes *c*.

[10] ventos quibus *scripsi* : viventis quibus *Sa* : viveras quibus *A* : vivenes(?) quibus *c*.

[11] ante erunt (runt *S*) constitui possit *SAac*.

[12] adhibeto *S* : adiuveto *Aac*.

[13] in *om. SAac*.

[14] trifurcam *ac* : trifurcamina *S* : trifurcam in *A*.

[15] extent *SAac*.

tall, free from excrescences and with sound bark. Such plants will take root well and quickly. If you take branches from trees, choose them from those which bear good and abundant fruit every year, taking them from the "shoulders" which face the rising sun. If you have set a plant with its root you will perceive that the growth will be quicker than in the other plants. A tree which is ingrafted is more fruitful than one which is not, that is, than one which is planted in the form of a branch or of a small plant. 7 But, before you transplant small trees, note what winds they had formerly faced, and afterwards get to work and transfer them from a sloping, dry position to moist soil. Preferably plant a tree which has three prongs, and let it project at least three feet from the ground. If you wish to put two or three small trees in the same trench, take care that they do not touch one another, since then they will be killed by worms. 8 When you set plants, lower right into the bottom of the trench, on the right and on the left hand side, bundles of twigs of the thickness of the arm in such a way that they project a little above the soil, so that in summer you may with little trouble

---

[16] voles *c* : volens *SAa*.
[17] constituere *SAac*.
[18] puter *SAac*.
[19] constringat *SAac*.
[20] nam ita vermibus interibunt *scripsi ex libro de Arboribus* 20, 2 : aut verbi ut interibunt *S* : aut verbi aut interibunt *Aac*.
[21] cum semina depones *ex libro de Arboribus l.c. Schneider* : depone *Aac* : depones *S*.
[22] usque *ac* : us *SA*.
[23] fasciculas *SA*.

per quos aestate parvo labore aquam radicibus subministres. Arbores ac semina cum radicibus autumno serito, hoc est circa idus Octob.; 9 taleas et ramos [1] primo vere,[2] antequam[3] germinent[4] arbores, deponito: ac ne tinea molesta sit seminibus ficulneis, in imum scrobem lentisci taleam inverso cacumine demittito.

Ficum frigoribus ne serito. Loca aprica, calculosa, glareosa, interdum et saxeta amat. Eiusmodi arbor cito convalescit, si scrobes amplos patentesque feceris. 10 Ficorum genera,[5] etsi sapore atque habitu distant, uno modo, sed pro differentia agri seruntur. Locis frigidis et autumni temporibus [6] aquosis praecoques ponito ut ante pluviam fructum deligas: locis calidis hibernas serito. At [7] si voles ficum quamvis non natura seram facere, tunc grossulos, prioremve fructum decutito,[8] ita alterum edet,[9] quem in hiemem [10] differet. Non nunquam etiam, cum frondere [11] coeperint arbores, cacumina fici ferro summa prodest amputare: sic firmiores arbores et feraciores sunt; ac semper conveniet, simulatque folia agere coeperit ficus, rubricam amurca [12] diluere, et cum stercore 11 humano ad radicem infundere. Ea res efficit uberiorem fructum, et farctum [13] fici pleniorem ac

[1] taleas et ramos *addidi ex libro de Arboribus*, 20, 3.
[2] primo vere *edd.*: removerere *S*: removeret *Aac.*
[3] antequam *edd.*: quã *S*: equam *A*: aquam *ac.*
[4] germinant *SAac.*
[5] fico generata *SAac.*
[6] temporis *SAac.*
[7] aut *SAac.*
[8] decutit *SAac.*
[9] det *SAac.*
[10] hieme *SAac.*
[11] frondere *Aac*: fronde *S.*
[12] amurgam *S.*

convey water through them to the roots. Set trees and seedlings with roots in autumn, that is, about October 15th, but plant cuttings and branches in 9 the early spring before the trees begin to bud; and, in order that the moth may not damage fig-tree seedlings, put in the bottom of the trench a slip from a mastic-tree with its top inverted.

Do not plant a fig-tree in cold weather. It likes sunny positions, where there are pebbles and gravel, and sometimes also rocky places. This kind of tree quickly gains strength if you make your trenches roomy and wide. The various kinds of fig-tree, al- 10 though they differ greatly in flavour and habit, are planted in the same manner, allowance being made for the difference of soil. In cold places and where the autumn season is wet, you should plant those whose fruits ripen early, so that you may gather the fruit before the rain comes; but plant winter figs in warm places. If, on the other hand, you wish to make a fig-tree bear late fruit, which it does not naturally do, shake down the unripe or early fruit, and it will then produce another crop which it will defer to the winter. Sometimes too, when the trees begin to bear leaves, it is beneficial to cut off the extreme tops of the fig-tree with a knife; the trees are then sturdier and more prolific. It will be always a good plan, as soon as the fig-tree begins to put forth leaves, to dissolve ruddle in lees of olive-oil and pour it together with human ordure over the roots. This makes the 11 fruit more abundant and the inner part of the fig fuller

[13] farctum *add. edd. ex libro de Arboribus* 21, 2: partum *SAac.*

meliorem. Serendae sunt autem praecipue Livianae,[1] Africanae, Chalcidicae, Fulcae, Lydiae, Callistruthiae,[2] Astropiae,[3] Rhodiae[4] Libycae, Tiburnae,[5] omnes etiam biferae et triferae flosculi.

12 Nucem Graecam serito circa cal. Febr., quia prima gemmascit: agrum durum, calidum, siccum desiderat. Nam in locis diversis nucem si deposueris, plerumque putrescit. Antequam nucem deponas, in aqua mulsa nec nimis dulci macerato. Ita iucundioris saporis fructum, cum adoleverit, 13 praebebit, et interim melius atque celerius frondebit. Ternas nuces in trigonum statuito, et nux a nuce minime palmo absit, et apex[6] ad Favonium spectet. Omnis autem nux unam radicem mittit, et simplici stilo prorepit. Cum ad scrobis solum radix pervenit, duritia humi coercita recurvatur, et ex se in modum ramorum alias radices emittit.

14 Nucem Graecam et Avellanam Tarentinam facere hoc modo poteris. In quo scrobe destinaveris nuces serere,[7] in eo terram minutam in modo[8] semipedis ponito, ibique semen ferulae repangito.[9] Cum ferula fuerit enata, eam findito, et in medulla eius sine

[1] libianae *S*.
[2] callistrustiae *S* : callistrustiae *A* : callistrusitae *c* : callistrusneae *a*.
[3] astopiae *SA* : asthopie *c* : stopie *a*.
[4] rhodie *ac* : rohiae *SA*.
[5] Tiburnae *scripsi* : tybernae *Aa* : tiberne *S* : thiberne *c*.
[6] apex *scripsi* : anceps *SAac*.
[7] necesse rere *SA* : nec esse serere *a*.
[8] in modum *S* : pro modum *Aac*.
[9] repangito *A* : pangito *S* : repaginato *c* : *om. a*.

[a] Pliny, *N.H.* XV. § 70. It is said to have been called after Livia, the wife of Augustus.
[b] Pliny, *N.H.* XIV. § 69 : called after Chalcis in Euboea.

and better. You should chiefly plant the Livian,[a] African, Chalcidian,[b] Fulcan,[c] Lydian, Callistruthian,[d] Astropian,[e] Rhodian, Libyan and Tiburnian[f] fig-trees, also all those which bear a floweret twice or three times a year.

You should plant the almond-tree, since it is the first tree to put out buds, about February 1st. It requires hard, warm, dry ground; for if you plant a nut in places which have different qualities from these, it generally rots. Before you put the nut in the ground, soak it in honey-water, which should not be too sweet; it will then, when it comes to maturity, produce fruit of a pleasanter flavour, and meanwhile its foliage will grow better and quicker. Place three nuts so as to form a triangle and let them be at least a hand's breadth away from one another, and let one apex of the triangle face towards the West. Every nut sends out one root and creeps out of the ground with a single stem. When the root has reached the bottom of the planting-hole, it is checked by the hardness of the soil and bent back and puts forth from itself other roots like the branches of a tree. 12 13

You will be able to make an almond and a filbert into a Tarentine nut in the following manner. In the planting-hole in which you intend to sow the nuts place fine soil to a depth of half a foot and set in it a fennel-root. When the fennel has grown up, split it and secrete in the pith of it an almond or a 14

[c] This kind is not otherwise mentioned and the name is perhaps corrupt.

[d] Book X. line 416: so called because sparrows (στρουθοί) were fond of it. It was also called *passeraria*.

[e] This kind is not otherwise mentioned and the name is perhaps corrupt.

[f] From Tibur in Latium, the modern Tivoli.

putamine nucem Graecam aut Avellanam abscondito, et ita adobruito. Hoc ante calend. Martias facito, vel etiam inter nonas et idus Mart. Eodem tempore iuglandem et pineam et castaneam serere oportet.

15 Malum Punicum vere[1] usque in cal. Aprilis recte seritur. Quod si acidum aut minus dulcem fructum feret, hoc modo emendabitur. Stercore suillo et humano urinaque vetere radices rigato. Ea res et fertilem arborem reddet, et primis annis fructum vinosum; post quinquennium dulcem, et apyrenum[2] facit. Nos exiguum admodum laseris vino diluimus, et ita cacumina arboris summa oblevimus. Ea res

16 emendavit acorem malorum. Mala Punica ne in arbore hient,[3] remedio sunt[4] lapides tres, si, cum seres[5] arborem, ad radicem ipsam collocaveris.[6] At si iam arborem satam[7] habueris, scillam secundum radicem arboris serito. Alio modo, cum iam matura mala fuerint, antequam rumpantur, ramulos, quibus dependent, intorqueto. Eodem modo servabuntur incorrupta etiam toto anno.

17 Pyrum autumno ante brumam serito, ita ut minime dies xxv ad brumam[8] supersint. Quae ut sit ferax, cum adoleverit, alte eam ablaqueato, et iuxta ipsam radicem truncum findito, et in[9] fissuram cuneum[10] tedae pineae adicito, et ibi relinquito: deinde adobruta ablaqueatione cinerem supra terram

18 inicito. Curandum est autem, ut quam generosis-

[1] vere *edd. ex libro de Arboribus* 23, 1: habere *SAac.*
[2] aprinum *SAac.*
[3] hient *a*: ient *SA*: ventre medio *c.*
[4] erit *SAac.*
[5] seres *Aa*: res *Sc.*
[6] collocaveris *S*: colueris *Aac.*
[7] sitam *SAac.*
[8] ad brumam *ac*: a brumam *A*: abruma *S.*

filbert without its shell, and then cover it over with earth. Do this before March 1st or between March 7th and 15th. You should at the same time plant the walnut, the pinenut and the chestnut.

15 It is correct to plant the pomegranate in the spring up to April 1st. But if it bears fruit which is bitter and not sweet, this will be remedied by the following method: moisten the roots with sow-dung and human ordure and stale urine. This will both render the tree fertile and during the first years cause the fruit to have a vinous taste; after five years it makes it sweet and its kernels soft. We ourselves have mixed just a little juice of alexanders with wine and smeared the uppermost tops of the tree. This has remedied the tartness of the fruit. 16 To prevent pomegranates from bursting on the tree, the remedy is to place three stones at the very root of the tree when you plant it; if, however, you have already planted it sow a squill near the root of the tree. According to another method, when the fruit is already ripe and before it bursts, you should twist the little boughs on which it hangs. By the same method the fruit will keep without decaying for a whole year.

17 Plant the pear-tree in the autumn before winter comes, so that at least twenty-five days remain before mid-winter. In order that the tree may be fruitful when it has come to maturity, trench deeply round it and split the trunk close to the very root and into the fissure insert a wedge of pitch-pine and leave it there; then, when the loosened soil has been filled in, throw ashes over the ground. 18 We must take care to plant our orchards with the most excellent

---

[9] in *om. SAac.* [10] cuneo *SAac.*

simis pyris pomaria conseramus. Ea sunt Crustumina, regia, Signina, Tarentina, quae Syria dicuntur, purpurea, superba, hordeacea, Aniciana, Naeviana, Favoniana, Lateritana, Dolabelliana, Turraniana, volaema, mulsa,[1] praecocia, venerea, et quaedam alia, 19 quorum enumeratio nunc longa est. Praeterea malorum genera exquirenda maxime Scaudiana,[2] Matiana, orbiculata, Cestina, Pelusiana,[3] Amerina, Syrica, melimela, Cydonia: quorum genera tria sunt, struthia,[4] chrysomelina, mustea. Quae omnia non solum voluptatem, sed etiam salubritatem afferunt. Sorbi quoque et Armeniaci atque Persici non minima est gratia. Mala, sorba, pruna, post mediam 20 hiemem [5] usque in idus Feb. serito. Mororum [6] ab idib. Feb. usque ad aequinoctium vernum satio est.

[1] mulsa *ac*: mulsia *SA*.
[2] Scaudiana *scripsi*: Scaidianam *S*: Gaudiana *Aac*.
[3] pedusiana *SAac*.
[4] struthia *Aac*: struti *S*.
[5] hiemem *edd. ex libro de Arboribus* 25, 1: essem *SA*: messem *ac*.
[6] mororum *edd. ex libro de Arboribus l.c.*: malorum *SA*.

[a] From Crustumium in Etruria.
[b] From Signia in Latium.
[c] So called, according to Pliny (*N.H.* XV. § 55) because they are ripe at the time of the barley-harvest.
[d] So called from the person who introduced it (Pliny, *l.c.*; Cato, *R.R.*, VII. 3).
[e] Probably called after a member of the *gens Naevia*, who perhaps also introduced the Naevian olive (Book XII. 50. 1).
[f] Called after M. Favonius, an imitator of Cato (Cicero, *Att.* I. 14. 5).
[g] From Laterium near Arpinum, where Q. Cicero had a villa (Cicero, *Att.* XI. 1).

pear-trees that we can find. These are the Crustuminian,[a] the Royal, the Signine,[b] the Tarentine, which are called Syrian, the Purple, the Superb, the Barley-pear,[c] the Anician,[d] the Naevian,[e] the Favonian,[f] the Lateritan,[g] the Dolabellian,[h] the Turranian,[i] the Warden-pear,[j] the Honey-pear, the Early-ripe, the Venus-pear and certain others, which it is a long task to enumerate now. Moreover, the 19 following kinds of apple should be especially sought after, the Scaudian,[k] the Matian,[l] the Globe-apple, the Cestine,[m] the Pelusian,[n] the Amerian,[o] the Syrian, the Honey-apple and the Cydonian[p] (of which there are three kinds, the Sparrow-apple, the Golden apple and the Must-apple[q]). All these cause not only pleasure but also good health. Service-apples also and apricots and peaches have no small charm. You should plant apple-trees, service-trees and plum trees after the middle of winter and until February 13th. The time for planting mul- 20 berries is from February 13th to the spring equinox.

[h] Called after an unknown member of the Dolabella family.

[i] Called after D. Turranius Niger, the friend of Varro (Varro, *R.R.* II. Introd. 6).

[j] Vergil, *Georg.* 88. Servius derives the name from *vola* and says it means "hand-filler."

[k] Called after a certain Scaudius (Pliny, *N.H.* XV. § 49).

[l] Called after C. Matius, a favourite of Augustus (Book XII. 46. 1).

[m] Called after a certain Cestius (Pliny, *loc. cit.*).

[n] From Pelusium in north Egypt.

[o] From Ameria, a town of Umbria.

[p] *Malum Cydonium* is the quince. Cydonia is a town in Crete.

[q] So called according to Pliny (*N.H.* XV. § 51) because it ripens quickly.

Siliquam Graecam, quam quidam κεράτιον vocant, et Persicum ante brumam per autumnum serito. Amygdala, si parum ferax erit, forata arbore lapidem adicito, et ita librum arboris inolescere sinito.

21 Omnium autem generum ramos [1] circa cal. Martias in hortis subacta [2] et stercorata terra super pulvinos arearum disponere convenit. Danda est opera, ut dum teneros ramulos habent, veluti pampinentur, et ad unum stilum primo anno semina redigantur. Et cum autumnus incesserit, ante quam frigus cacumina

22 adurat, omnia folia decerpere expedit, et ita crassis arundinibus, quae ab una parte nodos integros habeant, velut pileis [3] induere, atque a frigore et gelicidiis teneras adhuc [4] virgas tueri. Post viginti quattuor deinde menses sive transferre et disponere in ordinem voles, sive inserere, satis tuto utrumque [5] facere poteris.[6]

XI. Sed omnis surculus omni arbori inseri potest, si non est ei, cui inseritur, cortice dissimilis. Si vero etiam similem fructum et eodem tempore affert, sine scrupulo egregie inseritur. Tria genera porro insitionum antiqui tradiderunt. Unum, quo resecta et fissa arbor resectos [7] surculos accipit. Alterum, quo

[1] ramos *S* : ramis *A* : ramus *ac*.

[2] subacta *edd. ex libro de Arboribus l.c.* : in hostis tritta *SA* : trita *a* : truta *c*.

[3] pileis *edd.* : tiliae sic *SAc* : taliae sic *a*.

[4] adhuc *edd. ex libro de Arboribus l.c.* : adit ut *SAc* : adit et *a*.

[5] utrumque *edd. ex libro de Arboribus l.c.* : utrius *SAa* : utriusque *c*.

[6] poteris *om. SAac*.

[7] resectos *Ac* : sextos *S* : reseptos *a*.

The carob-tree, which some people called *Ceration*,[a] and the peach-tree you should plant during the autumn before winter comes. If an almond is not productive enough, make a hole in the tree and drive in a stone and so allow the bark of the tree to grow over.

21 It is proper to plant out the branches of all kinds of fruit trees about March 1st in gardens on raised beds after the soil has been well worked and manured. Care must be taken to trim them while the little branches are young and tender and in the first year the seedlings should be reduced to a single stem. When autumn has come on, before the cold nips the tops, it is well to strip off all the foliage and to cover

22 the trees with caps, as it were, of thick reeds which have their knots intact on one side, and thus protect the still tender rods from cold and frosts. Then after twenty-four months you will be able quite safely to do whichever you wish of two things—either to transplant and arrange them in rows or else to engraft them.

The grafting of fruit trees.

XI. Any kind of scion can be grafted on any tree, if it is not dissimilar in respect of bark to the tree in which it is grafted; indeed if it also bears similar fruit and at the same season, it can perfectly well be grafted without any scruple. Further, the ancients have handed down to us three kinds of grafting; one in which the tree, which has been cut and cleft, receives the scions which have been cut; the second, in which the tree having been cut admits

[a] κεράτιον, which is found in the same sense as here in an inscription at Abydos (*O.G.I.*, 5. 21. 27), is used in *Luke* XV. 16 of the "husks" eaten by the Prodigal Son. The name is no doubt due to the shape of carob-nuts, which Pliny (*N.H.* XV. § 95) describes as "sometimes curved like a sickle."

resecta inter librum et materiam semina admittit. Quae utraque genera veris temporis sunt. Tertium, quo ipsas gemmas cum exiguo cortice in partem sui delibratam recipit, quam vocant agricolae emplastrationem; vel, ut quidam, inoculationem. Hoc genus 2 insitionis aestivo tempore optime usurpatur. Quarum insitionum rationem cum tradiderimus, a nobis repertam quoque docebimus.

Omnes arbores simulatque gemmas agere coeperint,[1] luna crescente inserito; olivam autem circa 3 aequinoctium vernum usque in idus Aprilis. Ex qua arbore inserere voles, et surculos ad insitionem sumes, videto ut sit tenera et ferax nodisque crebris: et cum primum germina tumebunt,[2] de ramulis anniculis, qui solis ortum spectabunt, et integri erunt, eos legito crassitudine digiti minimi. Surculi sint bifurci vel trifurci. Arborem, quam inserere voles, serra diligenter exsecato[3] ea parte, qua maxime nitida et sine cicatrice est:[4] dabisque operam, ne librum 4 laedas. Cum deinde truncum recideris, acuto ferramento plagam levato. Deinde quasi cuneum tenuem ferreum vel osseum inter corticem et materiam ne minus digitos tres, sed considerate, demittito, ne laedas aut rumpas corticem. Postea surculos quos inserere voles falce acuta ex ima parte deradito

[1] coeperit *SAa* : coepit *c*.
[2] tumibunt *SAac*.
[3] excato *S* : exsecato *Aa* : excecato *c*.
[4] et *S*.

[a] So called from the plaster of clay or wax used in this method.

grafts between the bark and the hard wood (both these methods belong to the season of spring); and the third, when the tree receives actual buds with a little bark into a part of it which has been stripped of the bark. The last kind the husbandmen call emplastration[a] or, according to some, inoculation.[b] This type of grafting is best employed in the summer. When we have imparted the method of these graftings, we will also set forth another which we have discovered. 2

You should engraft all other trees as soon as they begin to put forth buds and when the moon is waxing, but the olive-tree about the spring equinox and until April 13th. See that the tree from which you intend to graft and are going to take scions for insertion is young and fruitful and has frequent knots and, as soon as the buds begin to swell, choose from among the small branches which are a year old those which face the sun's rising and are sound and have the thickness of the little finger. The scions should have two or three points. You should cut the tree into which you wish to insert the scion carefully with a saw in the part which is most healthy and free from scars, and you will take care not to damage the bark. Then, when you have cut away part of the trunk, smooth over the wound with a sharp iron instrument; then put a kind of thin wedge of iron or bone between the bark and the firm-wood to a depth of not less than three inches, but do so carefully so as not to damage or break the bark. Afterwards with a sharp pruning-knife pare down the scions which you wish to insert, at their bottom end 3 4

[b] Because an "eye" or bud is taken from one tree and inserted in another.

tantum, quantum cuneus demissus [1] spatii dabit, atque ita, ne medullam [2] neve alterius partis corticem
5 laedas. Ubi surculos [3] paratos habueris,[4] cuneum vellito, statimque surculos dimittito [5] in ea foramina,[6] quae cuneo adacto inter corticem et materiam feceris. Ea autem fine, qua adraseris, surculos sic inserito, ut semipede [7] vel amplius de arbore extent. In una arbore duos, vel si truncus vastior est, plures calamos recte inseres, dum ne minus quattuor digitorum inter eos sit spatium. Pro arboris magnitudine et
6 corticis bonitate haec facito. Cum omnes surculos, quos arbor ea patietur, demiseris, libro ulmi vel iunco aut vimine arborem constringito: postea paleato luto bene subacto oblinito totam plagam, et spatium quod est inter surculos, usque eo dum [8] minime quattuor digitis insita extent.[9] Supra deinde muscum [10] imponito, et ita ligato, ne pluvia dilabatur. Quosdam tamen magis delectat in trunco arboris locum seminibus serra facere, insectasque partes
7 tenui scalpello levare, atque ita surculos aptare. Si pusillam arborem inserere [11] voles, imam abscindito, ita ut sesquipede e terra [12] extet. Cum deinde praecideris, plagam diligenter levato: et medium truncum acuto scalpello modice findito, ita ut fissura digitorum

[1] demissis *c* : dimissus *SAa.*
[2] medullis *SAac.*
[3] *post* surculos *add. Aac* dimittito.
[4] straveris *SAac.*
[5] dimittito *addidi ex libro de Arboribus l.c.*
[6] foramina *edd. ex libro de Arboribus l.c.* : forma *SAac.*
[7] semipedem *SAac.*
[8] usque ad eodem *S.*
[9] in una—§ 6, insita extent *S* : om. *Aac.*
[10] muscum *edd. ex libro de Arboribus* 26, 6 : ramuscula *Sac*: ramicula(?) *A.*

to such a size as will fill the space given by a wedge which has been thrust in, in such a way as not to damage the cambium or the bark on the other side. When you have got the scions ready, pull out the wedge and immediately push down the scions into the holes which you made by driving in the wedge between the bark and the firm-wood. Put in the scions by inserting the end where you have pared them down in such a way that they stand out half-a-foot or more from the tree. You will be correct in inserting two grafts in one tree, or more if the trunk is larger, provided that the space between them is not less than four inches. In doing so take into account the size of the tree and the quality of the bark. When you have put in all the scions that the tree will stand, bind the tree with elm-bark or reeds or osiers; next with well-worked clay mixed with straw daub the whole of the wound and the space between the grafts to the point at which the scions still project at least four inches. Then put moss over the clay and bind it on so that the rain may not seep through. Some people, however, prefer to make a place for the slips in the trunk of the tree with a saw and then smooth the parts in which cuts have been made with a thin surgical-knife and then fit in the grafts. If the tree which you wish to engraft is small, cut it off low down so that it projects a foot and a half from the ground; then, after cutting it down, carefully smooth the wound and split the stock in the middle a little way with a sharp knife,

5

6

7

[11] serere *SAac*.

[12] sesquipedam e terra *A* : sequipedamen terra *S* : sexquipedam e terra *a* : sexquipedem e terra *c*.

trium sit in ea. Deinde cuneum, quo[1] diducatur, inserito, et surculos ex utraque parte derasos demittito, sic ut librum seminis libro arboris aequalem facias. 8 Cum surculos diligenter aptaveris, cuneum eximito, et arborem, ut supra dixi, alligato: deinde terram circa arborem adaggerato usque ad ipsum insitum. Ea res a vento et calore maxime tuebitur.

Nos tertium genus insitionis invenimus, quod[2] cum sit subtilissimum, non omni generi arborum[3] idoneum est, sed fere recipiunt talem insitionem, quae humidum succosumque et validum librum habent, sicut ficus. 9 Nam et lactis plurimum mittit, et corticem robustum habet. Optime itaque[4] inseritur tali ratione.[5] Ex arbore, de qua inserere voles, novellos et nitidos ramos eligito, in iisdemque quaerito[6] gemmam, quae bene apparebit, certamque[7] spem germinis habebit: eam duobus digitis quadratis circumsignato, ut gemma media sit: et ita acuto scalpello circumcisam diligenter, ne eam laedas, delibrato. 10 Item alterius arboris, quam emplastraturus es, nitidissimum ramum eligito, et eiusdem spatii corticem circumcidito, et materiam delibrato. Deinde in eam partem, quam nudaveris, praepara-

[1] quo *Aac*: quod(?) *S*.
[2] invenimus quod *add. edd.*: *om. SAac.*
[3] generi arboris *a*: generi arborum *c*: generiem arbori *A*: geriem arbori *S*.
[4] ea *add. S*.
[5] tali ratione *scripsi ex libro de Arboribus* 26, 7: carifici ratione *S*: caprifici raneus *Ac*: caprifici ramos *a*.
[6] querito *SA*: serito *ac*.

so that there is a cleft of three inches in it. Then insert a wedge by which the cleft may be kept open, and thrust down into it scions which have been pared away on both sides, in such a way as to make the bark of the scion exactly meet the bark of the tree. When 8 you have carefully fitted in the scions, pull out the wedge and bind the tree in the manner described above; then heap the earth round the tree right up to the graft. This will give the best protection from wind and heat.

A third kind of grafting is our own invention; being a very delicate operation, it is not suited to every kind of tree. Generally speaking those trees admit of this kind of grafting which have moist, juicy and strong bark, like the fig-tree; for this both yields 9 a great abundance of milk and has a stout bark, and so a graft can be very successfully inserted by the following method. On the tree from which you wish to take your grafts, you should seek out young and healthy branches, and you should look out on them for a bud which has a good appearance and gives sure promise of producing a sprout. Make a mark round it enclosing two square inches, so that the bud is in the middle, and then make an incision all round it with a sharp knife and remove the bark carefully so as not to damage the bud. Also choose 10 the healthiest branch of the other tree, which you are going to inoculate, and cut out a part of the bark of the same dimensions as before and strip the bark off the firm-wood. Then fit the scutcheon which you have prepared to the part which you have

[7] certamque *edd. ex libro de Arboribus* 26, 8: certaminis *SAac.*

tum emplastrum aptato,[1] ita ut alterius [2] delibratae
11 parti conveniat. Ubi ita haec feceris, circa gemmam bene alligato, cavetoque ne laedas ipsum germen. Deinde commissuras et vincula luto oblinito, spatio relicto, ut gemma libera vinculo non urgeatur. Arboris autem insitae sobolem et ramos superiores praecidito, ne quid sit, quo [3] possit succus [4] avocari,[5] aut ne cui [6] magis quam insito serviat. Post XXI diem solvito emplastrum. Et hoc genere optime etiam olea inseritur.

12 Quartum illud genus insitionis iam docuimus, cum de vitibus disputavimus. Itaque supervacuum est hoc loco repetere traditam rationem terebrationis.

Sed cum antiqui negaverint posse omne genus surculorum in omnem arborem inseri, et ex illa quasi finitione, qua nos ante paulo usi sumus, veluti quandam [7] legem sanxerint, eos tantum [8] surculos posse coalescere, qui sint cortice ac libro et fructu consimiles iis arboribus, quibus inseruntur, existimavimus errorem huius opinionis discutiendum, tradendamque posteris, rationem, qua possit omne genus surculi
13 omni generi [9] arboris inseri. Quo ne longiore exordio legentem fatigemus, unum quasi exemplum subiciemus, quo possit omne genus surculi dissimilibus [10] arboribus inseri.

[1] aptato *Sc* : apto *Aa*.

[2] alterius *scripsi* : altere *Aac* : alte *S*.

[3] quo *edd. ex libro de Arboribus* 26, 9 : quod *SAac*.

[4] succus *add. edd. ex libro de Arboribus l.c.* : *om. SAac*.

[5] avocari *S* : vocari *Aac*.

[6] ne cui *edd. ex libro de Arboribus l.c.* : necuim *SA* : nec humi *a* : ne vim *c*.

[7] veluti quandam *edd. ex libro de Arboribus l.c.* : vel antequam *SAa* : ut antiquum *c*.

[8] tantum *c* : tantos *SAa*.

bared, so that it exactly corresponds to the area on the other tree from which the bark has been stripped. Having done this, bind the bud well all round and be careful not to damage the sprout itself. Then daub the joints of the wound and the ties round them with mud, leaving a space, so that the bud may be free and not be constricted by the binding. Cut away the shoot and upper branches of the tree into which you have inserted the graft, so that there may be nothing to which the sap can be drawn off or benefit from the sap to another part rather than the graft. After the twenty-first day unbind the scutcheon. This kind of grafting is very successful with the olive also. 11

The fourth method of grafting we have already explained when we treated of vines; so it is superfluous to repeat here the method of "terebration" already described.[a] 12

But since the ancients denied that *any* kind of scion could be grafted on *any* kind of tree and, according to the limitation which we made use of just now,[b] established as a hard and fast rule that only those scions can unite which resemble the trees in which they are inserted in bark and rind and fruit, we have thought it advisable to destroy this erroneous opinion and to hand down to posterity a method by which any kind of scion can be grafted upon any kind of tree. That we may not weary the reader with too long a discourse, we will submit a single example by following which any kind of scion can be grafted upon a different kind of tree. 13

[a] IV. 29, 13: V. 9. 16. [b] § 1 of this chapter.

[9] omni generi *c* : omni genere *A* : omne genere *S* : *om. a.*
[10] dissimilibus *scripsi* : dissimilis *S* : dimissis *Aac.*

Scrobem[1] quoquoversus pedum[2] quattuor ab arbore olivae tam longe fodito, ut extremi rami oleae possint eam contingere.[3] In[4] scrobem deinde fici arbusculam deponito, diligentiamque adhibeto, ut 14 robusta et nitida fiat. Post biennium, cum iam satis amplum incrementum ceperit, ramum olivae, qui videtur nitidissimus, deflecte, et ad crus arboris ficulneae religa: atque ita amputatis ceteris ramis, ea tantum cacumina,[5] quae inserere voles,[6] relinque; tum arborem fici detrunca, plagamque leva, et me- 15 diam cuneo finde. Cacumina deinde olivae, sicuti matri cohaerent, ex utraque parte adrade, et ita fissurae fici insere, cuneumque exime, diligenterque ramulos colliga, ne qua vi revellantur.[7] Sic interposito triennio coalescet ficus cum olea, et tum demum quarto anno, cum bene[8] coierint,[9] velut propagines, ramulos olivae a matre resecabis. Hoc modo omne genus in omnem arborem inseres. At prius quam finem[10] libri faciamus, quoniam fere species[11] surculorum omnes persequimur prioribus[12] libris, de cytiso praecipere[13] nunc parvum[14] ac tempestivum est.[15, 16]

[1] scrobe *SAac.*

[2] pedes *SAac.*

[3] possit ea contingere *S*: positae contingere *ac*: positae tangere *A*.

[4] in *om. SAac.*

[5] cacumine *ac*: cacumineris *SA*.

[6] quae inserere voles *add. edd. ex libro de Arboribus* 27. 3: *om. SAac.*

[7] revellantur *ac*: revellatur *SA*.

[8] bene *edd. ex libro de Arboribus* 27. 4: ene *S*: eno *Ac*: eo *a*.

[9] coierit *SAac.*

[10] finem *Aac*: fine *S*.

[11] species *a*: specimus *S*: spes *c*.

Dig a trench measuring four feet each way at such a distance from an olive-tree that the ends of the branches can reach it. Then plant a small fig-tree in the trench, and be careful that it grows strong and healthy. After two years, when it has made enough 14 growth, bend down the branch of the olive-tree which seems to be the healthiest and bind it to the stock of the fig-tree. Then lop off the rest of the branches and leave only the tops which you wish to engraft; then cut through the trunk of the fig-tree and smooth off the wound and split it in the middle with a wedge. Then pare the tops of the olive-tree, 15 still adhering to the mother-tree, on both sides, and then insert[a] them in the cleft in the fig-tree, and take away the wedge and carefully tie the little branches so that no force may tear them away. Then after an interval of three years the fig-tree will coalesce with the olive-tree, and finally, in the fourth year, when they have become properly united, you will cut off the little olive branches from the mother-tree, just as if they were layers. This is the way in which you will graft any kind of scion on any kind of tree. But before we make an end of this book, since in the earlier books we treat of almost every kind of small tree, I regard it as a brief and opportune task to give instructions about the shrub-trefoil.[b]

[a] By bending them over, not cutting them off.
[b] The text here is doubtful: one MS. seems to contain two sets of words expressing the same thing.

---

[12] prioribus *a* : priores *SAc*. [13] incipere *SAac*.
[14] parvum *Aac* : pravum *S*. [15] puto *A* : fuit *Sa* : *om. c*.
[16] *post* puto *add.* nunc (hunc *c*) arboris praecipientes opportune eius meminerimus *SAac*.

XII. Cytisum in agro esse quam plurimum maxime refert, quod gallinis, apibus, ovibus, capris, bubus quoque et omni generi pecudum utilissimus est: quod ex eo cito pinguescit, et lactis plurimum praebet ovibus,[1] tum [2] etiam [3] octo mensibus viridi eo pabulo uti et postea arido possis. Praeterea in quolibet agro quamvis macerrimo celeriter comprehendit:
2 omnem iniuriam sine noxa patitur. Mulieres quidem si lactis inopia premuntur, cytisum aridum in aqua macerare oportet, et cum tota nocte permaduerit, postero die expressi succi ternas heminas permiscere modico vino, atque ita potandum dare: sic et ipsae [4] valebunt, ct pucri abundantia lactis confirmabuntur. Satio autem cytisi vel autumno circa idus Octobris, vel vere fieri potest.

3 Cum terram bene subegeris, areolas facito, ibique velut ocimi semen cytisi autumno serito. Plantas deinde vere disponito, ita ut inter se quoquoversus quattuor pedum spatio distent. Si semen non habueris, cacumina cytisorum vere deponito, et ster-
4 coratam terram circumaggerato. Si pluvia non incesserit, rigato quindecim proximis diebus: simulatque novam frondem agere coeperit, sarrito, et post triennium deinde caedito, et pecori praebeto. Equo [5] abunde est viridis pondo xv, bubus pondo vicena, ceterisque pecoribus pro portione virium.[6] Potest [7]

[1] ovibus *add. edd. ex libro de Arboribus* 28. 1 : *om. SAac.*
[2] cum *SAc*: tum *a*.
[3] ovis *post* etiam *add. SAc*: iovis quod *a*.
[4] ipsa *SAa* : ipse *c*.
[5] equo *edd.* : aeque *SA* : eque *ac*.
[6] virium *Aac* : virum *S*. [7] potest *Ac* : potes *S* : *om. a*.

[a] Presumably at one feeding.

XII. It is very important to have as much shrub-trefoil as possible on your land, because it is most useful for chickens, bees, sheep, goats, oxen and cattle of every kind, which quickly grow fat upon it and it makes ewes yield a very large quantity of milk; moreover you could also use it for eight months of the year as green fodder and afterwards as dry. Furthermore, on any ground whatsoever, even if it be very lean, it quickly takes root, and it bears any ill-treatment without taking harm. Indeed if women suffer from lack of milk, dry shrub-trefoil ought to be steeped in water and, after it has soaked for a whole night, on the following day three *heminae* of the juice squeezed out of it should be mixed with a little wine and given them to drink; in this way they themselves will enjoy good health, and the children will grow strong on the abundance of milk provided for them. Shrub-trefoil can be sown either in the autumn about October 15th or in the spring.

Of the Cytisus or Shrub-Trefoil.

2

When you have worked the soil thoroughly, make little beds and in the autumn sow there the seed of the shrub-trefoil as you would that of basil. Then in the spring set out the plants so that they are distant four feet each way from one another. If you have no seed, plant out tops of shrub-trefoil in the spring and heap well-manured soil round them. If rain has not come on, water them on the fifteen following days. As soon as a plant begins to put forth young foliage, hoe the ground. Then after three years cut down the plants and give them to the cattle. Fifteen pounds of shrub-trefoil when it is green is quite enough [a] for a horse, and twenty pounds for an ox, and it should be given to the other animals according to their strength. Shrub-trefoil can also

3

4

etiam circa sepem agri satis commode ramis cytisus
seri, quoniam facile comprehendit et iniuriam sustinet. Aridum si dabis, parcius praebeto, quoniam
vires maiores habet, priusque aqua macerato, et
5 exemptum paleis permisceto. Cytisum cum aridum
facere voles, circa mensem Septembrem, ubi semen
eius grandescere incipiet, caedito, paucisque horis,
dum flaccescat, in sole habeto: deinde in umbra ex-
siccato, et ita condito.

Hactenus de arboribus praecepisse abunde est,
redditur o pecoris curam et remedia sequenti volumine.

be quite conveniently propagated by planting boughs round the fence of a field, since it easily takes root and stands up to rough usage. If you give it dry, give it rather sparingly, since it has more strength, and soak it first in water and after taking it out of the water, mix it with chaff. When you wish 5 to dry it, cut shrub-trefoil about the month of September, when its seed begins to grow large, and keep it in the sun for a few hours until it withers; then dry it in the shade and store it.

In what has gone before I have given ample instruction about trees; in the next book I intend to deal with the care of cattle and the remedies for their diseases.

BOOK VI

LIBER VI

PRAEFATIO

Scio quosdam, Publi Silvine, prudentes agricolas pecoris abnuisse curam, gregariorumque pastorum velut inimicam suae professionis disciplinam constantissime repudiasse. Neque infitior id eos aliqua 1 ratione fecisse, quasi [1] sit agricolae contrarium pastoris propositum: cum ille quam maxime subacto et puro solo gaudeat, hic novali graminosoque; ille fructum e terra speret, hic e pecore; ideoque arator abominetur, at contra pastor optet herbarum proventum. 2 Sed in his tam [2] discordantibus votis est tamen quaedam societas atque coniunctio: quoniam et pabulum e [3] fundo plerumque domesticis pecudibus magis quam alienis depascere ex usu est, et [4] copiosa stercoratione, quae contingit e gregibus, terrestres 3 fructus exuberant. Nec tamen ulla regio est, in qua modo frumenta gignantur, quae non ut hominum [5] ita armentorum adiutorio colatur.[6] Unde etiam iumenta et armenta nomina a re [7] traxere, quod

---

[1] qua *SA*¹. [2] in his tam *R*: in ista *SA*¹.

[3] ex *Lundström*: et *SA*¹: e *A*²: est *c*.

[4] ex usu est et *Schneider*: exueste *S*: exuestet *AR*.

[5] hominum *S*: omnium *AR*.

[6] adiutorio colatur *Schneider*: adiuratorio colatur *S* adiurator inculator *A*: adiurator inculatur *A*².

[7] nomina a re *Lundström*: nominare *S*: nom̄ are *A*.

BOOK VI

PREFACE

I am well aware, Publius Silvinus, that there are some intelligent farmers who have refused to keep cattle and have consistently rejected the pursuit of the master of a flock as harmful to their profession. I do not deny that they have some reason for so doing on the ground that the aim of the farmer is contrary to that of the shepherd, since the former rejoices in land which is tilled and cleared to the greatest possible extent, while the latter takes pleasure in ground which is fallow and grassy; the one hopes for the fruits of the earth, the other for the produce of his cattle, and so the cultivator detests while on the other hand the grazier longs for a rich yield of grass. But, in spite of these irreconcilable 2 desires, there exists a sort of alliance and union between them, because, firstly, it is generally better to use the food provided by one's own farm in feeding one's own cattle rather than those of other people, and, secondly, because it is owing to the plentiful use of manure, which is derived from flocks, that the fruits of the earth abound. Nor indeed is there any 3 region in which nothing but cereals is grown and which is not cultivated quite as much by the aid of cattle as of men. Hence also draught-animals (*iumenta*) and animals which draw the plough (*armenta*)

nostrum laborem, vel onera subvectando [1] vel arando iuvarent.

Itaque sicut veteres Romani praeceperunt, ipse quoque censeo tam pecorum quam agrorum cultum 4 pernoscere. Nam in rusticatione vel antiquissima est ratio pascendi eademque [2] quaestuosissima, propter quod nomina quoque pecuniae et peculii tracta videntur a pecore: quoniam et solum id [3] veteres possederunt, et adhuc apud quasdam gentes unum hoc usurpatur divitiarum genus: sed [4] ne apud nostros quidem colonos alia res uberior.[5] Ut etiam M. Cato prodidit,[6] qui consulenti, quam partem rei rusticae exercendo celeriter locupletari posset [7] respondit: Si bene is pasceret; [8] rursusque interroganti, quid deinde faciendo satis uberes fructus percepturus esset, affirmavit: Si mediocriter pasceret. 5 Ceterum de tam sapiente viro piget dicere, quod eum quidam auctores memorant eidem quaerenti, quidnam [9] tertium in agricolatione quaestuosum esset, asseverasse, si quis vel male pasceret; cum praesertim maius dispendium sequatur inertem et inscium [10] pastorem, quam prudentem [11] diligentemque compendium.[12] De secundo tamen responso dubium

[1] subvectando *R* : subiectando *S* : subtectando *A*.
[2] eandemque *SA*.
[3] id *Lundström* : in *SA*.
[4] sed *S* : et *AR*.
[5] re superior *SAR*.
[6] prodidit *S* : reddidit *AR*.
[7] posset *S* : possit *AR*.
[8] is pasceret *Lundström* : ipsasciret *S* : ipsas geret *A*.
[9] quidam *SA*.
[10] scium *SAR*.
[11] prudenti *SAR*.
[12] conpendium S^2 : conprendium S^1A.

[a] The author here derived *iumentum* from *iuvare*, to aid, and *armentum* from *arare*, to plough. In the latter case the

derive their names from the fact that they aid our labour either by carrying burdens or by ploughing.[a]

Therefore, as the ancient Romans taught, I myself am also of the opinion that we should thoroughly understand the management of cattle as well as the cultivation of the fields. For in the history of farming the system of grazing is certainly very ancient and at the same time very profitable, and it is on this account also that the names for money (*pecunia*) and private property (*peculium*) seem to have been derived from the word for cattle (*pecus*), because this was the only possession which the men of old time had, and, even at the present day, amongst some peoples, this is the only kind of wealth in general use, and even among our farmers there is nothing which yields a richer increase. This was the opinion of Marcus Cato among others, who, when someone seeking advice asked him what department of agriculture he should practise in order to get rich quickly, replied that he would get rich if he were a competent grazier. When the same person went on to ask him what is the second best thing to do in order to obtain a sufficiently rich return, Cato insisted that he could achieve this by being a moderately good grazier. I feel some hesitation in relating about so wise a man the reply, which some authors attribute to him, when the same person enquired what was the third most lucrative practice in agriculture, namely, for a man to be even a bad grazier; since certainly the losses which attend a lazy and ignorant grazier are greater than the profits which attend one who is prudent and careful. As for Cato's second answer, there is no doubt that the

4

5

derivation is correct, but *iumentum* is derived from *iugum*, a yoke.

non est, quin mediocrem negligentiam domini fructus pecoris exsuperet.

6 Quam ob causam nos hanc quoque partem rei rusticae, Silvine, quanta valuimus industria, maiorum secuti[1] praecepta posteritati mandavimus. Igitur cum sint duo genera quadrupedum, quorum alterum paramus in consortium operum, sicut bovem, mulam, equum, asinum; alterum voluptatis ac reditus et custodiae causa, ut ovem, capellam, suem, canem:[2] de eo genere primum dicemus, cuius usus[3] 7 nostri laboris est particeps. Nec dubium, quin, ut ait Varro, ceteras pecudes bos honore superare debeat, praesertim in Italia, quae ab hoc nuncupationem traxisse creditur, quod olim Graeci tauros italos vocabant,[4] et in ea urbe, cuius moenibus condendis mas[5] et femina boves aratro terminum signaverunt, vel, ut antiquiora repetam,[6] quod item Atticis[7] Athenis Cereris et Triptolemi fertur minister: quod inter fulgentissima sidera particeps caeli sit:[8] quod deinde laboriosissimus adhuc hominis socius in agricultura: cuius tanta fuit apud antiquos[9] veneratio, ut tam capitale esset[10] bovem necuisse, quam civem. Ab hoc igitur promissi operis capiamus exordium.

[1] secuti *S*: sicuti *AR*.

[2] tamen *SAR*.

[3] usum *SAR*.

[4] vocant *SAR*.

[5] condendis mas *S*: condendissimas *A*.

[6] vel, ut antiquiora repetam *Lundström ex cit. Mulomedicina Chironis* (*Ed. Oder*): petam *SA*.

[7] atticus *SAR*.

[8] caelis *SA*: celi *R*.

[9] aputanti quis S^1: apud antiquis A^1: apud antiquos S^2A^2.

[10] capitales set *S*: capitales et *AR*.

profit from cattle more than makes up for a moderate amount of carelessness on the part of their owner.[a]

6 It is on this account, Silvinus, that, following the precepts of our forefathers, we have taken all the pains which we can to hand on to posterity an account of this department of agriculture also. There are, then, two classes of fourfooted animals, one of which we procure to share our labours, such as the ox, the mule, the horse and the ass, and the other which we keep for our pleasure and the profit which they bring us or for keeping watch, such as the sheep, the goat, the pig and the dog. We will deal first with the class which we employ to take part in our work. 7 There is no doubt, as Varro says, that the ox ought to be ranked above all other cattle, especially in Italy, which is believed to have derived its name from this animal, which the Greeks formerly called *italos*,[b] and in that city[c] at the founding of whose walls an ox and a cow drew the plough which marked its boundaries; also because, to go still further back, at Athens in Attica the ox too is said to have been the attendant of Ceres (Demeter) and Triptolemus, and because it has its place in the heavens, among the most brilliant constellations, and, lastly, because it is still man's most hardworking associate in agriculture, and so great was the respect in which it was held among the ancients that it was equally a capital crime to have killed an ox and to have killed a fellow-citizen. Let us, therefore, begin the task before us with the ox.

[a] Cicero, *de Off.* II, § 89, gives a fourth way of getting rich, by tilling the soil.
[b] Or, more usually, *vitulus*, calf. [c] *I.e.* Rome.

I. Quae in emendis bubus sequenda quaeque vitanda sint, non ex facili dixerim; quoniam pecudes pro regionis caelique statu et habitum [1] corporis et ingenium animi et pili colorem gerant. Aliae formae sunt Asiaticis, aliae Gallicis, Epiroticis aliae. Nec tantum diversitas provinciarum, sed ipsa quoque Italia partibus suis discrepat. Campania plerumque boves progenerat albos et exiles, labori tamen 2 et culturae patrii soli non inhabiles. Umbria vastos et albos; eademque rubios; [2] nec minus probabiles animis quam corporibus. Etruria et Latium compactos, sed ad opera fortes. Apenninus durissimos omnemque difficultatem tolerantes, nec ab aspectu decoros. Quae cum tam varia et diversa [3] sint, tamen quaedam quasi communia et certa praecepta in emendis iuvencis arator sequi debet; eaque Mago Carthaginiensis ita prodidit, ut nos deinceps memo-3 rabimus. Parandi sunt boves novelli, quadrati, grandibus [4] membris, cornibus proceris ac nigrantibus et robustis, fronte lata et crispa, hirtis auribus, oculis et labris nigris, naribus resimis patulisque, cervice longa et torosa, palearibus amplis et pene ad genua promissis, pectore magno, armis vastis, [5] capaci et tanquam implente utero, lateribus [6] porrectis, lumbis latis, dorso recto planoque vel etiam subsidente, [7]

[1] habitum *S* : habitu *AR*.
[2] rubios *A* : rabios *S*[1] : robios *S*[2] : rubeos *a*.
[3] diversa *a* : versa *SAR*.
[4] grandis *SA*.
[5] vasti *SAR*.
[6] lateribus *S*[2]*R* : latibus *S*[1]*A*.
[7] susidente *S*[2] : subidente *S*[1]*AR*.

---

[a] His work on agriculture was translated into Latin by order of the senate (I. 1. 13; Varro *R.R.* I. 1. 10: Cicero, *Or.* I. 58, 249).

I. I should find it far from easy to say what are the points to be looked for and what to be avoided in the purchase of oxen; for cattle show variation in bodily form and disposition and the colour of their hair according to the nature of the district and climate in which they live. Those of Asia and of Gaul and of Epirus are different in form, and not only are there diversities in the various provinces, but Italy itself shows varieties in its different parts. Campania generally produces small, white oxen, which are, however, well suited for their work and for the cultivation of their native soil. Umbria breeds huge white oxen, but it also produces red oxen, esteemed not less for their spirit than for their bodily strength. Etruria and Latium breed oxen which are thick-set but powerful as workers. The oxen bred in the Apennines are very tough and able to endure every kind of hardship but not comely to look upon. Though there is so much variety and diversity, yet there are certain as it were universal and fixed principles which the farmer of arable land ought to follow in buying bullocks. Mago[a] the Carthaginian has laid down these principles in the form which we will now detail. The bullocks which should be purchased are those which are young, squarely built, with large limbs and horns which are long and blackish and strong; the forehead should be wide and covered with curly hair, the ears shaggy, the eyes and lips dark in colour, the nostrils bent back and wide spreading, the neck long and muscular, the dewlap ample and falling almost to the knees, the chest broad, the shoulders huge; the belly should be capacious and have the appearance of pregnancy, the flanks extended, the loins wide, the back straight

Points to be looked for in oxen.

2

3

clunibus rotundis, cruribus compactis ac rectis, sed brevioribus potius quam longis, nec genibus improbis, ungulis[1] magnis, caudis longissimis et setosis, pilo totius[2] corporis denso brevique, coloris rubii vel fusci, tactu corporis mollissimo.

II. Talis notae vitulos oportet, cum adhuc teneri sunt, consuescere manu tractari, ad praesepia religari, ut exiguus in domitura labor eorum et minus sit periculi. Verum nec ante tertium neque post quintum annum iuvencos domari placet, quoniam illa aetas adhuc tenera est, haec iam praedura. Eos autem, qui de grege feri comprehenduntur, sic subigi convenit. 2 Primum omnium spatiosum stabulum praeparetur, ubi domitor facile versari, et unde egredi sine periculo possit. Ante stabulum nullae angustiae sint, sed aut campus aut via late patens, ut cum producentur[3] iuvenci, liberum habeant excursum; ne pavidi aut arboribus aut obiacenti cuilibet rei se implicent[4] noxamque capiant. 3 In stabulo sint ampla praesepia, superque transversi asseres in modum iugorum a terra septem pedibus elati configantur, ad quos religari possint iuvenci. Diem deinde, quo domituram auspiceris, bonum a tempestatibus et a religionibus matutinum eligito, cannabinisque funibus cornua iuvencorum ligato. 4 Sed laquei, quibus capulantur, lanatis pellibus involuti sint, ne tenerae frontes[5] sub cornua laedantur.

[1] ungulis *S* : vinculis *Aa*.
[2] pilo totius *Pontedera ex cit. Palladii* : pilosius *SA*.
[3] producentur *SA²a* : producerentur *A¹R*.
[4] implicent *Sa* : impluent *c* : inplicent *AS²* : inplicet *S¹*.
[5] tenere frontet *S¹A¹* : frontes *S²* : tenera fronte *R*.

and flat or even sinking slightly, the buttocks round, the legs compact and straight but short rather than long and the knees not ill-shaped, the hoofs large, the tail very long and bristly, the hair all over the body thick and short and of a red or brindle colour and the body very soft to the touch.

How to train oxen.

II. Calves of such a strain, you must accustom, while they are still young, to allow themselves to be handled and fastened to their mangers, so that there may be little trouble and less danger in breaking them in. The general opinion is that bullocks should not be broken in before their third or after their fifth year, since the former age is as yet too tender and the latter too hard. Those which are taken wild from the herd ought to be tamed in the following manner. 2 First of all a spacious shed should be got ready, where the trainer may be able to move about easily and from which he can withdraw without danger. There should be no narrow spaces in front of the shed but either open country or a wide road, so that, when the bullocks are driven forth, they may have room to escape and that they may not, in their alarm, become entangled in trees or anything else which gets in their way and hurt themselves. 3 In the shed there should be roomy stalls, and overhead horizontal beams should be fixed shaped like yokes, raised seven feet above the ground to which the bullocks can be tied. Then, to inaugurate the training, choose the morning of a day which is free from storms and not the occasion of any religious ceremony and fasten the horns of the bullocks with hempen cords. 4 The nooses with which they are caught should be wrapped round with woolly skins, so that the tender part of the forehead

Cum deinde buculos comprehenderis, perducito ad stabulum, et ad stipites religato ita, ut exiguum laxamenti habeant, distentque inter se aliquanto spatio, ne in colluctatione alter alteri noceat. Si nimis asperi erunt, patere unum diem noctemque desaeviant. Simulatque iras contuderint,[1] mane producantur, ita ut et a tergo complures, qui sequuntur, retinaculis eos contineant, et unus cum clava salignea procedens modicis ictibus subinde impetus eorum coerceat.

5 Sin autem placidi et quieti boves erunt, vel eodem die, quo alligaveris, ante vesperum licebit producere, et docere per mille passus composite [2] ac sine pavore ambulare: cum domum reduxeris,[3] arcte ad stipites religato, ita ne capite moveri possint. Tum demum ad alligatos boves neque a posteriore parte neque a latere, sed adversus, placide et cum quadam vocis adulatione venito, ut accedentem consuescant aspicere. Deinde nares perfricato, ut hominem discant 6 odorari. Mox etiam convenit tota tergora et tractare et respergere mero, quo familiariores bubulco fiant: uteris quoque et sub femina manum subicere, ne ad eius modi tactum postmodum pavescant, et ut ricini [4]

[1] contulerint *SA*.
[2] composite *R*: conposita *SA*.
[3] preduxeris *S*[1]: pro- *AR*.
[4] riclini *SA*.

below the horns may not be injured. Then when you have captured the steers, you should lead them to the shed and attach them to the posts in such a way that their ropes give very little play and that they are a little distance apart from one another, so that they may not hurt each other in their struggles. If they are too savage, allow them a day and a night to expend their fury, and as soon as the edge of their anger is blunted, they should be driven forth early in the morning, care being taken that several persons follow them behind also and hold them back by their tethers while one man, going in front of them with a club of willow wood in his hand, from time to time restrains their onrush with light blows.

If, however, the cattle are placid and quiet, it will 5 be possible for you to drive them out even before the evening of the day on which you have tied them up and train them to walk for a thousand paces in an orderly manner and without fear. When you have conducted them home again, you should bind them very closely to the posts, so that they cannot move their heads. Then is the time to approach the oxen, when they are tied, not from behind or from the side but from straight in front, quietly and by using a soothing tone of voice, in order that they may become accustomed to see you approaching them, and next rub their noses so that they may learn to know a man by his odour. Soon after this it is also a good plan 6 both to stroke their hides all over and to sprinkle them with unmixed wine, so that they may become on more familiar terms with their oxherd; it is well also to put the hand on the belly and under the thighs, so that they may not be alarmed if they are touched in this way afterwards, and also so that

qui plerumque feminibus inhaerent, eximantur. Idque cum fit, a latere domitor stare debet, ne calce 7 contingi possit. Post haec diductis malis educito linguam, totumque os et[1] palatum sale defricato, libralesque offas in praesulsae adipis liquamine tinctas in gulam[2] demittito, ac vini singulos sextarios per cornu faucibus infundito: nam per haec blandimenta triduo fere mansuescunt, iugumque quarto die accipiunt, cui ramus illigatus temonis vice traicitur: interdum et pondus aliquod iniungitur, ut maiore nisu 8 laboris exploretur patientia. Post eiusmodi experimenta vacuo plaustrum subiungendi, et paulatim longius cum oneribus producendi sunt. Sic perdomiti mox ad aratrum instituantur, sed in subacto agro, ne statim difficultatem operis reformident, neve adhuc tenera colla dura proscissione terrae contundant.[3] Quemadmodum autem bubulcus in arando bovem instituat, primo praecepi volumine. Curandum[4] ne in domitura bos calce aut cornu quemquam contingat. Nam nisi haec caventur, nunquam eiusmodi vitia quamvis subacto[5] eximi poterunt.

9 Verum ista sic agenda praecipimus, si veteranum[6] pecus non aderit; alioqui[7] expeditior tutiorque ratio

[1] os et *Lundström ex cit. Palladii*: eo sed *S*: eo sub *AR*.
[2] gulam *Palladius*: singula *SAR*.
[3] contundunt *Ald.*: condant *SAR*.
[4] Curanda *SAa*: Curandum *c*.
[5] subacta *SAR*. [6] veranum *SA*.
[7] alioqui *Lundström*: adeoque *SAR*.

---

[a] Which it must become used to later.
[b] These instructions occur in Book II. 2. 22 ff.

ticks, which generally fasten on the thighs, may be removed. In doing this the trainer ought to stand at the side, so that the animal may not reach him with its hoof. After this you should pull the jaws 7 apart and draw out the tongue and rub the whole mouth and palate with salt and put down the animal's throat cakes of a pound's weight of meal moistened with well-salted drippings of fat, and pour into their jaws a *sextarius* of wine at a time by means of a horn; for by blandishments of this kind they generally become tame in three days and allow themselves to be yoked on the fourth day. This yoke has the bough of a tree tied to it instead of a pole;[a] sometimes too a weight is attached, so that the capacity of the animal for enduring toil may be tested by the greater effort which is involved. After experiments of this kind the bullocks should 8 be yoked to an empty wagon and gradually be made to go longer journeys with loads. Soon after they have been thus broken in, they should be set to draw the plough, but over land already tilled, so that they may not be frightened at first by the difficulty of their task and that their still tender necks may not be bruised by the tough first breaking of the ground. I have already in my first book[b] given instructions how the ploughman is to train the ox in ploughing. Care must be taken that the ox does not strike anyone with his hoof or his horn while he is being trained; for, unless precautions are taken against this, it will never be possible to get rid of faults of this kind, though the animal has been broken in.

The method which we are prescribing should be 9 followed only if no ox is available which has already done service; otherwise the system of training which

domandi est, quam nos in nostris agris sequimur. Nam ubi plaustro aut aratro iuvencum consuescimus, ex domitis bubus valentissimum eundemque placidissimum cum indomito iungimus. Is et procurrentem 10 retrahit, et cunctantem producit. Si vero non pigeat iugum fabricare, quo tres iungantur, per [1] hanc machinationem consequemur, ut etiam contumaces boves gravissima opera non recusent. Nam ubi piger iuvencus medius inter duos veteranos iungitur, aratroque iniecto terram moliri [2] cogitur, nulla est imperium respuendi facultas. Sive enim efferatus prosilit, duorum arbitrio inhibetur: seu consistit, duobus gradientibus etiam invitus obsequitur: seu conatur decumbere, a valentioribus sublevatus trahitur: propter quae undique necessitate contumaciam deponit, et ad patientiam laboris paucissimis verberibus perducitur.

11 Est etiam post domituram mollioris generis bos, qui decumbit in sulco: eum non saevitia, sed ratione [3] censeo emendandum. Nam qui stimulis aut ignibus aliisque tormentis id vitium eximi melius iudicant, verae rationis ignari sunt: quoniam pervicax contumacia plerumque saevientem fatigat. Propter quod utilius est citra [4] corporis vexationem fame potius et siti cubitorem bovem emendare. Nam eum vehementius afficiunt naturalia desideria, quam

[1] per *add. Lundström.*
[2] terra molli *codd.*
[3] rationem *SA*: sed ratione *ac.*
[4] citra *S*: circa *AR.*

we follow on our own farm is more expeditious and safer. For when we are accustoming the young bullock to the wagon or plough, we yoke with the untrained animal the strongest and at the same time quietest of the trained oxen, which both keeps it back if it rushes forward and makes it advance if it lags behind. Indeed, if we have no objection to 10 constructing a yoke to which three animals can be fastened, we shall by this device achieve the result that even obstinate oxen do not refuse the heaviest tasks. For when an idle bullock is yoked between two veteran oxen and forced to till the ground with the plough which is put upon them, he has no opportunity of refusing to obey the order which has been given him; for, if he has become savage and rushes forward, he is checked by the controlling power of the other two; or, if he stands still when the other two pace along, he also follows even against his will; or, if he tries to lie down, he is upheld and dragged along by his more powerful companions. Hence he is forced from all sides to lay aside his obstinacy, and it takes very few blows to induce him to submit to hard work.

There is also an ox of a softer kind after it has been 11 broken in, which lies down in the furrow; in my opinion he should be made to mend his ways by reasoning rather than by cruelty. Those who think that the vice is better eradicated by means of goads, fire or other forms of torture, do not know how to reason aright; for the animal's stubborn obstinacy usually wears out the angry ploughman. Hence it is more expedient to cure the ox which has the habit of lying down by hunger and thirst without having recourse to doing it bodily hurt; for its natural desires

12 plagae. Itaque si bos decubuit, utilissimum est pedes eius sic cingulis [1] obligari, ne aut insistere aut progredi aut pasci possit. Quo facto inedia et siti [2] compulsus deponit ignaviam; quae tamen rarissima [3] est in pecore vernaculo: longeque omnis bos indigena melior est quam peregrinus. Nam neque aquae nec pabuli nec caeli mutatione [4] tentatur, neque infestatur conditione regionis, sicut ille, qui ex planis et campestribus locis in montana et aspera perductus 13 est, vel ex montanis in campestria. Itaque etiam, cum [5] cogimur ex longinquo boves arcessere, curandum est, ut in similia patriis locis traducantur. Item custodiendum est, ne in comparatione vel statura vel viribus impar cum valentiore iungatur. Nam utraque res inferiori celeriter affert exitum.

14 Mores huius pecudis probabiles habentur, qui sunt propiores placidis quam concitatis, sed non inertes: qui sunt verentes plagarum et acclamationum, sed fiducia virium nec auditu nec visu pavidi, nec ad ingredienda flumina aut pontes formidolosi, multi cibi edaces, verum in eo conficiendo lenti. Nam hi melius concoquunt, ideoque robora corporum citra maciem conservant, qui ex commodo,[6] quam qui 15 festinanter mandunt. Sed tam vitium est bubulci pinguem quam exilem bovem reddere: habilis enim et modica corporatura pecoris operarii debet esse,

[1] singulis *SA*[1]*R*.
[2] siti *R*: sitis *SA*.
[3] quam et amarissima *S*: qua et amarissima *AR*.
[4] mutationem *SA*: mutatione *c*.
[5] cui *SAR*.
[6] commodo *S*: commoda *AR*.

affect it more deeply than blows. So, if an ox has 12 lain down, the best plan is for its feet to be fastened together with straps in such a way that it can neither stand up nor walk nor feed. As a result, under the compulsion of starvation and thirst, it lays aside its sloth, which, however, is very rarely found amongst our home-grown cattle. Indeed a native ox is far superior to one which comes from elsewhere; for it is not disturbed by change of water or food or climate and is not troubled by the local conditions, as an ox would be which has been brought from flat plain-lands to a rough mountainous country or *vice versa*. When, therefore, we are obliged to bring oxen 13 from a distance, care must be taken that they are transferred to country which resembles that in which they were born. You must also be on your guard when pairing oxen together not to yoke one which is inferior in height or strength with one which is more powerful; for either of these circumstances quickly proves fatal to the weaker of the two.

Characteristics which are esteemed in oxen are 14 possessed by those which are placid rather than excitable and at the same time not lazy, and which are afraid of blows or shouts, but, being confident in their own strength, are not alarmed by anything which they hear or see, and which are not nervous at having to cross rivers or bridges, and which can eat plenty of food but are slow in finishing it; for leisurely chewers digest better and therefore preserve their bodily strength without becoming thin better than those which eat their food hurriedly. But it is quite 15 as much a fault in an oxherd to make his oxen fat as to make them thin; for the bodily form of a working ox ought to be active and moderate in bulk, with

nervisque et musculis robusta, non adipibus obesa, ut nec sui tergoris mole nec labore operis degravetur. Sed quoniam quae sequenda sunt in emendis domandisque bubus tradidimus, tutelam eorum praecipiemus.

III. Boves calore sub divo,[1] frigoribus intra tectum, manere oportet. Itaque hibernae stabulationi[2] eorum praeparanda sunt stramenta, quae mense Augusto intra dies triginta sublatae messis[3] praecisa[4] in acervum extrui debent. Horum desectio cum pecori tum agro est utilis: liberantur arva sentibus, qui aestivo[5] tempore per Caniculae ortum recisi plerumque radicitus intereunt, et stramenta pecori[6] subiecta plurimum stercoris efficiunt.

Haec cum ita curaverimus, tum et omne genus pabuli praeparabimus, dabimusque operam, ne penuria cibi macrescat pecus. Boves autem recte 2 pascendi non una ratio est. Nam si ubertas regionis viride pabulum subministrat nemo dubitat quin id genus cibi ceteris praeponendum sit: quod tamen nisi riguis aut roscidis locis non contingit. Itaque in iis ipsis vel maximum commodum est, quod sufficit una opera duobus iugis, quae eodem die alterna 3 temporum vice vel arant vel pascuntur. Siccioribus agris ad praesepia boves alendi sunt, quibus pro conditione regionum cibi praebentur: eosque nemo dubitat, quin optimi sint[7] vicia[8] in fascem ligata, et

[1] divo *R*: dio *SA*²: diu *A*¹.
[2] stabulationi *Ald.*: stabulati *S*: stabulatio *AR*.
[3] messis *a*: mensis *SAR*.
[4] precisas *SAR*.
[5] quaestivo *A*: qui aestivo *A*²: questivo *S*.
[6] stramenta pecori *Ursinus*: stramentis pecoris *SAR*.
[7] sunt *AR*: sint *S*.
[8] vitia *SA*: vicia *c*.

strong sinews and muscles and not encumbered by fat, so that it may not be wearied either by the weight of its own body or by the exertion necessary for its work. But since we have now set down the principles which must be followed in buying oxen and in breaking them in, we will next give directions for the care of them.

III. Oxen should remain out of doors when it is warm and under cover when it is cold; therefore, for their winter stabling, straw must be prepared, which ought to be cut and heaped up in stacks in August within thirty days of the gathering of the harvest. The cutting of the straw is beneficial both to the cattle and to the ground; for the fields are thus freed from briers, which, if they are cut back in the summer at the time of the rising of the Dogstar, usually die off at the roots, and also, if straw is put down as litter for the cattle, it produces a very large quantity of dung.

The care and feeding of oxen.

When we have arranged for this, we shall make provision also for every kind of fodder and ensure that the cattle will not be thin for want of food. 2 There is more than one system of feeding cattle properly. If the fertility of the district supplies green fodder, there is no doubt that this kind of food is to be preferred to all others; but this is only to be found in well-watered or dewy places. In these circumstances there is the very great advantage that one farm-labourer is enough to look after two yoke of oxen, which on the same day either plough or graze alternately. 3 On drier farms the oxen must be fed at their stalls, the fodder provided varying according to the nature of the district. There can be no doubt that the best foods are vetches

cicercula itemque pratense faenum. Minus commode tuemur armenta[1] paleis, quae ubique et quibusdam regionibus solae praesidio sunt. Eae[2] probantur maxime ex milio, tum ex hordeo, mox etiam ex tritico. Sed iumentis iusta operum reddentibus hordeum praeter has praebetur.

4 Bubus autem pro temporibus anni pabula dispensantur. Ianuario mense singulis fresi et aqua macerati ervi quaternos sextarios mixtos paleis dare convenit, vel lupini macerati modios, vel cicerculae maceratae semodios, et super haec affatim paleas.[3] Licet etiam, si est leguminum inopia, et eluta et siccata vinacia, quae de lora eximuntur, cum paleis 5 miscere. Nec dubium quin ea longe melius cum suis folliculis, ante quam eluantur, praeberi possint. Nam et cibi et vini vires habent, nitidumque et hilare et corpulentum pecus faciunt. Si grano abstinemus, frondis aridae corbis pabulatorius[4] modiorum viginti sufficit, vel faeni pondo triginta, vel sine modo viridis laurea et ilignea frondes. Et his, si regionis vis[5] permittit, glans adicitur: quae nisi ad satietatem detur, scabiem parit. Potest etiam si proventus[6] vilita- 6 tem[7] facit, semodius fabae fresae praeberi. Mense Februario plerumque eadem sunt cibaria. Martio et Aprili debet ad faeni pondus adici, qua terra proscinditur: sat autem erit pondo quadragena singulis dari. Ab idibus tamen mensis Aprilis usque in idus

[1] armenta *S*: armento *AR*.
[2] eae *S*: ea *AR*.
[3] paleas *S*: paleis *AR*.
[4] pabulatoribus *SAR*.
[5] vis *om. SAR*.
[6] proventus *S*: pro ventu *A*.
[7] vilitatem *R*: vilitem *SA*.

tied up in bundles and chickpea and also meadow-hay. We are not looking after our cattle so well if we feed them on chaff, which is a universal, and in some districts the only, resource. The chaff which is most highly esteemed comes from millet, the next best from barley, and the third best from wheat; beasts of burden which are rendering regular terms of labour are given barley as well as chaff.

4 The diet of oxen is regulated according to the time of year. In January it is a good plan to give them four *sextarii* each of bitter-vetch crushed and soaked in water and mixed with chaff, or a *modius* of soaked lupines, or half a *modius* of soaked chickling-pea, as well as chaff in abundance. If there is a lack of pulse, it is allowable to mix with chaff grape skins taken from the after-wine which have been washed and 5 dried, but there is no doubt that it is far better to give them the grapemash, skins and all, before they have been washed, for they contain the strength both of food and of wine and make the cattle sleek and of good cheer and plump. If we abstain from giving them grain, it is enough to supply a fodder-basket holding twenty *modii* of dry leaves or thirty pounds of hay, or green bay-leaves or the foliage of the holm-oak in unlimited quantities. To these mast is added, if the resources of the district permit, but, unless enough is provided to cause satiety, it causes the scab. A half-*modius* of crushed beans may also be provided, if a good crop makes it cheap enough 6 to do so. In February the food is usually the same. In March and April an addition should be made to the weight of hay in places where the ground is being broken up for the first time; forty pounds, however, will be enough to give to each animal. From April

Iunias viride pabulum recte secatur: potest etiam in calend. Iulias frigidioribus locis idem praestari: a quo tempore in calend. Novemb. tota aestate et deinde autumno satientur fronde; quae tamen non ante est utilis, quam cum maturaverit [1] vel imbribus vel assiduis roribus: probaturque maxime ulmea,
7 post fraxinea, et ab hac populnea. Ultimae sunt ilignea et quernea et laurea: sed eae post aestatem necessariae deficientibus ceteris. Possunt etiam folia ficulnea probe dari, si sit ea copia, aut stringere arbores expediat. Ilignea tamen vel melior est quernea, sed eius generis, quod spinas non habet; nam id quoque, ut iuniperus,[2] respuitur a pecore
8 propter aculeos. Novembri mense ac Decembri per sementem quantum appetit bos, tantum praebendum est: plerumque tamen sufficiunt singulis modii glandis et paleae ad satietatem datae, vel lupini macerati modii, vel ervi aqua conspersi sextarii VII permixti paleis, vel cicerculae similiter conspersae sextarii XII mixti paleis, vel singuli modii vinaceorum, si iis, ut supra dixi, large paleae adiciantur; vel si nihil horum est, per se faeni pondo quadraginta.

IV. Sed non proderit cibis [3] satiari pecora, nisi omnis adhibeatur [4] diligentia, ut salubri sint corpore, viresque conservent: quae utraque [5] custodiuntur large dato per triduum medicamento, quod com-

[1] maturuit *SAR*.
[2] iuniperus *R*: -erius *S*: imperius *A*.
[3] cibos *S*: cibus *AR*.
[4] adhibeatur S^2: adhibetur S^1: adiuvetur *AR*.
[5] utroque *SAR*.

13th to June 15th it is proper to cut green forage for them; supply of it can even be continued until July 1st in cooler regions. From then through the whole summer and the following autumn up to November 1st, they should be given their fill of leaves, which, however, are not fit for use until matured either by rain or by continual dew. The most highly esteemed is the foliage of the elm, next comes that of the ash, and, thirdly, that of the poplar; the least 7 satisfactory is that of the holm-oak, the oak and the bay-tree, but these may have to be used after the summer, if all other kinds fail. It is also proper to give them fig-leaves, if there is abundance of them or if it is expedient to strip the trees. Holm-oak-leaves are better than oak-leaves, but they should not be of the kind that has spines, for this is refused by cattle because of the prickles, as also are juniper-leaves. In November and December, during the 8 period of sowing, an ox should be given all the food which it wants; but a *modius* of mast a head is generally enough and as much chaff as they can eat, or a *modius* of soaked lupines or seven *sextarii* of bitter-vetch sprinkled with water and mixed with chaff, or twelve *sextarii* of chickpeas similarly sprinkled and mixed with chaff, or a *modius* of grape-skins each, provided that, as I have said above, chaff is generously added to them; if none of these foods is available, forty pounds of hay should be given by itself.

The diseases of oxen and their remedies.

IV. It will be no use to give cattle a satisfying diet unless every care is taken that they are healthy in body and that they keep up their strength. Both these objects are secured by administering on three consecutive days a generous dose of medicine com-

ponitur pari pondere triti lupini[1] cupressique, et cum aqua nocte una sub divo habetur; idque quater anno fieri debet ultimis temporibus veris, aestatis, 2 autumni, hiemis. Saepe etiam languor et nausea discutitur, si integrum gallinaceum crudum ovum ieiuni faucibus inseras, ac postero die spicas ulpici vel alii cum vino conteras, et in naribus infundas: neque haec tantum remedia salubritatem faciunt. Multi[2] et largo sale miscent pabula; quidam marrubium deterunt[3] cum oleo et vino; quidam porri fibras, alii grana thuris, alii sabinam herbam rutamque pinsitam[4] mero diluunt, eaque medicamenta 3 potanda praebent. Multi caulibus vitis[5] albae et valvulis ervi bubus medentur: nonnulli pellem serpentis obtritam cum vino miscent. Est[6] etiam remedio cum dulci vino tritum serpyllum, et concisa et in aqua macerata scilla. Quae omnes praedictae potiones trium heminarum singulis diebus per triduum datae alvum purgant, depulsisque vitiis re- 4 creant vires. Maxime tamen habetur salutaris amurca, si tantundem aquae misceas, et ea pecus insuescas; quae protinus dari non potest, sed primo cibi asperguntur;[7] deinde exigua portione medicatur aqua, mox pari mensura mixta datur ad saturitatem.

[1] lupini *a*: *om. SAR.*
[2] multo *SAR.*
[3] deterunt *S*: dederunt *AR.*
[4] rutamque pinsitam *Lundström*: putaque vinitam *SA.*
[5] vitibus *SA*: vitis *ac.*
[6] est *S*: sed *AR.*
[7] adspergunt *SA.*

pounded of equal weights of the crushed leaves of the lupine and of the cypress, which is mixed with water and left out of doors for a night. This should be done four times a year—at the end of the spring, of the summer, of the autumn and of the winter. Lassitude and nausea also can often be dispelled if you force a whole raw hen's egg down the animal's throat when it has eaten nothing; then on the following day you should crush together spikes of leek or garlic in wine and pour it into its nostrils. Nor are these the only remedies which make for health. Many people mix also a generous quantity of salt with the fodder; some grate white horehound in oil and wine; some infuse fibres of leek, others grains of frankincense, others savin [a] and crushed rue in unmixed wine and give them these medicaments to drink. Many people use the stalks of white-vine (bryony) and the shells of bitter-vetch as a medicine for oxen; some crush a snake's skin and mix it with wine. Thyme crushed in sweet wine and squill cut up and soaked in water are also used as remedies. All the above-mentioned potions in doses of three *heminae* given daily for three days purge the bowel and renew the animal's strength by driving away its maladies. But lees of olive-oil are regarded as particularly salutary if you mix them with an equal portion of water and accustom the cattle to them; this remedy cannot be administered all at once, but at first is sprinkled on the food, next a small portion is infused in the water, and then the animal is given as much as it can take mixed in equal portions of both ingredients.

2

3

4

[a] A kind of juniper which yields a volatile oil.

V. Nullo autem tempore et minime aestate utile est boves in cursum concitari: nam ea res aut alvum movet, aut [1] frequenter [2] febrem. Cavendum quoque est, ne ad praesepia sus aut gallina perrepat. Nam hoc [3] quod decidit [4] immixtum pabulo, bubus affert necem.

Sus aegra pestilentiam facere [5] valet. Quae cum in gregem [6] incidit, confestim mutandus [7] est caeli status, et in plures partes distributo pecore longinquae regiones petendae sunt, atque ita segregandi a sanis morbidi, ne quis interveniat, qui contagione 2 ceteros labefaciat. Itaque cum ablegabuntur, in ea loca perducendi sunt, quibus nullum impascitur pecus, ne adventu suo etiam illi tabem afferant. Evincendi sunt autem quamvis pestiferi morbi, et exquisitis remediis propulsandi. Tunc panacis et eryngii radices faeniculi seminibus miscendae, et cum fricti [8] ac moliti tritici farina candenti aqua conspergendae, eoque medicamine salivandum aegrotum 3 pecus. Tunc paribus casiae myrrhaeque et thuris ponderibus, ac tantundem sanguinis marinae testudinis miscetur potio cum vini veteris sextariis tribus, et ita per nares infunditur. Sed ipsum medicamentum ponderis sescunciae divisum portione aequa per triduum cum vino dedisse sat erit. Praesens etiam remedium cognovimus radiculae, quam pastores consiliginem vocant. Ea in Marsis montibus plurima nascitur, omnique pecori maxime est salutaris. Laeva

[1] meta ut *SAR*.
[2] frequenter *Lundström*: frequen (*sequ. vac. sp.*) *SAR*.
[3] haec *S*: hec *A*: hoc *c*.
[4] desidit *SA*: decidit *ac*.
[5] facere S^2: face S^1AR.
[6] grege *SAR*.
[7] mutandus *S*: mutatus *AR*.
[8] fricti *S*: defruti *AR*.

Causes and remedies of plague in oxen.

V. At no season of the year and least of all in the summer is it beneficial to incite oxen to run; for this either relaxes the bowels or else often gives rise to fever. Care must also be taken that no pig or chicken slips into their stalls, for the excrement which falls from them, mixed with their food, is fatal to oxen. A diseased sow may cause plague. If this falls upon a herd, a change of climate must immediately be made, and the cattle must be divided up, in a number of groups, and sent to distant places and those which are infected segregated from the healthy, that no infected animal may come into contact with the rest and destroy them with the contagion. 2 When they are thus isolated, they have to be taken to places where no herd is pastured, so that they may not by their arrival bring the plague there also. Diseases, however pestilential, must be overcome and expelled by carefully sought-out remedies. Sometimes roots of all-heal and sea-holly should be mixed with fennel-seeds and, together with flour of crushed and ground wheat, should be sprinkled with boiling water, and the suffering herd given a drench with this medicament. 3 Sometimes a potion consisting of equal weights of cinnamon, myrrh and frankincense and a like quantity of the blood of a sea-tortoise is mixed with three *sextarii* of old wine and poured through the animal's nostrils. It will suffice to have given the medicine itself divided into equal doses of one and a half ounces together with wine for three days. We have also found a sovereign remedy in the root which the shepherds call *consiligo*.[a] It grows in large quantities in the Marsian mountains and is very salutary for all cattle; it is dug up with

[a] *Pulmonaria officinalis*, lungwort.

manu effoditur ante solis ortum. Sic enim lecta
4 maiorem vim creditur habere. Usus eius traditur
talis. Aenea fibula pars auriculae latissima circumscribitur, ita ut manante sanguine tanquam O literae
ductus[1] appareat. Hoc et intrinsecus et ex superiore
parte auriculae cum factum est, media pars descripti
orbiculi eadem fibula transuitur, et facto foramini
praedicta radicula inseritur; quam cum recens plaga
comprehendit, ita continet, ut elabi[2] non possit: in
eam deinde auriculam omnis vis morbi pestilensque
virus elicitur,[3] donec pars, quae fibula circumscripta
est, demortua excidit, et minimae partis iactura caput
5 conservatur. Cornelius Celsus etiam visci folia cum
vino trita per nares infundere iubet. Haec facienda,
si gregatim pecora laborant: illa deinceps, si singula.

VI. Cruditatis signa sunt crebri ructus ac ventris
sonitus, fastidia cibi, nervorum intentio, hebetes oculi.
Propter quae bos neque ruminat neque lingua se
deterget. Remedio erunt aquae calidae duo congii,
et mox triginta brassicae modicae caules cocti et ex
2 aceto dati. Sed uno die abstinendum est alio cibo.
Quidam clausum intra tecta continent, ne pasci
possit. Tum lentisci oleastrique cacuminum pondo
IV, et libram[4] mellis una trita permiscent aquae

[1] littere ductum *SA*.
[2] ea labi *SAR*.
[3] eligitur *S*: efficitur *A*[1]: elicit *c*: elicitur *A*[2]*a*.
[4] libra *SAR*.

---

[a] Aulus Cornelius Celsus, a contemporary of Columella, besides his book on medicine which has survived, also wrote on agriculture.

the left hand before sunrise, for it is believed to have greater potency if it is picked in this way. 4 The following is the traditional manner of using it. A line is drawn round the widest part of the ear-lap with brazen pin in such a way that a figure resembling the letter O appears where the blood flows. When this operation has been performed both inside and on the upper part of the ear, the middle of the circle which has been described is pierced with the same pin and the root mentioned above is inserted in the hole thus made, and, when the newly made wound has closed on it, it holds the root so tightly that it cannot slip out. Then all the virulence of the disease and the poison of the plague is attracted to this ear, until the part round which the line was described by the pin mortifies and comes away. Thus the head is saved by the sacrifice of a very small portion of it. 5 Cornelius Celsus[a] also recommends the pouring into the nostrils of wine in which the leaves of mistletoe have been crushed. The latter course must be adopted if the cattle are suffering as a herd, the former if individual animals are affected.

Remedies for indigestion in cattle.

VI. Signs of indigestion are frequent eructations, rumblings of the belly, distaste for food, tension of the sinews and dimness of the sight, with the result that the ox neither ruminates nor cleanses himself by licking. The appropriate remedy will be two *congii* of hot water, followed by thirty moderate-sized stalks of cabbage cooked and dipped in vinegar; but the animal must abstain from other food for one day. 2 Some people keep the animal shut up indoors, so that it cannot graze; they then mix four pounds of the tops of mastic and wild olive crushed up with a pound of honey in a *congius* of water,

congio, quam nocte una[1] sub dio habent, atque ita faucibus infundunt. Deinde interposita hora macerati ervi quattuor libras obiciunt, aliaque potione 3 prohibent. Hoc per triduum fieri debet, ut omnis causa languoris discutiatur. Nam si neglecta cruditas est, et inflatio ventris et intestinorum maior dolor insequitur,[2] qui nec capere cibos sinit, gemitus exprimit, locoque stare non patitur, saepe decumbere, et agitare caput, caudamque crebrius agere cogit. Manifestum remedium est proximam clunibus partem caudae vinculo vehementer obstringere, vinique sextarium cum olei hemina faucibus infundere atque ita citatum per mille et quingentos passus agere. 4 Si dolor permanet, ungulas circumsecare, et uncta[3] manu per anum inserta fimum extrahere, rursusque agere currentem. Si nec hoc profuit, tres caprifici aridi conteruntur, et cum dodrante aquae calidae dantur. Ubi nec haec medicina processit, myrti silvestris foliorum duae librae laevigantur, totidemque sextarii calidae aquae mixti per vas ligneum faucibus infunduntur. Atque ita sub cauda sanguis emittitur. Qui cum satis profluxit, inhibetur papyri ligamine. Tum concitate agitur pecus eousque dum 5 anhelet. Sunt et ante detractionem sanguinis illa remedia: tribus heminis vini triens[4] pinsiti alii[5] permiscetur, et post eam potionem currere cogitur, vel salis sextans cum cepis decem conteritur, et ad-

[1] nocte unam *SA* : noctem unam *R*.
[2] dolor sequitur *S*²: dolori sequitur *SA*¹*R*.
[3] cuncta *SAR*.
[4] triens *Svennung* : tribus *SAR*.
[5] pinsiti alii *ed. pr.* : pinsitiali *SAR*.

which they keep for one night in the open air, and then pour it down the animal's throat. Then after an interval of an hour they put before it four pounds of soaked bitter-vetch and keep it away from any other drink. This should be continued for three days, 3 so that every cause of lassitude is dissipated. If indigestion is neglected, inflation of the belly and more severe pain in the intestines follow, which does not allow the animal to take its food, causes it to bellow, does not suffer it to remain in one place, and makes it lie down frequently, toss its head and lash its tail continually. An obvious remedy is to bind down tightly the part of the tail nearest to the haunches and to pour down its throat a *sextarius* of wine and a *hemina* of oil and then drive it for a mile and a half at a quick pace. If the pain persists, you 4 should cut the hoof all round, draw off the excrement by greasing the hand and inserting it into the anus, and again drive the animal at a running pace. If this also has done no good, three dried wild figs are crushed and administered with a *dodrans* of hot water. If this remedy has also been unsuccessful, two pounds of the leaves of wild myrtle are pulverized and mixed with the same number of *sextarii* of hot water and poured down the throat by means of a wooden vessel; then the animal is bled under the tail and, when enough blood has flowed, it is checked by a bandage of papyrus; then the animal is driven at a quick speed until it is out of breath. The following 5 remedies are applied before drawing off any blood: a *triens* of pounded garlic is mixed with three *heminae* of wine, and, after drinking this, the animal is compelled to run; or else a *sextans* of salt is pounded up with ten onions, and after being mixed with boiled-

mixto melle dococto collyria immittuntur alvo,[1] atque ita citatus bos agitur.

VII. Ventris quoque et intestinorum dolor sedatur visu nantium et maxime anatis. Quam si conspexerit, cui intestinum dolet, celeriter tormento liberatur. Eadem anas maiore profectu mulos [2] et equinum genus conspectu suo sanat. Sed interdum nulla prodest medicina. Sequitur torminum [3] vitium, quorum signum est cruenta et mucosa ventris 2 proluvies. Remedio sunt cupressini quindecim coni, totidemque gallae, et utrorumque [4] ponderis vetustissimus caseus.[5] Quibus in unum tunsis admiscentur austeri vini quattuor sextarii, qui pari mensura per quatriduum dispensati dantur: nec desint lentisci myrtique [6] et oleastri cacumina. Viridis alvus [7] corpus ac vires carpit, operique inutilem reddit. Quae cum accident, prohibendus erit bos potione per triduum, 3 primoque die cibo abstinendus. Sed mox cacumina oleastri et arundinis, item baccae lentisci myrtique [8] dandae; nec potestas aquae nisi quam parcissime facienda est. Sunt qui tenerorum lauri foliorum libram [9] et abrotonum ceraticum,[10] pari portione deterant [11] cum aquae calidae duobus sextariis, atque ita faucibus infundant, eademque pabula, ut supra diximus, obiciant. 4 Quidam vinaceorum duas libras torre-

[1] albo *SAR*.
[2] mulos *R* : mulus *SA*.
[3] terminum *S* : minũ *A* : minus *R*.
[4] utrorumque *S* : vivorumque *AR*.
[5] caseus *S* : caeses *A* : ceses *R*.
[6] myrtiq; *ed. pr.* : murtisq; *SAa*.
[7] albos *SAR*.
[8] myrti *ed. pr.* : multi *SAR*.
[9] tenerorum lauri foliorum libram *scripsi* : teneram laurum coloni libram *SAR*.
[10] ceraticum *SAR* : cepaticum *Lundström*.
[11] deterant *S* : dederant *AR*.

down honey is introduced as a suppository into the bowel and the ox is driven at a quick pace.

Remedy for internal pains in oxen.

VII. Pain in the belly and intestines is assuaged by the sight of swimming birds, especially a duck. If an ox which has a pain in its intestines sees a duck, it is quickly delivered from its torment. The sight of a duck is also even more successful in curing mules and the race of horses. Sometimes, however, no remedy is of any avail and colic follows, the sign of which is a flux of blood and mucous matter from the belly. 2 The cure for this consists of fifteen cypress-cones and the same number of oak-apples and very old cheese equal in weight to the other two ingredients. When these have been pounded up together, four *sextarii* of rough wine are mixed with them, and the mixture is administered in equal doses for four days; nor should tops of mastic and myrtle and wild olive be lacking. Diarrhoea [a] wastes the body and the strength and renders an animal useless for work. When this happens, the ox will have to be kept from drinking for two days and on the first day must be kept from eating; 3 but soon thereafter tops of wild olive and of reeds must be given, also berries of mastic and myrtle, but no opportunity of drinking water must be allowed except as sparingly as possible. Some people crush a pound of tender leaves of bay and the same quantity of horned southernwood [b] in two *sextarii* of hot water and pour it down the animal's throat and put before it the same food as I mentioned above. 4 Some people heat two pounds of

[a] That *viridis* agrees with *alvus* ("green bowel") and does not belong to the previous sentence is clear from Vegetius, who writes, *si venter coeperit fluere viridis* (quoted by Schneider).

[b] Probably *Artemisia abrotonum*.

faciunt, et ita conterunt cum totidem sextariis vini austeri, potandumque medicamentum praebent, omni alio humore subtracto,[1] nec minus cacumina praedictarum arborum obiciunt. Quod si neque ventris restiterit[2] citata proluvies, neque intestinorum ac ventris dolor, cibosque respuet, et praegravato capite saepius coniverit,[3] lacrimaeque ab oculis et pituita a naribus profluent, usque ad ossa frons media uratur, auresque ferro descindantur.[4] Sed vulnera facta igne dum[5] sanescunt, defricare bubula urina convenit; at ferro rescissa melius pice et oleo curantur.

VIII. Solent etiam fastidia ciborum afferre vitiosa incrementa linguae, quas ranas veterinarii vocant. Haec ferro reciduntur, et sale cum alio pariter trito vulnera defricantur, donec lacessita pituita profluat.[6] Tum vino perluitur os, et interposito unius horae spatio virides herbae vel frondes dantur, dum facta

2 ulcera cicatrices ducant. Si neque ranae fuerint, neque alvus citata, et nihilo minus cibos non appetet, proderit alium pinsitum cum oleo per nares infundere, vel sale et cunila defricare fauces, vel eandem partem alio tunso et alecula linire. Sed haec si solum fastidium est.

[1] omni alio humore subtracto *Svennung ex cit. Palladii*: omnia in umores supra dixi *SA*.

[2] restiterit *edd.*: eriserit *S*: crescerit *AR*.

[3] coniverit *Svennung*: consuevit *SAR*.

[4] decidantur *SA*.

[5] indũ *S*: interdum *AR*.

grape-skins and crush them in two *sextarii* of rough wine and then give them to be drunk as a medicine, keeping any other liquid away from them, but nevertheless putting before them the tops of the trees mentioned above. But if neither the violent flux from the belly nor the pain in the intestine and stomach has ceased and the animal refuses his food, and its head is very heavy and it frequently blinks and tears flow from its eyes and slime from its nostrils, the middle of its forehead should be burnt down to the bone and its ears cleft with a knife. It is in fact a good plan to rub with ox-urine the wounds caused by the fire while they are healing; but those which are due to cuts with the knife are better treated with pitch and oil.

Treatment for swollen tongue in oxen.

VIII. Aversion to food is often caused by morbid swellings of the tongue which veterinary surgeons call "frogs." They are cut back with a knife and the wounds rubbed with salt and garlic crushed together in equal quantities, until a viscous discharge thus provoked flows forth. The mouth is then washed out with wine and after the interval of one hour a diet of green herbs or leaves is administered until the sores which had formed are scarred over. 2 If no "frogs" have formed and the bowel is not disturbed but nevertheless the animal has no appetite for its food, it will be beneficial to pour a mixture of pounded garlic and oil through its nostrils or to rub the throat with salt and marjoram, or to smear the same part with crushed garlic and fish-sauce. But this remedy should be used if aversion to food is the only symptom.

[6] proluat *S*.

IX. Febricitanti bovi convenit abstineri cibo uno die, postero deinde exiguum sanguinem ieiuno sub cauda emitti, atque interposita hora modicae magnitudinis coctos brassicae coliculos triginta ex oleo et garo[1] salivati more demitti, eamque escam per quinque dies ieiuno dari. Praeterea cacumina lentisci aut oleae, vel tenerrimam quamque frondem, ac pampinos vitis obici; tum etiam spongia labra detergeri, et aquam frigidam ter die praeberi potandam.
2 Quae medicina sub tecto fieri debet, nec ante sanitatem bos emitti. Signa febricitantis manantes lacrimae, gravatum caput, oculi compressi, fluidum salivis os, longior et cum quodam impedimento tractus spiritus, interdum et cum gemitu.

X. Recens tussis optime salivato farinae hordeaceae discutitur. Interdum magis prosunt gramina concisa, et his admixta fresa faba. Lentis quoque valvulis exemptae et minute molitae miscentur aquae calidae sextarii duo, factaque sorbitio per cornu infunditur. Veterem tussim sanant duae librae hyssopi macerati sextariis aquae tribus. Nam id medicamentum teritur, et cum lentis minute, ut dixi, molitae sextariis quattuor more salivati datur, ac
2 postea aqua hyssopi per cornu infunditur. Porri

[1] garo *ed. pr.*: caro *SAR.*

IX. When an ox suffers from fever, it is a good plan that it should go without food for a day, and that on the following day a little blood should be drawn off under the tail before it eats anything, and that after an interval of an hour it should be made to swallow thirty cooked stalks of cabbage of moderate size which have been dipped in oil and pickled fish in the manner of drench. This food should be given for five days on an empty stomach. Furthermore, tops of mastic or olive or any other very tender foliage and vine-shoots should be placed before it, also its lips should be wiped with a sponge and cold water given it to drink three times a day. This treatment should be carried out under cover and the animal should not be allowed to go out until it is cured. The symptoms of a state of fever are running at the eyes, a heavy head, contracted eyes, a flow of saliva from the mouth, an unusually slow and a somehow obstructed respiration, accompanied also at times by lowing.

Fever in oxen.

2

X. A cough, if treated early, is best dispelled by a medicine which causes salivation made of barley-flour. Sometimes grass cut up small and crushed beans mixed with it are more beneficial; also two *sextarii* of lentils removed from their pods and ground up small are mixed with hot water and the draught thus formed is poured down the throat through a horn. A cough of long standing is cured with two pounds of hyssop infused in three *sextarii* of water. Now this medicament is crushed up and administered with four *sextarii* of lentils ground small, in the manner I have described, and given to cause salivation, and the hyssop-water is afterwards poured in through a horn. The juice of a leek together with

Coughs of oxen.

2

enim succus cum[1] oleo, vel ipsa fibra cum hordeacea farina contrita remedium[2] est. Eiusdem radices diligenter lotae, et cum farre triticeo pinsitae[3] ieiunoque datae vetustissimam tussim discutiunt. Facit idem pari mensura ervum sine valvulis cum torrido[4] hordeo molitum et salivati more in fauces demissum.

XI. Suppuratio melius ferro rescinditur quam medicamento. Expressa deinde sanie sinus[5] ipse, qui eam continebat, calida bubula urina eluitur, atque ita linamentis pice liquida et oleo imbutis colligatur: vel si colligari ea pars non potest, lamina candenti sebum caprinum aut bubulum instillatur. 2 Quidam, cum vitiosam partem inusserunt, urina vetere humana eluunt,[6] atque ita aequis ponderibus incocta pice liquida cum vetere axungia linunt.

XII. Sanguis demissus in pedes claudicationem affert. Quod cum accidit, statim ungulam inspicito.[7] Tactus autem fervorem demonstrat: nec bos vitiatam partem vehementius premi patitur. Sed si sanguis adhuc supra ungulas in cruribus est, frictione assidua discutitur; vel, cum ea nihil profuit, scarificatione emittitur.[8] At si iam in ungulis est, inter duos 2 ungues cultello leviter aperies.[9] Postea linamenta

[1] succus cum *Lundström*: sucum *SA*.
[2] remedium *S*: remedia *A*[1]: remedio *A*[2]*R*.
[3] pinsite *R*: pinsita *SA*.
[4] torreo *SAR*.
[5] sasinus *SAR*.
[6] partem—eluunt *ex cit. Palladii edd.* partẽminus seruntur inhabeture hurane luunt *S*: parte minus seruntur inhabeturae rane luunt *A*.
[7] ungula inspicit *SAR*.
[8] hemitur *SAR*: emititur *c*.

oil, or the fibre itself of the leeks crushed up with barley-flour, is also used as a cure; the roots too of the same plant carefully washed and pounded up with wheaten flour, given to the animal when it is fasting, dispel the most inveterate cough. The same effect is produced by bitter-vetch without its husk pounded up with an equal portion of toasted barley and poured down the throat in the manner of a drench.

Remedies for suppuration.

XI. It is better to get rid of suppuration by the surgeon's knife than with medicine. Then, when the pus has been squeezed out, the sinus itself which contained it is washed out with warm ox-urine and then bound up with linen bandages soaked in liquid pitch and oil, or, if the part affected cannot be bound up, goat's or ox's tallow is dripped upon it by means of a red hot plate of iron. 2 Some people, when they have cauterized the part affected, wash it with stale human urine and then anoint it with raw liquid pitch and stale axle-grease in equal quantities.

Remedy for lameness in oxen.

XII. Down-flow of blood into the animal's feet gives rise to lameness. When this happens, the first thing that you should do is to inspect the hoof; merely touching it proves the presence of inflammation, and the animal cannot bear any at all violent pressure on the affected part. But if the blood is still in the legs above the hoofs, it can be dissipated by continual friction, or, if that has no effect, it can be removed by scarification. But if it has already reached the hoofs, you will make a slight incision with a lancet between the two halves of the hoof; 2 then bandages dipped in salt and vinegar are

[9] aperiet *SAR*.

sale atque aceto imbuta[1] applicantur, ac solea spartea pes induitur,[2] maximeque datur opera, ne bos in aquam pedem mittat et ut sicce stabuletur. Hic idem sanguis nisi emissus fuerit, famicem creabit, qui si suppuraverit, tarde percurabitur: primum ferro circumcisus et expurgatus, deinde pannis aceto et sale et oleo madentibus inculcatis, mox axungia vetere et sebo hircino pari pondere decoctis, ad

3 sanitatem perducitur. Si sanguis in inferiore parte ungulae est, extrema pars ipsius unguis ad vivum resecatur, et ita emittitur, ac linamentis pes involutus spartea munitur. Mediam ungulam ab inferiore parte non expedit aperire, nisi eo loco iam suppuratio facta est. Si dolore nervorum claudicat, oleo et sale genua poplitesque [3] et crura confricanda sunt, donec sanetur.

4 Si genua intumuerint, calido aceto fovenda sunt, et lini semen aut milium detritum conspersumque aqua mulsa imponendum: spongiae [4] quoque ferventi aqua imbutae [5] et expressae [6] litaeque [7] melle recte genibus applicantur, ac fasciis circumdantur. Quod si tumori subest aliquis humor, fermentum vel farina hordeacea ex passo aut aqua mulsa decocta imponitur: et cum maturuit suppuratio, rescinditur ferro, eaque emissa, ut supra docuimus, linamentis

5 curatur. Possunt etiam, ut Cornelius Celsus prae-

[1] imbuta *R*: inbuto *SA*.
[2] inducitur *SAR*.
[3] popliteque *SA*: poplitesque *c*.
[4] sphongia *S*: spongia *R*: phongio *A*.
[5] inbuta *SA*.
[6] expressa *SAR*.
[7] litae quae *A*: lita que *S*.

applied and the foot is covered with a "slipper" of broom and the greatest care is taken to prevent the ox from putting his foot in water and that it keep dry in its stall. This same blood, unless it is drawn off, will give rise to a bruise, and, if this suppurates, it will take a long time to heal. First a cut must be made round it with a knife and it must be cleaned, then it is brought to a healthy condition by having rags pressed against it soaked in vinegar, salt and oil, and afterwards by treatment with stale axle-grease and goat's tallow boiled in equal quantity. 3 If the blood is in the lower part of the hoof, the extremity of the hoof itself is cut to the quick and the blood thus discharged, and the foot is wrapped in bandages and protected with a "slipper" of broom. It is not advisable to open the middle of the hoof from below, unless suppuration has already taken place in that part. If the lameness is due to pain in the sinews, the knees, the ham and the legs should be rubbed with oil and salt until it is cured.

4 If the knees are swollen, they must be fomented with warm vinegar and poulticed with linseed or millet which has been ground up and sprinkled with honey-water; also sponges soaked in boiling water and then wrung out and smeared with honey are correctly applied to the knees and wrapped round with bandages. But if there is any liquid matter under the swelling, some yeast or barley-flour boiled in raisin-wine or honey-water is placed upon it; and when the suppuration has come to a head, it is cut with the surgeon's knife, and, when the pus has been extracted, it is treated with bandages in the manner described above. 5 Incisions made with the knife can also be treated, as Cornelius Celsus taught, by means

cipit, lilii radix aut scilla cum sale, vel sanguinalis herba, quam poligonum Graeci appellant, vel marrubium ferro reclusa sanare. Fere autem omnis dolor corporis, si sine vulnere est, recens melius fomentis discutitur; vetus uritur, et supra ustum butyrum vel caprina instillatur adeps.

XIII. Scabies extenuatur trito alio defricta;[1] eodemque remedio curatur rabiosae canis vel lupi morsus, qui tamen et ipse imposito vulneri vetere salsamento aeque bene sanatur. Et ad scabiem praesentior alia medicina est. Cunila bubula,[2] et sulphur conteruntur, admixtaque amurca cum oleo atque [3] aceto incoquuntur; deinde tepefactis scissum alumen tritum spargitur. Id medicamentum can- 2 dente [4] sole illitum maxime prodest. Ulceribus gallae tritae remedio sunt; nec minus succus marrubii cum fuligine.

Est et infesta pestis bubulo pecori, coriaginem rustici appellant, cum pellis ita tergori adhaeret, ut apprehensa manibus diduci a costis non possit. Ea res non aliter accidit,[5] quam si bos aut ex languore aliquo ad maciem perductus est, aut sudans in opere faciendo refrixit, aut si sub onere pluvia madefactus 3 est. Quae quoniam perniciosa sunt, custodiendum est, ut cum ab opere boves redierint [6] adhuc aestuantes anhelantesque, vino aspergantur, et offae adipis faucibus eorum inserantur. Quod si praedictum vitium inhaeserit, proderit [7] decoquere laurum, et ea

[1] defricta *S* : defricto *AR*.
[2] cunicula bubula *S* : cuniculabula *AR*.
[3] aqua *SAR*.
[4] cantendente *SAR*.
[5] accedit *SAR*.
[6] redideriht *SA* : redierint *ac*.
[7] proderit *S*²*c* : prodiderit *S*¹*AR*.

of lily-roots or squills mixed with salt, or the staunching plant which the Greeks call *polygonum*,[a] or horehound. Almost all bodily pains, if there is no wound, can in their early stages be better dissipated by fomentation; in the advanced stage they are treated by cauterizations and the dropping of burnt butter or goat's fat upon the place.

Remedies for the scab, ulcers, etc.

XIII. The scab is alleviated if it is rubbed with bruised garlic, and the same remedy is used for the bite of a mad dog or wolf, which, however, is also quite as easily cured by placing stale pickled fish upon the wound. There is also a still more efficacious remedy for the scab; ox-marjoram and sulphur are pounded up together and cooked in lees of olives mixed with oil and vinegar; then, when the mixture is hot, split alum is ground up and sprinkled upon it. This remedy is most efficacious if it is smeared on when the sun is hot. 2 Ground oak-galls are a cure for ulcers, likewise the juice of horehound together with soot.

There is also a dangerous plague which affects cattle, called by the farmers "hide-binding," when the skin adheres so closely to the back that, if it is taken hold of with the hands, it cannot be drawn away from the ribs. It occurs only when the ox is either reduced to a lean condition as the result of some illness or has become chilled when sweating in the course of its labours, or if it has been drenched by rain when it is carrying a load. 3 Since these conditions are dangerous, care must be taken that the oxen, when they have returned from work still hot and panting, are sprinkled with wine and that balls of fat are thrust down their throats. If, however, the above-mentioned malady has already taken hold of

[a] Knotgrass (*Polygonum aviculare*).

calda fovere terga, multoque oleo et vino confestim subigere, ac per omnes partes apprehendere et attrahere pellem.[1] Idque optime fit sub dio, sole fervente. Quidam faeces vino et adipe commiscent, eoque medicamento post fomenta praedicta utuntur.

XIV. Est etiam illa gravis pernicies, cum pulmones exulcerantur. Inde tussis et macies, et ad ultimum phthisis invadit. Quae ne mortem afferant, radix consiliginis ita, ut supra docuimus, perforatae auriculae inseritur, tum porri succus instar heminae pari olei mensurae miscetur, et cum vini sextario 2 potandus datur diebus compluribus. Interdum et tumor palati cibos respuit, crebrumque suspirium facit, et hanc speciem praebet, ut bos in latus[2] pendere videatur. Ferro palatum prodest[3] et sauciare, ut sanguis profluat, et exemptum valvulis ervum maceratum, viridemque frondem, vel aliud molle pabulum, dum sanetur, praebere.

3 Si in opere collum contuderit,[4] praesentissimum est remedium sanguis de aure emissus: aut si id factum non erit, herba, quae vocatur avia,[5] cum sale trita et imposita. Si cervix mota et deiecta est, considerabimus quam in partem declinet, et ex diversa auricula sanguinem detrahemus. Ea porro vena, quae in aure videtur esse amplissima, sarmento prius

[1] pelle *SAR*.
[2] bos in latus *Lundström*: bos lotus *SAR*.
[3] prodest *add. Schneider ex Vegetio* iv. 14.
[4] contuderit *S*: contunderit *AR*.
[5] avia *Aldus*: habia *SAR*.

them, it will be beneficial to make a concoction of bay-leaves and foment their backs with it while they are still warm and immediately after to massage them with a large quantity of oil and wine and to take hold of the hide all over the animal and draw it away. This is best done in the open air in burning sunshine. Some people mix dregs of oil with wine and fat and use it as a remedy after the fomentations mentioned above.

Remedies for ulcerated lungs and swellings of the palate and neck of an ox.

XIV. It is also a serious distemper when the lungs become ulcerated; it results in coughing and emaciation and finally in phthisis. To prevent these conditions from causing death, a root of lungwort, as we prescribed above, is inserted in a hole made in the ear and then about a *hemina* of the juice of leek is mixed with a like quantity of oil and given as a potion for several days with a *sextarius* of wine. 2 Sometimes too a swelling of the palate causes the animal to refuse its food and heave frequent sighs, and an impression is caused that it is hanging over towards one side.[a] It is beneficial also to make a wound in the palate with a knife, so that the blood may flow, and to administer bitter-vetch without its husk and soaked and green leaves or some other soft fodder, until the wound heals.

3 If in the course of its work the ox has his neck bruised, the most efficacious remedy is to draw blood from the ear, or, if that is not done, the herb called groundsel is crushed up with salt and placed on the part affected. If the neck is moved in a certain direction and hangs down, we shall examine and see to which side it declines and draw blood from the ear on the other side; moreover, what appears to be the largest vein in the ear is first beaten with a twig, and

[a] The text, however, seems to be in need of further correction.

verberatur. Deinde cum ad ictum intumuit, cultello solvitur; et postero die iterum ex eodem loco sanguis emittitur, ac biduo ab opere datur vacatio. Tertio deinde die levis iniungitur labor, et paulatim ad iusta 4 perducitur. Quod si cervix in neutram partem deiecta est, mediaque intumuit, ex utraque auricula sanguis emittitur. Qui cum intra triduum, cum bos vitium cepit, emissus non est, intumescit collum, nervique tenduntur, et inde nata durities iugum non 5 patitur. Tali vitio comperimus aptum[1] esse medicamentum ex pice liquida et bubula medulla et hircino sebo et vetere oleo aequis ponderibus compositum atque incoctum. Hac compositione sic utendum est. Cum disiungitur ab opere, in ea piscina, ex qua bibit, tumor cervicis aqua madefactus subigitur, praedictoque medicamento defricatur et 6 illinitur. Si ex toto propter cervicis tumorem iugum recuset, paucis diebus requies ab opere danda est. Tum cervix aqua frigida defricanda, et spuma argenti illinenda est. Celsus quidem tumenti cervici herbam, quae vocatur avia, ut supra dixi, contundi et imponi iubet. Clavorum, qui fere cervicem infestant, minor molestia est: nam facile oleo[2] per ardentem 7 lucernam instillato sanantur. Potior tamen ratio est custodiendi, ne nascantur,[3] neve colla calvescant, quae non aliter glabra fiunt, nisi cum sudore aut pluvia cervix in opere madefacta est. Itaque cum

[1] aptum *ex Vegetio l.c.*: autem *SAR*.
[2] oleo *et* sanantur *om. SAR*.
[3] nascantur *S*: nascatur *AR*.

then when it has swollen up as a result of the blows, it is opened with a lancet, and on the following day blood is again drawn from the same spot and the animal is given two days' rest from work. Then on the third day a light task is enjoined upon it, which is gradually increased until it does a full day's work. 4 If, however, the neck does not incline to either side but is swollen in the middle, blood is let from both ears. If bleeding is not performed within three days after the ox has got the disease, the neck swells up and the sinews become taut and as a result a hard lump is formed which cannot endure the pressure of the yoke. 5 For this kind of malady we have discovered a suitable remedy composed of liquid pitch, beef-marrow, goat's fat, and stale oil in equal quantities and cooked together. This compound should be used in the following manner: when the ox is unharnessed after its work, the swelling on its neck is moistened with water in the trough from which it drinks and then massaged and rubbed and smeared with the medicament described above. 6 If the animal absolutely refuses the yoke because of the swelling on its neck, it must be given a few days' rest from work; then the neck must be rubbed with cold water and anointed with litharge of silver. Celsus indeed recommends that to a swollen neck the herb called groundsel should, as I have already said, be crushed and applied. The warts which generally infest the neck constitute only a minor malady; for they can easily be cured with oil dripped on them from a burning lamp. 7 A better plan, however, is to take care that they do not form and that the necks of the oxen do not become bald, for they only become hairless when the neck is moistened by sweat or rain during work.

id accidit, pulveri [1] lateritio trito priusquam disiungantur, colla conspergi oportet: deinde cum assiccuerint,[2] subinde oleo imbui.

XV. Si talum aut ungulam vomer laeserit, picem duram et axungiam cum sulfure et lana succida involvito [3] candente ferro supra vulnus inurito. Quod idem remedium optime facit exempta stirpe, si forte surculum calcaverit, aut acuta testa vel lapide ungulam pertuderit [4]; quae tamen si altius vulnerata est, latius ferro circumciditur, et ita inuritur, ut supra praecepi: deinde spartea calceata per triduum suffuso

2 aceto curatur. Item si vomer crus sauciarit, marina lactuca, quam Graeci tithymallum vocant, admixto sale imponitur. Subtriti pedes eluuntur calefacta bubula urina: deinde fasce [5] sarmentorum incenso, cum iam ignis in favillam recidit, ferventi cineri [6] bos cogitur insistere, ac pice liquida cum oleo vel axungia cornua eius linuntur. Minus tamen claudicabunt armenta, si opere disiunctis multa frigida laventur pedes; et deinde suffragines coronaeque ac discrimen ipsum, quo divisa est bovis ungula, vetere axungia defricentur.[7]

[1] pulveri (pulvere) *Richter* (*Hermes* LXXX, 201): veteri *prior. edd.*
[2] ad siccum erit *S*: ad siccum erint *A*.
[3] involvito *Svennung*: involuta *SAR*.
[4] pertulerit *SAR*.
[5] fasce *R*: fasces *SA*.
[6] ferveti cineribus *A*: ferventi cineribus *SR*.
[7] defricentur *S*: defricetur *AR*.

When this happens, therefore, their necks ought to be sprinkled with dust made by grinding brick-work before they are unyoked; then, when their necks have dried, they ought to be moistened from time to time with oil.

Remedies for damage to pasterns and hoofs.

XV. If the pastern or hoof has been injured by the ploughshare, wrap round it hard pitch and axle-grease, bind it with sulphur and greasy wool and make a burn above the wound with a piece of red-hot iron. The same remedy has an excellent effect after the removal of a piece of wood from the hoof, if the ox has by chance trodden on a shoot or pierced its hoof with a sharp tile or stone. If, however, the wound is rather deep, a wider cut is made round it with a knife and it is then cauterized according to the method which I have described above; next the hoof is covered with a "slipper" made of broom and treated for three days with a suffusion of vinegar. 2 Also if an ox has damaged its leg on the ploughshare, sea-spurge,[a] which the Greeks call *tithymallus*, mixed with salt, is applied to the wound. The feet are rubbed underneath and are washed with warmed ox-urine; then a bundle of twigs is burnt and when now the fire has sunk to embers, the animal is made to stand on the glowing ashes and the horny parts of the hoof are anointed with liquid pitch mixed with oil or axle-grease. Cattle, however, will be less likely to go lame, if their feet are washed in plenty of cold water when they are unyoked after work, and if their hocks, the crowns of their hoofs and the division itself between the two halves of the hoofs are rubbed with stale axle-grease.

[a] *Euphorbia paralius.*

XVI. Saepe etiam vel gravitate longioris itineris,[1] vel cum in proscindendo aut duriori solo aut obviae radici obluctatur,[2] convellit armos. Quod cum accidit, e prioribus cruribus sanguis mittendus est: si dextrum armum laesit, in sinistro; si laevum, in dextro; si vehementius utrumque vitiavit, item in posterioribus cruribus venae [3] solventur. Praefractis 2 cornibus linteola sale atque aceto et oleo imbuta superponuntur, ligatisque per triduum eadem infunduntur. Quarto demum axungia pari pondere cum pice liquida, et cortice pineo levigata [4] imponitur. Et ad ultimum cum iam cicatricem ducunt, fuligo infricatur.

Solent etiam neglecta ulcera scatere verminibus: qui si mane perfunduntur aqua frigida, rigore contracti decidunt. Vel si hac ratione non possunt eximi, marrubium aut porrum conteritur, et admixto sale imponitur. Id celerrime necat praedicta ani-3 malia. Sed expurgatis [5] ulceribus confestim adhibenda sunt linamenta cum pice et oleo vetereque axungia, et extra vulnera eodem medicamento circumlinenda, ne infestentur a muscis, quae ubi ulceribus insederunt, vermes creant.

XVII. Est etiam mortiferus serpentis ictus, est et minorum [6] animalium noxium virus. Nam et vipera et caecilia saepe cum in pascuo bos improvide super-

[1] itineris *om. SAR.*
[2] obluctatur *S*: obluctatus *AR.*
[3] vene *S* : bene *AR.*
[4] pineo levigata *S* : pineolo vigata *A* : pineolo iugata *R.*
[5] expurgatis *S* : -i *AR.*
[6] minorum *ex Vegetio* : magnorum *codd.*

XVI. It often happens that an ox wrenches its shoulders either owing to the weight of its load on a somewhat prolonged journey or when, in breaking up the ground, it has to struggle against an unusually hard patch or a root which gets in its way. When this happens, blood must be drawn from its front legs—from the left leg if it has injured its right shoulder and from the right leg if the left shoulder is affected. If it has injured both shoulders rather seriously, veins will have to be opened in the hind legs as well. If the horns are broken, pieces of linen soaked in salt and vinegar and oil are put upon them and the same things poured over them for three days after they have been bound up; next on the fourth day axle-grease and liquid pitch in equal portions and pulverized pine-bark are applied, and, finally, when they are already beginning to scar over, they are rubbed with soot.

Remedies for sprained shoulders and damaged horns.

2

Ulcers, too, if they are neglected, generally swarm with worms. If they are drenched in the morning with cold water, they shrivel up with the cold and die. If they cannot be got rid of by this method, horehound or leek is pounded up and applied with a mixture of salt; this promptly kills these creatures. After the ulcers have been cleaned out, linen bandages must be immediately applied with pitch, oil and stale axle-grease, and the wounds must be anointed outside with the same medicament, so that they may not become infested by flies which, when they settle on the ulcers, breed worms.

3

XVII. The bite of a snake is also fatal to oxen, and the poison of certain lesser animals is also hurtful. For an ox while grazing often lies down unawares

Remedies for bites of snakes and other animals and for diseases of the eye.

cubuit, lacessita onere morsum imprimit. Musque araneus, quem[1] Graeci μυγαλῆν appellant, quamvis exiguis dentibus non exiguam pestem molitur. Venena viperae depellit super scarificationem ferro factam[2] herba, quam vocant personatam,[3] trita et 2 cum sale imposita. Plus etiam eiusdem radix contusa prodest, vel si montanum trifolium invenitur, quod confragosis locis efficacissimum nascitur, odoris gravis, neque absimilis bitumini, et idcirco Graeci eam ἀσφάλτειον appellant; nostri autem propter figuram vocant acutum trifolium: nam[4] longis et hirsutis foliis viret, caulemque robustiorem facit, 3 quam pratense. Huius herbae succus vino mixtus infunditur faucibus, atque ipsa folia cum sale trita malagmatis in vicem cedunt.[5] Vel si hanc herbam viridem tempus anni negat, semina eius collecta et levigata cum vino dantur potanda, radicesque cum suo caule tritae atque hordeaceae farinae et sali commixtae ex aqua mulsa scarificationi superponuntur. 4 Est etiam praesens remedium, si conteras fraxini tenera cacumina quinque librarum, cum totidem vini et duobus sextariis olei, expressumque[6] succum faucibus infundas; itemque cacumina eiusdem arboris cum sale trita laesae parti superponas.

Caeciliae[7] morsus tumorem suppurationemque molitur. Idem facit etiam muris aranei. Sed illius sanatur noxa subula aenea, si locum laesum com-

[1] quem *R* : quae *S* : que *A*.
[2] captam *SAR*.
[3] personatam (cf. Pliny, *N.H.* XXV. § 104): persona *SAac*.
[4] non *SAR*.
[5] in vicem cedunt *scripsi* : vicedunt *SA* : incendunt *R*.
[6] expressus quae *S* : expressusque *A*.
[7] celi *S* : caeli *A*.

upon vipers and lizards, which, provoked by its weight, inflict a bite upon it. The shrew-mouse, which the Greeks call *mygale*, though its teeth are small, gives rise to a malady which is far from being slight. A viper's poison can be expelled by scarifying with a knife the part affected and applying to it the herb called burdock, pounded up and mixed with salt. The crushed root of the same plant is 2 even more beneficial, or the mountain trefoil, which grows in rugged places and is most efficacious, if it can be found; it has a strong odour like that of bitumen, whence the Greeks call it *asphalteion*, but our countryfolk call it "sharp trefoil" from its shape, for it grows long, hairy leaves and forms a stouter stalk than the meadow trefoil. The juice of this herb 3 mixed with wine is poured down the throat, and the leaves themselves are pounded up with salt to form a poultice. If the season of the year makes it impossible to obtain this herb in a green state, its seeds are collected and pulverized and given with wine as a potion, while the roots are pounded up with their stalks and mixed with barley-flour and salt and, after being dipped in honey-water, are applied to the scarified part. A sovereign remedy is also provided 4 by crushing five pounds of tender tops of ash with the same number of *sextarii* of wine and two of oil and by pouring the juice which you have squeezed out down the animal's throat. You should also apply the tops of the same tree pounded up with salt to the part affected.

The bite of a lizard causes swelling and suppuration, as also does that of a shrew-mouse, but the injury caused by the former is cured if you puncture the part affected with a brazen awl and anoint it with

5 pungas, cretaque cimolia ex aceto linas. Mus perniciem, quam intulit, suo corpore luit: nam animal ipsum oleo mersum necatur, et cum imputruit, conteritur, eoque medicamine morsus muris aranei linitur. Vel si id non adest, tumorque[1] ostendit iniuriam dentium, cuminum conteritur, eique adicitur exiguum picis liquidae et axungiae, ut lentorem 6 malagmatis habeat. Id impositum perniciem commovet. Vel si antequam tumor discuteretur, in suppurationem convertitur, optimum est ignea lamina collectionem[2] resecare, et quicquid vitiosi est, inurere, atque ita liquida pice cum oleo linire. Solet etiam ipsum animal vivum creta figulari circumdari; quae cum siccata est, collo boum suspenditur. Ea res innoxium pecus a morsu muris aranei praebet.

7 Oculorum vitia plerumque melle sanantur. Nam sive intumuerunt, aqua mulsa[3] triticea farina conspergitur et imponitur: sive album in oculo est, montanus sal Hispanus vel Ammoniacus vel etiam Cappadocus, minute tritus et immixtus melli vitium extenuat. Facit idem trita sepiae testa, et per fistulam ter die oculo inspirata. Facit et radix, quam Graeci σίλφιον vocant, vulgus autem nostra 8 consuetudine[4] laserpitium appellant. Huius quantocunque ponderi decem partes salis ammoniaci adiciuntur, eaque pariter trita oculo similiter in-

[1] umorque *A*: umorquae *S*.
[2] collectionem *ex Vegetio*: convertionem *SA*.
[3] mulsa *A*²*R*: mulsae *S*.
[4] consuetudinem *SA*: consuetudine *ac*.

---

[a] Fuller's earth from Cimolus, an island in the Cyclades.
[b] From Ammon in the Libyan desert.

Cimolian chalk [a] dipped in vinegar. The shrew-mouse atones with its own body for the harm which it has inflicted; for the animal itself is killed by being drowned in oil, and, when it has putrefied, it is crushed and the bite inflicted by the shrew-mouse is anointed with it as a remedy. If this is not available and the swelling shows teeth-marks, cumin is crushed up and a little liquid pitch and axle-grease is added to it, so that it may have the soft consistency of a poultice. The application of this gets rid of the mischief. If the swelling turns into a suppuration before it is dispersed, it is best to cut away the abscess with a hot iron plate and burn away any harmful matter and then anoint the place with liquid pitch and oil. There is also a practice of encasing the shrew-mouse itself while still alive in potter's clay and, when the clay is dry, hanging it round the ox's neck. This renders the animal immune from the bite of a shrew-mouse. 5 6

Maladies of the eyes are generally cured with honey. If they have swollen up, wheaten flour is sprinkled with honey water and applied to the eyes; or, if there is a white film on the eye, Spanish or Ammoniac [b] or even Cappadocian rock-salt, pounded small and mixed with honey, lessens the malady. The shell of a cuttle-fish ground up and blown into the eye three times a day through a pipe has the same effect, as also has the root which the Greeks call *silphion* and of which the common name in our language is *laserpitium*.[c] To any quantity of this ten parts of Ammoniac salt are added; and both are poured similarly into the eye after being ground up in the same manner, or else the root of the same plant crushed up and 7 8

[c] Laserwort, *Ferula tingitana*.

funduntur, vel eadem radix tunsa[1] et cum oleo lentisci inuncta vitium expurgat. Epiphoram supprimit polenta conspersa mulsa aqua, et in supercilia genasque imposita; pastinacae quoque agrestis semina, et succus armoraceae, cum melle conlevata 9 oculorum sedant dolorem. Sed quotiensque mel aliusve succus remediis adhibetur, circumlinendus erit oculus pice liquida cum oleo, ne a muscis infestetur. Nam et ad dulcedinem et odorem[2] mellis aliorumque medicamentorum non hae solae, sed et apes advolant.

XVIII. Magnam etiam perniciem saepe affert hirudo hausta cum aqua. Ea adhaerens faucibus sanguinem ducit, et incremento suo transitum cibis praecludit. Si tam difficili loco est, ut manu trahi non possit, fistulam vel arundinem inserito, et ita calidum oleum infundito: nam eo contactum animal 2 confestim decidit. Potest etiam per fistulam deusti cimicis nidor immitti: qui ubi superpositus[3] igni fumum emisit, conceptum nidorem fistula usque ad hirudinem perfert; isque nidor depellit haerentem. Si tamen vel stomachum vel intestinum tenet, calido aceto per cornu infuso necatur. Has medicinas quamvis bubus adhibendas praeceperim, posse tamen ex eis[4] plurima etiam omni maiori pecori convenire nihil dubium est.

XIX. Sed et machina fabricanda est, qua clausa iumenta bovesque curentur, ut et propior[5] accessus

[1] tunsa *S* : contunsa *A*¹.
[2] oculorum *SAR*.
[3] superpositus *S* : superponuntur *AR*.
[4] is *S* : his *AR*.
[5] proprior *SAR*.

mixed with oil of mastic is used to anoint the eye and purges away the malady. Running at the eyes is stopped by pearl-barley sprinkled with honey-water and applied to the eyebrows and cheeks; wild parsnip seeds and the juice of the horse-radish diluted with smooth honey assuage pain in the eyes. But whenever honey or any other juice is introduced into the remedies employed, the eye will have to be anointed all round with liquid pitch and oil to prevent its being infested with flies; for not only flies but also bees are attracted to the sweetness and odour of honey and other medicaments. 9

Remedies for oxen which have swallowed leeches.

XVIII. Much harm too is often caused by a leech swallowed with the drinking-water, which, fastening on the throat, sucks the blood and blocks the passage of food with its own added bulk. If the leech is in such a difficult place that it cannot be removed by hand, you should insert a pipe or reed and then pour in warm oil; for if this touches it, the leech immediately falls off. The odour from a burnt bug may also be introduced through a pipe (for when a bug is put upon the fire and has produced smoke, the vapour given off reaches the leech through a pipe) and this vapour dislodges the leech from its clinging hold. If, however, it is attached to the stomach or intestine, it can be killed by pouring hot vinegar through a horn. Though I have prescribed these remedies to be used for oxen, most of them are certainly suitable also for all the larger kinds of cattle. 2

How to construct a machine for holding cattle when they are being treated.

XIX. It is necessary also to construct a machine in which one can enclose beasts of burden and oxen and treat them, in order that those who are applying remedies may have readier access to their patients

ad pecudem medentibus sit, nec in ipsa curatione quadrupes reluctando remedia respuat. Est autem talis machinae forma: roboreis axibus compingitur solum, quod habet in longitudinem pedes novem, et in latitudinem pars prior dipundium semissem, pars 2 posterior quattuor pedes. Huic solo septenum pedum stipites recti ab utroque latere quaterni applicantur. Ii autem in ipsis quattuor angulis affixi sunt, omnesque transversis sex temonibus quasi vacerrae inter se ligantur,[1] ita ut a posteriore parte, quae latior[2] est, velut in caveam quadrupes possit induci, nec exire alia parte prohibentibus adversis axiculis. Primis autem duobus statuminibus imponitur firmum iugum, ad quod iumenta capistrantur, vel boum cornua religantur. Ubi potest etiam numella[3] fabricari, ut inserto capite descendentibus per foramina regulis 3 cervix catenetur. Ceterum corpus laqueatum et distentum temonibus obligatur, immotumque medentis arbitrio est expositum. Haec ipsa machina communis erit omnium maiorum quadrupedum.

XX. Quoniam de bubus satis praecepimus, opportune de tauris vaccisque dicemus. Tauros maxime membris amplissimis, moribus placidis, media aetate probandos censeo. Cetera fere eadem omnia in his observabimus, quae in bubus eligendis. Neque enim alio distat bonus taurus a castrato, nisi quod huic torva facies est, vegetior aspectus, breviora cornua, torosior cervix, et ita vasta, ut sit maxima portio

[1] ligantur *S* : ligatur *AR*.
[2] latior *c ed. pr.* : laterior *SAR*.
[3] numella *S* : numelli *AR*.

[a] The details of the construction are not altogether clear, and the text appears in need of emendation.

and that these quadrupeds, while they are actually being doctored, may not struggle and reject the remedies. The shape of this machine is as follows: a piece of ground nine feet long and two and a half feet wide in front and four feet wide at the back is floored with boards of oak. In this space four upright 2 posts seven feet high are placed on the right and left sides; they are set upright in the four corners and are all bound to each other with six cross-poles [a] to form a kind of railing, so that the animal can be driven in from the back, which is broader, as into a cage, but cannot get out on any of the other sides, because the bars get in his way and prevent him. On the two front posts a stout yoke is placed, to which beasts of burden are fastened with halters and oxen tied by their horns, and you can also contrive here stocks, so that, when the animal's head has been inserted, bars may descend and pass through holes and the neck thus be held tight. The rest of the body, secured 3 with nooses and stretched out, is bound to the cross-poles and is subject to the will of the person who is doctoring the animal. This machine will serve alike for all the greater quadrupeds.

Bulls.

XX. Now that we have given enough instruction about oxen, it will be proper to deal next with bulls and cows. In my opinion we ought to esteem most highly bulls which have very large limbs and a calm temperament and are not too young or too old. In other respects we shall look for much the same qualities as we sought when choosing oxen. For a good bull does not differ from a gelded ox except that its expression is fierce, its appearance more animated, its horns shorter, its neck more brawny and so huge as to form the greatest part of its body; its belly is

corporis, venter[1] paulo subtruncior, qui magis rectus[2] et ad ineundas feminas habilis sit.

XXI. Vaccae quoque probantur altissimae formae longaeque, maximis uteris, frontibus latissimis, oculis nigris et patentissimis, cornibus venustis et levibus et nigrantibus, pilosis auribus, compressis malis, palearibus et caudis amplissimis, ungulis modicis, et cruribus parvis.[3] Cetera quoque fere eadem in feminis, quae et in maribus, desiderantur, et praecipue ut sint novellae: quoniam, cum excesserunt 2 annos decem, fetibus inutiles sunt. Rursus minores bimis iniri non oportet. Si ante tamen conceperint, partum earum removeri placet, ac per triduum, ne laborent, ubera exprimi, postea mulctra prohiberi.

XXII. Sed et curandum est omnibus annis aeque ac in reliquis gregibus pecoris, ut delectus habeatur. Nam et enixae[4] et vetustae,[5] quod gignere desierunt, summovendae sunt, et utique taurae, quae locum fecundarum occupant, ablegandae vel aratro domandae; quoniam laboris et operis non minus quam iuvenci propter uteri sterilitatem patientes sunt. Eiusmodi armentum maritima et aprica 2 hiberna desiderat; aestate[6] opacissima nemorum et montium,[7] elata[8] magis quam plana pascua. Nam

[1] venter *Schneider*: ventre *SAR*.
[2] rectus *ed. pr.*: treus *S*: reus *AR*.
[3] parvis *dubitanter add. Lundström.*
[4] enixae *Ald.*: et visae *S*: et vise *AR*.
[5] vetustate *SAR*.
[6] e statim *S*: aestatim *A*.
[7] opacissima nemorum et montium *Lundström*: opicis morum omnium *SAR*: opacis nemorum omnium *a*.

rather less developed underneath, so that it forms a straighter line and is more convenient for coupling with the female.

Cows.

XXI. Cows also are most highly esteemed which are very tall and long in shape, with large bellies, very broad foreheads, eyes black and very wide-open, horns elegant, smooth and inclined to blackness, hairy ears, compressed cheek-bones, very large dewlaps and tails, hoofs of moderate size, and small legs. In other respects almost the same qualities are desirable in the females as in the males; above all things they should be young, since, when they have passed ten years, they are useless for breeding. On the other hand they should not be covered by the bulls when they are less than two years old; if, how- 2 ever, they conceive before reaching two years, it is thought proper that their young should be taken from them and their udders emptied for three days that they may not feel pain, and that after that they should be kept away from the milk-pail.

Annual review of the herd.

XXII. You should also take care to hold an examination of your cows, as of all herds of cattle, every year; for those which have done with calf-bearing and are old, since they have ceased bearing, should be removed, and barren cows in particular, which are occupying the place of the fertile, must be got rid of or broken in to the plough; for on account of their sterility they can endure toil and work quite as well as bullocks. This kind of cattle requires sunny 2 pasture-ground near the sea in the winter; but in summer they like the shadiest parts of the woods or mountains and pasturage on high ground rather than

---

[8] elata *Heinsius*: ac laeta *S*: ac leta *AR*: alta *a*.

melius nemoribus herbidis et frutectis [1] et carectis [2] pascitur,[3] quoniam siccis ac lapidosis locis durantur ungulae. Nec tam fluvios [4] rivosque desiderat, quam lacus [5] manu factos; quoniam et fluvialis [6] aqua, quae fere frigidior est, partum abigit, et caelestis iucundior est. Omnis tamen externi frigoris tolerantior equino armento vacca est, ideoque facile sub dio hibernat.

XXIII. Sed laxo spatio consepta facienda sunt, ne in angustiis conceptum altera alterius elidat, et ut invalida fortioris ictus effugiat. Stabula sunt optima saxo aut glarea strata, non incommoda tamen etiam sabulosa, illa, quod imbres respuant, haec, quod celeriter exsorbent transmittuntque. Sed utraque devexa sint, ut humorem effundant; spectentque ad meridiem, ut facile siccentur, et frigidis ventis non 2 sint [7] obnoxia. Levis autem cura pascui est. Nam ut laetior herba consurgat, fere ultimo tempore aestatis inceditur. Ea res et teneriora pabula recreat, incensis sentibus duris [8] et fruticem surrecturum in altitudinem compescit. Ipsis vero corporibus affert salubritatem iuxta conseptum saxis et canalibus sal superiectus, ad quem saturae pabulo libenter recurrunt, cum pastorali signo quasi receptui canitur. 3 Nam id quoque semper crepusculo fieri debet, ut ad sonum buccinae pecus, si quod in silvis substiterit,

[1] frutetis *ed. pr.* : fructibus *SAR.*
[2] curetis *SAR* : caretis *ed. pr.*
[3] pascitur,[1] *add. Schneider.*
[4] pluvios *SAR* : fluvios *a.*
[5] lacus *R* : lacu *SA.*
[6] pluvialis *SAR* : fluvialis *a.*
[7] sit *SAR.*
[8] incensis sentibus duris et *Lundström* : dentis durib; *S* : dentibus duribus *A*[1] : dentibus duris *R* : sentibus duris, *ed. pr.* : incensis aridis *Palladius,* IX. 4.

in the plain; for it is better for them to feed in grassy woods and places covered with bushes and sedge-beds, since in dry, stony places their hoofs become hard. They do not require rivers and streams so much as artificial ponds, since river-water, which is generally colder, causes abortion, while rain-water is pleasanter to the taste. Cows, however, endure every outdoor cold better than horses and so can easily pass the winter under the open sky.

Enclosures and cow-sheds.

XXIII. Enclosures must be constructed which allow ample space, so that one cow may not in narrow quarters cause abortion in another and that a feeble cow may avoid the blows of a stronger. The best cow-sheds are floored with stone or gravel, though sandy floors are also suitable, the former because they keep out rainwater, the latter because they quickly absorb it and drain it away. In either case they must be shelving, so as to make the moisture flow away, and they should face the south that they may dry easily and not be exposed to the cold winds. 2 The care of the pasturage is a small matter; for, in order that the grass may grow more abundantly, it is usually burnt in the last part of the summer. This makes the fodder more tender when it grows again, since the hard briers are burnt, and it keeps down the bushes which would grow to a great height. Salt sprinkled on the stones and water-courses near the enclosures contributes to the good bodily health of the cattle and they gladly have recourse to it after they have eaten their fill, when what may be called the cowherd's signal for retreat is sounded; 3 for this too ought always to be given at dusk, so that any cattle which have remained in the woods may be accustomed, when the horn

saepta repetere consuescat. Hic enim recognosci grex poterit, numerusque[1] constare si velut ex militari disciplina intra stabularii[2] castra manserint. Sed non eadem in tauros exercentur imperia, qui freti viribus per nemora vagantur, liberosque egressus et reditus habent, nec revocantur nisi ad coitus feminarum.

XXIV. Ex eis,[3] qui quadrimis minores sunt maioresque quam[4] duodecim annorum, prohibentur admissura: illi,[5] quoniam quasi puerili aetate seminandis armentis parum idonei habentur; hi, quia senio sunt effeti.[6] Mense Iulio feminae maribus plerumque permittendae, ut eo tempore conceptos 2 proximo vere adultis iam pabulis edant.[7] Nam decem mensibus ventrem perferunt, neque ex imperio magistri, sed sua sponte marem[8] patiuntur.[9] Atque in id fere[10] quod dixi tempus, naturalia congruunt desideria, quoniam satietate verni pabuli pecudes exhilaratae lasciviunt in venerem, quam si aut femina recusat, aut non appetit taurus, eadem ratione, qua fastidientibus equis mox praecipiemus, elicitur cupiditas odore genitalium admoto naribus. 3 Sed et pabulum circa tempus admissurae subtrahitur feminis, ne eas steriles reddat nimia corporis obesitas;

[1] numerumque *SAR*.
[2] stabularii *ed. pr.*: stabularum *SAR*.
[3] is *S*: his *AR*.
[4] quam *Ald.*: cum *SAR*.
[5] illi *ed. pr.*: illa *SAR*.
[6] effecti *A*[1]: effeti *SA*[2].
[7] edant *ed. pr.*: edat *SAR*.
[8] -que *post* marem *AR*; quae *S*.
[9] patitur *SAR*.
[10] ferre *SA*: fere *ac*.

sounds, to seek their enclosures. Here it will be possible to pass the herd in review and its numbers can be verified, if, as though under military discipline, they occupy the quarters assigned to them by the keeper of the stalls. But the same strict rules are not imposed upon the bulls, which, relying on their strength, wander about in the woods and have free exit and return and are only recalled when they are required to cover the females.

The breeding of cattle.

XXIV. Bulls which are less than four years old and more than twelve are prevented from mounting the females, the former because, being as it were in their infancy, they are regarded as hardly suitable for breeding purposes, the latter because they are worn out with old age. The females are generally allowed to consort with the males in the month of July, in order that they may give birth to the young which are conceived at this time in the following spring, when the fodder has already come to perfection; for the 2 period of gestation is ten months. The cows do not admit the male at their owner's command but of their own accord and their natural desires coincide generally with the time of year which I have mentioned, since exhilarated by the abundance of food which the spring provides they become wanton and desire intercourse. If the female refuses intercourse or the bull feels no desire for her, the same method is employed as we shall presently prescribe for the stallion who shows distaste for the mare, namely desire is stimulated by bringing to the nostrils the odour of the genital parts. Also towards the time 3 when the females are to be covered their food is reduced, so that excessive fatness may not render them barren, while the diet of the bulls is increased, so that

et tauris adicitur, quo fortius ineant. Unumque marem quindecim vaccis sufficere abunde est. Qui ubi iuvencam supervenit, certis signis comprehendere licet, quem sexum generaverit: quoniam si parte dextra desiluit, marem seminasse manifestum est; si laeva, feminam. Id tamen[1] verum esse non aliter apparet, quam si post unum coitum forda non ad-
4 mittit taurum: quod et ipsum raro accidit. Nam quamvis plena fetu non expletur libidine: adeo ultra naturae terminos etiam in pecudibus plurimum pollent blandae voluptatis illecebrae.

Sed non dubium est, ubi pabuli sit laetitia, posse omnibus annis partum educari; at ubi penuria est, alternis submitti: quod maxime in operariis vaccis fieri placet, ut et vituli annui temporis spatio lacte satientur, nec forda simul operis et uteri gravetur[2] onere. Quae cum partum edidit, nisi cibis fulta est, quamvis bona nutrix, labore fatigata nato subtrahit
5 alimentum. Itaque et fetae cytisus viridis[3] et torrefactum hordeum,[4] maceratumque ervum praebetur, et tener vitulus[5] torrido molitoque milio et permixto cum lacte salivatur. Melius etiam in hos usus Altinae vaccae parantur, quas eius regionis incolae cevas[6] appellant. Eae sunt humilis staturae,

[1] tam *SAR*.
[2] gravetur *S* : graventur *AR*.
[3] viridis *ed. pr.* : viri *SAR*.
[4] in horreum *SAR*.
[5] tener vitulus *Pontedera* : tenuervitolus *S*[1] : tenuē vitulus *S*[2] : teneruit olus *A*.
[6] gevas *S* : cevas *Aac*.

---

[a] *I.e.* from the point of view of nursing their young.
[b] A town near Venice.
[c] This word is probably the origin of the Low German *Keue*.

they may put more energy into the sexual act. One bull is quite enough for fifteen cows; and, when it has covered a heifer there are definite signs by which you can tell what is the sex of the offspring which it has begotten; since, if he uncouples towards the right side, it is clear that he has begotten a male, if towards the left, a female. But whether this is really true is only apparent when after one copulation the pregnant cow refuses to admit the bull again, and this actually happens only rarely; for although 4 the cow may have conceived, she is not satisfied in her desires; so true is it that the seductive allurements of pleasure exercise the greatest power even over cattle beyond the bounds prescribed by nature.

There is no doubt that where there is a great luxuriance of fodder, a calf can be reared from the same cow every year, but, where food is scarce, the cow must be used for breeding only every other year. This rule is particularly observed where cows are employed for work, in order that, firstly, the calves may have abundance of milk for the space of a year, and, secondly, that a breeding cow may not have to bear the burden of work and pregnancy at the same time. When she has given birth to a calf, however good a mother she may be, if she is worn out by work, she denies the calf its due nourishment if her diet does not give her enough support. That 5 is why green shrub-trefoil and toasted barley and sodden bitter-vetch are given to a cow which has borne a calf, and her tender young is given a drench of grilled millet ground up and mixed with milk. For these purposes [a] too it is better to procure cows from Altina,[b] which the inhabitants of that region call *cevae*.[c] They are of low stature and produce an

lactis abundantes, propter quod remotis earum fetibus, generosum pecus alienis educatur uberibus: vel si hoc praesidium non adest, faba fresa et vinum recte tolerat, idque praecipue in magnis gregibus fieri oportet.

XXV. Solent autem vitulis nocere lumbrici, qui fere nascuntur cruditatibus. Itaque moderandum est, ut bene concoquant: aut si iam tali vitio laborant, lupini semicrudi conteruntur, et offae salivati more faucibus ingeruntur. Potest etiam cum arida fico et ervo conteri herba Santonica, et formata in offam, sicut salivatum demitti. Facit idem axungiae pars una tribus partibus hyssopi permixta. Marrubii quoque succus et porri valet eiusmodi necare animalia.

XXVI. Castrare vitulos Mago censet, dum adhuc teneri sunt; neque id ferro facere, sed fissa [1] ferula comprimere testiculos, et paulatim confringere. Idque optimum genus castrationum putat, quod 2 adhibetur aetati tenerae sine vulnere. Nam, ubi iam induruit, melius bimus quam anniculus castratur. Idque facere vere [2] vel autumno luna decrescente praecipit, vitulumque ad machinam deligare: deinde prius quam ferrum admoveas, duabus angustis [3]

[1] fissa *S* : ipsa *AR*.
[2] vere *S* : *om.* *AR*.
[3] angustis *S* : angustiis *Aac*.

---

[a] *Herba Santonica* according to Pliny (*N.H.* XXVII. § 28) was a kind of *absinthium* or wormwood found in the territory of the Santoni in the province of Aquitania: the name of the town of Saintes in the department of Charentes Inférieure is derived from this tribe.

[b] Described in Chapter XIX above.

abundance of milk, for which reason, if their own young are taken from them, excellent cattle can be reared at the udders of cows who are not their mothers; or if this resource is not available, the calf puts up quite well with crushed beans and wine. This plan should be adopted particularly in large herds.

Remedies for worms in calves.

XXV. Worms, which generally occur when indigestion is present, are often harmful to calves. Their feeding, therefore, must be so regulated that they digest properly; or, if they are already suffering from a malady of this kind, half-raw lupines are crushed and pellets of them thrust down their throats to serve as a drench. Wormwood [a] can also be ground up with dried figs and bitter-vetch and made up into pellets and thrust down their throats to act as a drench. The same effect is produced by one part of axle-grease mixed with three parts of hyssop; also the juice of horehound and of leek is effectual for killing creatures of this kind.

The castration of calves.

XXVI. Mago is in favour of castrating calves while they are still young and tender, and he advises that the operation should not be performed with a knife but that the testicles should be compressed with a piece of cleft fennel and gradually broken up. He considers this to be the best method of castration, because it is applied when the animal is still tender and causes no wound. 2 When the animal has grown tougher, it is better that it should be castrated as a two-year-old than as a one-year-old. He recommends that the operation should take place in the spring or in the autumn when the moon is waning, and that the calf should be bound in the machine [b]; then, before applying the knife, you should seize

ligneis regulis veluti forcipibus apprehendere testium nervos, quos Graeci κρεμαστῆρας ab eo appellant, quod ex illis genitales partes dependunt. Comprehensos deinde testes ferro reserare, et expressos ita recidere, ut extrema pars eorum adhaerens praedictis nervis 3 relinquatur. Nam hoc modo nec eruptione sanguinis periclitatur iuvencus, nec in totum effeminatur adempta omni virilitate; formamque servat maris cum generandi vim deposuit, quam tamen ipsam non protinus amittit. Nam si patiaris eum a recenti curatione feminam inire, constat ex eo posse generari. Sed minime id permittendum, ne profluvio sanguinis intereat. Verum vulnera eius sarmenticio cinere cum argenti spuma linenda sunt, abstinendusque eo 4 die ab humore, et exiguo cibo alendus. Sequenti[1] triduo velut aeger cacuminibus arborum et desecto viridi pabulo oblectandus, prohibendusque multa potione. Placet etiam pice liquida et cinere cum exiguo oleo ulcera ipsa post triduum linere, quo et celerius cicatricem ducant, nec a muscis infestentur. Hactenus de bubus dixisse abunde est.

XXVII. Quibus cordi est educatio generis equini, maxime convenit providere actorem[2] industrium et pabuli copiam: quae utraque vel mediocria possunt aliis[3] pecoribus adhiberi. Summam sedulitatem et largam satietatem desiderat equitium. Quod ipsum

[1] sequensei *S*: sequens *AR*.
[2] actorem *Gesner*: auctorem *SAR*.
[3] aliis *S*: alias *AR*.

between two narrow laths of wood, as in a forceps, the sinews of the testicles, which the Greeks call "hangers," because the genital parts hang from them, and then take hold of the testicles and lay them open with a knife and after pressing them out cut them off in such a way that their extremities are left adhering to the said sinews. By this method the steer 3 runs no danger from an eruption of blood, nor is it likely to lose its masculinity and become totally effeminate, and it keeps the form of a male when it has been deprived of generative power. This, however, it does not lose immediately; for, if you allow it to cover a cow directly after the operation, it is certain that it is possible for it to beget offspring; but it should by no means be allowed to do so, lest it die from a flux of blood. The wounds should be anointed with the ash of brushwood and litharge of silver, and the animal should be kept away from water for that day and be fed on only a little food. For the 4 three following days it should be treated as a sick animal and tempted to eat with the tops of trees and green fodder cut off for it and must not be allowed to drink much. It is thought right also to anoint the actual sores after three days with liquid pitch and ashes mixed with a little oil, so that they may scar over more quickly and that they may not be infested by flies. I have now said enough about oxen.

Horses.

XXVII. For those whose pleasure it is to rear horses it is of the utmost importance to provide a painstaking overseer and plenty of fodder; both these points can be neglected up to a certain point in dealing with other domestic animals. A stud of horses, however, requires the most assiduous attention and a generous diet. Horses themselves fall

tripartite [1] dividitur. Est enim generosa materies, quae circo sacrisque certaminibus equos praebet. Est mularis, quae pretio fetus sui comparatur generoso. Est et vulgaris, quae mediocres feminas maresque progenerat. Ut quaeque est praestantior, 2 ita ubere campo pascitur. Gregibus autem spatiosa et palustria montana pascua eligenda sunt, rigua nec unquam siccanea,[2] vacuaque [3] magis quam stirpibus impedita frequentibus,[4] mollibus [5] potius quam 3 proceris herbis abundantia. Vulgaribus equis passim maribus ac feminis [6] pasci permittitur, nec admissurae certa tempora servantur.[7] Generosis circa vernum aequinoctium mares iniungentur, ut eodem tempore, quo conceperint,[8] iam laetis et herbidis campis post annum [9] parvo cum [10] labore fetum [11] educent. Nam mense [12] duodecimo [13] partum edunt. Maxime itaque curandum est praedicto tempore anni, ut tam feminis quam admissariis desiderantibus coeundi fiat potestas, quoniam id praecipue armentum, si prohibeas, libidinis exstimulatur furiis, unde etiam veneno inditum est nomen ἱππομανές, quod equinae cupidini 4 similem mortalibus amorem accendit. Nec dubium quin aliquot regionibus tanto flagrent ardore coeundi

[1] tripartito *SAR*.
[2] siccana *Sa* : sicana *AR*.
[3] bacuane *SAR*.
[4] frequentibus *S* : frequenter *Aac*.
[5] mollibus *S* : mollis *AR*.
[6] feminis *R* : finibus *SA*.
[7] servantur *S* : serventur *AR*.
[8] conceperint *S*² : coeperint *A*.
[9] *post* annum *add*. mensem *SAR*.
[10] parvo cum *S*²*Aac* : per vocum *S*¹.
[11] feitum *S* : fittũ *A*.
[12] mense *a* : mensem *SA*.

into three classes. There is the noble stock which supplies horses for the circus and the Sacred Games; then there is the stock used for breeding mules which in the price which its offspring fetches is a match for the noble breed; and there is the common breed which produces ordinary mares and horses. The more excellent each class is, the richer must be 2 the pasturage assigned to it. The feeding-grounds chosen for herds of horses must be spacious and marshy, mountainous, well-watered and never dry, empty rather than encumbered by many tree-trunks, and producing an abundance of soft rather than tall grass. The stallions and mares of the 3 common stock are allowed to be pastured everywhere together, and no fixed seasons are observed for breeding. The stallions of the noble stock will be put to the mares about the time of the spring equinox, so that the mares may be able to rear their offspring with little trouble, when the pasture is rich and grassy, at the same season a year later as that at which they conceived them; for they give birth to their young in the twelfth month. The greatest care, therefore, must be taken that at the said time of year every opportunity is given equally to mares as to their stallions to couple if they desire to do so, because, if you prevent them from doing so, horses beyond all animals are excited by the fury of their lust. (Hence the term "horse-madness" is given to the poison which kindles in human beings a passion like the desire in horses.) Indeed, in some regions, there is 4 no doubt that the mares are affected by such a burning desire for intercourse, that, even though

---

[13] duodecimo *a* : duodecima *SA*.

feminae, ut etiam si marem non habeant, assidua et nimia cupiditate figurando[1] sibi ipsae venerem cohortalium more avium vento concipiant. Neque enim poeta licentius dicit:

5 Scilicet ante omnes furor est insignis equarum.[a]

Illas ducit amor trans Gargara,[b] transque sonantem
Ascanium;[c] superant montes et flumina tranant,
6 Continuoque avidis ubi subdita flamma medullis,
Vere magis, quia vere calor redit ossibus, illae
Ore omnes versae ad Zephyrum, stant rupibus altis,
Exceptantque leves auras, et saepe sine ullis
Coniugiis, vento gravidae (mirabile dictu).[d]

7 Cum sit notissimum etiam in Sacro monte Hispaniae, qui procurrit in occidentem iuxta Oceanum, frequenter equas sine coitu ventrem[2] pertulisse fetumque educasse, qui tamen inutilis est, quod triennio, prius quam adolescat, morte absumitur. Quare, ut dixi, dabimus operam, ne circa aequinoctium vernum
8 equae desideriis naturalibus angantur.[3] Equos autem pretiosos reliquo tempore anni removere oportet a feminis, ne aut cum volent ineant aut, si id facere

[1] figurando *S* : figurandus *AR*.
[2] ventrem *S* : ventē *A* : ventum *R*.
[3] aguntur *SAR*.

---

[a] Vergil, *Georg.* III. 266 and 269–275.
[b] The highest peak of the range of Mt. Ida.
[c] A river of Bithynia (Strabo, XIV. 681).
[d] The story of the impregnation of mares by the wind seems to be as old as Homer (*Il.* XVI. 150).

there is no stallion at hand, owing to their continuous and excessive passion, by imagining in their own minds the pleasures of love they become pregnant with wind, just as farmyard hens produce "wind-eggs." Indeed the poet is not indulging his fancy too much when he says:[a]

> But, beyond all furies, wondrous is the rage 5
> Of mares;
>
> Love leads them over Gargara[b]
> And o'er Ascanius'[c] loudly roaring stream;
> They scale the mountain and through rivers swim.
> Soon as the flame has reached their craving marrow 6
> (More so in spring, for then the heat returns
> And warms their bones) all on high rocks they stand
> Facing the west, and the light breezes catch,
> And oft with wind conceive, without the aid
> Of union—a wondrous tale to tell![d]

For it is also well-known that on the Holy Mountain 7 of Spain,[e] which runs westward near the Ocean, mares have often become pregnant without coition and reared their offspring, which, however, is of no use, because it is snatched away by death at three years of age, before it can come to maturity. Therefore, as I have said, we shall take care that the brood-mares are not tormented by their natural desires about the time of the spring equinox. But during 8 the rest of the year the valuable stallions should be kept away from the mares, so that they do not cover them whenever they wish, nor, if they are prevented

[e] Varro, *de Re Rustica* (II. 1. 9) says that this occurred in the district in which Olisipo, the modern Lisbon, was situated.

prohibeantur, cupidine sollicitati[1] noxam contrahant.[2] Itaque vel in longinqua pascua marem placet ablegari, vel ad praesepia contineri: eoque tempore, quo vocatur a feminis, roborandus est largo cibo, et appropinquante vere hordeo ervoque[3] saginandus, ut veneri supersit, quantoque fortior inierit, firmiora 9 semina praebeat futurae stirpi. Quidam etiam praecipiunt eodem ritu, quo mulos, admissarium saginare, ut hac sagina hilaris pluribus[4] feminis sufficiat. Verum tamen nec minus quam quindecim nec rursus plures quam viginti unus debet implere, isque admissurae post trimatum usque in annos 10 viginti plerumque idoneus est. Quod si admissarius iners in venerem est, odore proritatur, detersis spongia feminae locis, et admota naribus equi. Rursus si equa marem non patitur, detrita scilla naturalia eius linuntur, quae res accendit libidinem. Nonnunquam ignobilis quoque ac vulgaris elicit[5] cupidinem coeundi. Nam ubi admotus[6] fere tentavit obsequium feminae,[7] abducitur,[8] et iam patientiori generosior equus imponitur.

Inde maior praegnantibus adhibenda cura est, 11 largoque pascuo firmandae. Quod si frigore hiemis herbae defecerint, tecto contineantur, ac neque opere

[1] sollicitationis *S*: -i *A*.
[2] contrahant *S*: -unt *AR*.
[3] herboque *S*¹*A*: hervoq; *S*²: ervoque *c*.
[4] pluribus *S*: plurimis *AR*.
[5] elicit *S*, *ed. pr.*: eligit *AR*.
[6] admotu *SAR*.
[7] feminae *Ursinus*: femina *SR*: semina *A*.
[8] adducitur *SAR*.

from doing so, harm themselves through excitement due to their desires. It is better, therefore, either to banish a stallion in some distant pasture or else keep it shut up in the stables; then at the time when it is summoned by the mare, it should be fortified by a generous diet, and with the approach of spring should be fattened on barley and bitter-vetch, so that it may be equal to the fatigues of intercourse, and that, the stronger it is when it covers the mare, the greater may be the sexual vigour which it communicates to its future descendants. Some authorities also prescribe that one should 9 fatten up a stallion by the method used for mules, so that, exhilarated by this condition, it may suffice for a number of mares. However, one stallion ought to be able to impregnate not less than fifteen and on the other hand not more than twenty mares, and is generally suitable to breeding purposes from three years of age to twenty. But if a stallion is dis- 10 inclined for intercourse, he can be roused by the odour of a sponge, with which the parts of the mare have been wiped, applied to his nostrils. On the other hand, if the mare refuses to submit to the stallion, her parts are anointed with crushed squill, and this kindles her desire. Sometimes, too, a badly-bred ordinary horse is used to arouse in the mare a longing for copulation; for, when he has approached her and, so to speak, invited her compliance, he is led away and the better-bred horse is mated with the now more complaisant mare.

From the time when mares become pregnant they need special care and must be fortified by generous fodder. If the grass has failed owing to the cold of winter, they should be kept under cover and not be 11

neque cursu exerceantur, neque frigori committantur, nec in angusto clauso, ne aliae aliarum conceptus elidant: nam haec omnia incommoda fetum abigunt. Quod si tamen aut partu aut abortu equa laboravit, remedio erit felicula trita, et aqua tepida
12 permixta ac data per cornu. Sin autem prospere cessit, minime manu contingendus pullus erit.[1] Nam laeditur etiam levissimo contactu. Tantum[2] cura adhibebitur, ut et amplo et calido loco cum matre versetur, ne aut frigus adhuc infirmo noceat, aut mater in angustiis eum obterat. Paulatim deinde producendus erit, providendumque, ne stercore ungulas adurat. Mox cum firmior fuerit, in eadem pascua, in quibus mater est, dimittendus, ne
13 desiderio partus sui laboret equa. Nam id praecipue genus pecudis amore natorum, nisi fiat potestas, noxam trahit. Vulgari feminae solenne est omnibus annis parere, generosam convenit alternis continere, quo firmior pullus lacte materno laboribus certaminum praeparetur.

XXVIII. Marem putant minorem trimo non esse idoneum admissurae, posse vero usque ad vigesimum annum progenerare; feminam bimam recte concipere, ut post tertium annum enixa fetum educet:

[1] pullus erit *S* : polluerit *AR* : poluerit *c*.
[2] tanta *SAR*.

fatigued by work or journeys, and they should not be exposed to the cold nor enclosed in a narrow space lest they should cause one another to miscarry; for all these unfavourable conditions cause abortion. But if a mare has suffered either in producing its offspring or from abortion, polypody crushed and mixed with tepid water and administered through a horn will serve as a remedy. If, on the other hand, all 12 goes well, the foal must on no account be touched with the hand, for even the lightest contact is harmful. All that one will have to do is to take care that the foal lives with its mother in a place which is both roomy and warm, so that the cold may not hurt it while it is still weak and that its mother may not crush it because its quarters are narrow. Then gradually it will have to be made to leave the stable, and care must be taken that it does not burn its hoofs with dung. Soon, when it has become stronger, it must be sent out to the same pasture as its mother, so that the latter may not be afflicted through longing for its offspring; for this kind of animal especially 13 suffers through its love for its young, if it have not the opportunity for indulging it. An ordinary mare is in the habit of bearing a foal every year; but a well-bred mare ought to be pregnant in alternate years, in order that, receiving greater strength from its mother's milk, the foal may be prepared for the toil of the contests.

The age of a stallion.

XXVIII. It is generally thought that a stallion is not suitable for breeding purposes before it is three years old, and that it can continue to procreate until its twentieth year, but that it is all right for a mare to conceive at the age of two years, so that it is three years old when it bears and rears its young, and it is

eandemque post decimum non esse utilem, quod ex annosa matre tarda sit atque iners proles. Quae sive ut femina sive ut masculus concipiatur, nostri arbitrii fore Democritus affirmat, qui praecipit, ut, cum progenerari marem velimus, sinistrum testiculum admissarii lineo funiculo aliove quolibet obligemus; cum feminam, dextrum. Idemque in omnibus paene pecudibus faciendum censet.

XXIX. Cum vero natus est pullus, confestim licet indolem aestimare, si hilaris, si intrepidus, si neque conspectu novae rei neque [1] auditu [2] terretur, si ante gregem procurrit, si lascivia et alacritate interdum et cursu certans aequales [3] exsuperat,[4] si fossam sine cunctatione transilit, pontem flumenque transcendit, haec erunt honesti animi documenta.

2 Corporis vero forma constabit exiguo capite, nigris oculis, naribus apertis, brevibus auriculis et arrectis, cervice molli lataque nec longa, densa iuba [5] et per dextram partem profusa, lato et musculorum toris numeroso pectore, grandibus armis et rectis, lateribus inflexis, spina duplici, ventre substricto, testibus paribus et exiguis, latis lumbis et subsidentibus, 3 cauda longa et setosa crispaque, mollibus atque altis rectisque cruribus, tereti genu parvoque neque introrsus spectanti, rotundis clunibus, feminibus torosis ac numerosis, duris ungulis et altis et concavis rotundisque, quibus coronae mediocres superpositae sunt. Sic universum corpus compositum, ut sit grande,[6] sublime, erectum, ab aspectu quoque agile,

[1] nove rei neq. *S* : noveque rei *AR*.
[2] auditu *S* : audita ut *AR*.
[3] aequalis *S* : exequalis *AR*.
[4] exuperat *AR* : exuberat *S*.
[5] iuba *c* : iuva *S* : tuta *A*.
[6] glande *S*¹*A* : grande *ac*.

also considered to be of no use after the tenth year, because the offspring of an aged mother is slow and lazy. Democritus declares that it will rest with us whether a male or a female is conceived, since he directs us, if we wish that a male should be begotten, to tie up the stallion's left testicle with a flaxen cord or some other material, and the right testicle if we want a female offspring; and he thinks that the same method should be adopted with almost all other cattle.

The qualities of a horse.

XXIX. As soon as a foal is born, it is possible to judge its natural qualities immediately. If it is good-humoured, if it is courageous, if it is not alarmed by the sight or sound of something unfamiliar, if it runs in front of the herd, if it surpasses its age-mates in playfulness and activity on various occasions and when competing in a race, if it leaps over a ditch and crosses a bridge on a river without baulking—these are the signs of generous mettle.

2 Its physical form will consist of a small head, dark eyes, wide-open nostrils, short, upstanding ears, a neck which is soft and broad without being long, a thick mane which hangs down on the right side, a broad chest covered with well-proportioned muscles, the shoulders big and straight, the flanks arched, the back-bone double, the belly drawn in, the testicles well matched and small, the loins broad and sunken, 3 the tail long and covered with bristling, curly hair, the legs soft and tall and straight, the knee tapering and small and not turned inwards, the buttocks round, the haunches brawny and well-proportioned, the hoofs hard, high, hollow and round with moderately large crowns above them; the whole body must be so formed as to be large, tall and erect, and also active

et ex longo, quantum figura permittit, rotundum. 4 Mores autem laudantur, qui sunt ex placido concitati, et ex concitato mitissimi. Nam hi et ad obsequia reperiuntur habiles, et ad certamina laboremque promptissimi. Equus bimus ad usum domesticum recte domatur; certaminibus autem expleto triennio: sic tamen ut post quartum demum annum labori committatur.

5 Annorum notae cum corpore mutantur. Nam dum bimus et sex mensium est, medii dentes superiores et inferiores cadunt. Cum quartum annum[1] agit his, qui canini appellantur, deiectis, alios affert. Intra sextum deinde annum molares superiores et inferiores[2] cadunt. Sexto anno, quos primos mutavit, exaequat. Septimo omnes explentur aequaliter, et ex eo cavatos gerit. Nec postea quot annorum sit, manifesto comprehendi potest. Decimo tamen anno tempora cavari incipiunt, et supercilia nonnunquam canescere, et dentes prominere. Haec, quae ad animum et mores corpusque et aetatem pertinent, dixisse satis habeo. Nunc sequitur curam recte et minus valentium demonstrare.

XXX. Si sanis[3] est macies, celerius torrefacto tritico, quam hordeo reficitur. Sed et vini potio danda

[1] annum *om. AR.*
[2] et inferiores *S* : *om. AR.*
[3] satis *SAR.*

---

[a] *I.e.* it should only contest after a year's training.

in appearance and, in spite of its length, rounded as far as its shape allows. As regards character, those 4 horses are esteemed which are roused to activity after being quiet and become very mild again after being roused ; for such animals are found to be both amenable to discipline and very ready to take part in public contests and the effort which they require. At two years of age a horse is suitable to be trained for domestic purposes; but, if it is to be trained for racing, it should have completed three years, and provided that it is entered for this kind of effort only after its fourth year.[a]

The signs which mark a horse's age change with its 5 physical changes. For when it is two years and six months old, its middle teeth, both the upper and the lower, fall out. In the course of its fourth year the so-called canine teeth are shed and it grows new ones in their place; then, before the end of its sixth year the upper and lower molars fall out, and in the course of the sixth year it makes up the number of the first set of teeth which it has changed; in the seventh year the whole set is completed, and henceforward the animal has some hollow teeth; and, subsequently, it is impossible to ascertain with certainty what its age is. In its tenth year, however, its temples begin to sink and its eyebrows sometimes begin to turn white and its teeth to project. I think I have said enough on the subject of the horse's disposition, character, physique and age. My next business is to set forth the way to look after horses in health and sickness.

Medicines for horses.

XXX. If a horse is thin without being ill, it can be restored to condition more quickly with roasted wheat than with barley; but it must also be given wine to

est, ac deinde paulatim eiusmodi cibi subtrahendi immixtis hordeo furfuribus, dum consuescat faba et puro hordeo ali.[1] Nec minus quotidie corpora pecudum quam hominum defricanda sunt: ac saepe plus prodest pressa manu subegisse terga, quam si largissime cibos praebeas.[2] Paleae vero equis

2 stantibus substernendae.[3] Multum autem refert robur corporis ac pedum [4] conservare.[5] Quod utrumque custodiemus, si idoneis temporibus ad praesepia, ad aquam, ad exercitationem pecus duxerimus, curaeque fuerit ut stabulentur sicco loco, ne humore madescant ungulae. Quod facile evitabimus,[6] si aut stabula roboreis axibus constrata, aut diligenter subinde emundata fuerit [7] humus, et paleae superiectae.[8]

3 Plerumque iumenta morbos concipiunt lassitudine et aestu, nonnunquam et frigore, et cum suo tempore urinam non fecerint; vel si sudant, et a concitatione confestim biberint; vel si, cum diu steterint, subito ad cursum extimulata sunt. Lassitudini quies remedio est, ita ut in fauces oleum vel adeps vino mixta infundatur. Frigori fomenta adhibentur, et calefacto oleo lumbi rigantur,[9] caputque et spina

4 tepenti adipe vel uncto liniuntur. Si urinam non facit, eadem fere remedia sunt. Nam oleum immixtum vino supra ilia et renes infunditur: et si hoc parum profuit, melle decocto et sale collyrium tenue

[1] ali *A*²*R* : alii *SA*¹.
[2] praebeat *A* : prebeat *SR*.
[3] paleae—substernandae *om. SAR*.
[4] pecudum *SA*¹*R* : pedum *A*².
[5] conservare *S* : servare *AR*.
[6] evitabimus *A*²*R* : evitavimus *SA*¹.
[7] fuerint *SAR*.
[8] superiactae *S* : superiecta *AR*.
[9] et—rigantur *om. AR*.

drink, and then by degrees foods of this kind must be reduced by mixing bran with barley until it becomes accustomed to a diet of beans and pure barley. The bodies of horses require a daily rubbing down just as much as those of human beings, and often to massage a horse's back with the pressure of the hand does more good than if you were to provide it most generously with food. Chaff ought to be spread on the ground where horses stand. It is also very important to maintain the vigour in their bodies and feet; we shall secure both these objects if we conduct the herd at suitable times to their stable, to their watering-place and to exercise, and if care is taken that they are stabled in a dry place, so that their hoofs are not wetted. This we shall easily avoid if the stable is floored with boards of hard wood, or if the ground is carefully cleaned from time to time and chaff thrown over it. 2

Beasts of burden generally fall ill from fatigue or from the heat, and sometimes also from the cold and when they have not passed urine at the proper time, or if they sweat and then drink immediately after having been in violent motion, or when they are suddenly spurred into a gallop after they have stood for a long time. Rest is the cure for fatigue, provided that oil or fat mixed with wine is poured down the throat. For a chill, fomentations are applied, and the loins moistened with heated oil, and the head and spine soaked with tepid fat or ointment. If the animal does not pass urine, the remedies are almost the same; for oil mixed with wine is poured over the flanks and loins, and if this has not produced the desired effect, a small suppository made of boiled honey and salt is applied to 3 4

inditur foramini,[1] quo manat urina,[2] vel musca viva, vel turis mica, vel de bitumine collyrium inseritur naturalibus. Haec eadem remedia adhibentur, si 5 urina genitalia deusserit. Capitis dolorem indicant lacrimae, quae profluunt, auresque flaccidae; et cervix cum capite aggravata, et in terram summissa. Tum rescinditur vena, quae sub oculo est, et os calda fovetur, ciboque abstinetur primo die. Inde postero autem potio ieiuno tepidae aquae praebetur ac viride gramen, tum vetus faenum vel molle stramentum substernitur, crepusculoque aqua iterum datur, parumque hordei cum vicialibus, ut per exiguas 6 potiones [3] cibi ad iusta perducatur. Si equo maxillae dolent, calido aceto fovendae, et axungia vetere confricandae sunt, eademque medicina tumentibus adhibenda est. Si armos laeserit, aut sanguinem demiserit,[4] medio fere in utroque crure [5] venae solvantur, et thuris polline cum eo qui profluit sanguine immixto, armi linantur, et ne plus iusto exanimetur, stercus ipsius iumenti fluentibus venis admotum [6] fasciis obligetur. Postero quoque die ex iisdem locis [7] sanguis detrahatur, eodemque modo curetur, et [8] hordeo abstineatur exiguo faeno dato. 7 Post triduum deinde usque in diem sextum porri succus instar trium cyathorum mixtus cum olei

[1] foramini *S* : -a *AR*.
[2] manat urina *S* : maturina *AR*.
[3] exiguas potiones *S* : exigua potione (portione *A*²) *A*¹.
[4] demiserit *S* : di- *AR*.
[5] crure *A*²*c* : cruore *SA*¹*R*.
[6] admotum *A*²*R* : -am *SA*¹.
[7] locis *om. AR*. [8] et *om. AR*.

the orifice from which the urine flows, or a live fly or a grain of incense or a suppository of bitumen is inserted in the genital organs. The same remedies will be applied, if the urine has scalded these organs. Head-aches are indicated by tears which flow from 5 the eyes and the hanging down of the ears, and the neck and head which are weighed down and droop towards the ground. In these circumstances the vein under the eyes is opened and the mouth fomented with hot water and the animal is kept away from food for the first day. Then on the next day, before it has eaten anything, it is given a drink of tepid water and some green grass; then a litter of old hay or soft straw is spread under it and, at dusk, water is again given and a little barley with haulm of vetch, so that by means of small doses the animal may be brought back to regular forms of food. If a horse's jaws give it pain, they should be fomented with hot 6 vinegar and rubbed with old axle-grease, and the same remedy should be applied if the jaws are swollen. If it has damaged its shoulders or has had an extravasation of blood to these parts, the veins somewhere near the middle of each leg should be opened and the shoulders should be anointed with a mixture of incense-dust and the blood which flows from the wound, and, that the animal may not be unduly weakened, some of its own ordure should be applied to the bleeding veins and bound with bandages. On the following day blood should again be drawn from the same places and the same treatment given, and the animal should be kept away from barley and only given a little hay. After three days and until the 7 sixth day the juice of a leek to the quantity of about three *cyathi* mixed with a *hemina* of oil should be

hemina faucibus per cornu infundatur. Post sextum diem lente ingredi cogatur, et cum ambulaverit, in piscinam demitti eum conveniet, ita ut natet: sic paulatim firmioribus cibis adhibitis[1] ad iusta per-
8 ducetur. At si bilis molesta iumento est, venter intumescit, nec emittit ventos, manus uncta inseritur alvo, et obsessi naturales exitus adaperiuntur, exemptoque stercore postea[2] cunila bubula et herba pedicularis cum sale trita et decocto[3] melli miscentur, atque ita facta collyria subiciuntur, quae
9 ventrem movent, bilemque omnem deducunt. Quidam myrrhae tritae quadrantem cum hemina vini faucibus infundunt, et anum[4] liquida pice oblinunt. Alii marina aqua lavant alvum, alii recenti muria.

Solent etiam vermes atque[5] lumbrici nocere intestinis; quorum signa sunt, si iumenta cum dolore crebro volutantur, si admovent caput utero, si caudam saepius iactant. Praesens medicina est, ita ut supra scriptum est, inserere[6] manum, et fimum eximere; deinde alvum marina aqua vel muria dura lavare, postea radicem capparis tritam cum sextario aceti[7] faucibus infundere; nam hoc modo praedicta intereunt animalia.

XXXI. Omni autem imbecillo pecori alte substernendum est, quo mollius cubet. Recens tussis celeriter sanatur, pinsita lente et a valvulis separata minuteque molita. Quae cum ita facta sunt,

[1] adivitis S^1A^1 : adiutus *R* : adibitis S^2.
[2] posite acunila *S* : posita ea *AR*.
[3] decoctos *SA* : -a *R*.
[4] anum S^2a : annũ S^1 : annul *A*.
[5] in qua *SAR*.
[6] insere *SA*.
[7] aceti *S* : cum aceto *AR*.

poured down its throat through a horn. After the sixth day it should be made to walk slowly and, after it has taken this exercise, it will be a good plan to drive it into a pond so that it may swim; then, by the administration by degrees of a more solid diet, it will be brought back to normal conditions. If a horse is 8 troubled by bile and its belly swells and it cannot get rid of wind, the hand is greased and inserted into its bowel and the natural exits which have been blocked are opened up; afterwards, when the ordure has been removed, ox-marjoram and lousewort crushed up with salt are mixed with boiled-down honey, so as to form a suppository, and inserted from below; these move the belly and bring away all the bile. Some people pour down the throat a 9 *quadrans* of ground myrrh in a *hemina* of wine and anoint the anus with liquid pitch; others wash out the bowel with sea-water, still others with fresh brine.

Tape-worms and maw-worms, too, often do harm to the intestines. It is a sign of their presence when horses roll about on the ground in internal pain or bring heads near their bellies or frequently flick their tails. An efficacious remedy is that described above, namely, the insertion of the hand and the removal of ordure followed by the washing out of the bowel with salt water or hard brine, and afterwards the pouring down the throat of the root of the caper-tree ground up with a *sextarius* of vinegar; for by this method the animals mentioned above are killed.

Remedies for a cough.

XXXI. When any animal is sick, deep litter must be provided, so that it may have a softer resting-place. A cough which has only just begun is quickly cured with crushed lentils separated from the pods and pounded into minute fragments. When

sextarius aquae calidae in eandem mensuram lentis miscetur, et faucibus infunditur; similisque medicina triduo adhibetur, ac viridibus herbis cacuminibusque arborum recreatur aegrotum pecus. Vetus autem tussis discutitur porri succo trium cyathorum cum olei hemina compluribus diebus [1] infuso, iisdemque, ut supra monuimus, cibis praebitis.

2 Impetigines et quicquid scabiei est [2] aceto et alumine defricantur. Nonnunquam, si haec permanent, paribus ponderibus mixtis nitro et scisso alumine cum aceto linuntur. Papulae [3] ferventissimo sole usque eo strigile raduntur, quoad eliciatur sanguis. Tum ex aequo miscentur radices agrestis hederae,[4] sulfurque et pix liquida cum alumine. Eo medicamine praedicta vitia curantur.

XXXII. Intertrigo bis in die subluitur aqua calida. Mox decocto ac trito sale cum adipe defricatur, dum sanguis emanet. Scabies mortifera huic quadrupedi est, nisi celeriter succurritur: quae si levis est, inter initia candenti [5] sub sole vel cedro [6] vel oleo lentisci linitur vel urticae semine et oleo detritis vel unguine ceti, quod in lancibus salitus thynnus re-
2 mittit. Praecipue tamen huic noxae salutaris est adeps marini vituli. Sed si iam inveteraverit, vehementioribus opus est remediis. Propter quod bitumen, et sulfur,[7] et veratrum [8] pici liquidae axungiaeque vetere [9] mixta pari pondere incoquuntur, atque

[1] diebus *add. Lundström.*
[2] scabiei est *Lundström* : scabies *SAR.*
[3] pabulo SA^1R : papulae A^2.
[4] herbe *SAR.*
[5] candentis *SAR.*
[6] cedro *S* : cedre *AR.*
[7] sulpure S^1A : sulphure S^2R.
[8] veratro *SAR.* [9] veteri *R* : veterio *SA.*

this has been done, a *sextarius* of hot water is mixed with the same quantity of lentils and poured down the animal's throat; the same treatment is continued for three days and the sick animal is strengthened by a diet of green grass and tree-tops. A cough of long standing can be dispelled by pouring down the throat on several days three *cyathi* of leek-juice in a *hemina* of oil and providing the same diet as we have prescribed above.

2 Remedy for skin diseases.

Skin-eruptions and any form of scab are rubbed with vinegar and alum. Sometimes, if these sores persist, they are anointed with equal quantities of soda and split alum mixed together in vinegar. Pustules are scraped with a curry-comb in very hot sunlight until blood is made to flow, then equal portions of the root of wild ivy, sulphur and liquid pitch are mixed with alum. The aforesaid ailments are treated with this medicament.

Remedies for chafing and scabies.

XXXII. Sores due to chafing are washed twice a day with hot water, and then they are rubbed with salt powdered and boiled with fat until the blood flows. Scabies is fatal to this kind of quadruped, unless help is speedily given. If the attack is only slight, in the first stages the sores should be anointed in burning sunlight with cedar-oil or mastic-gum or nettle seed and oil crushed together or the fish-oil which is deposited on dishes by salted tunnies. 2 The fat of the sea-calf is particularly efficacious against this malady. If, however, the trouble is of long standing, more violent remedies are needed; and so bitumen and sulphur and hellebore mixed with liquid pitch and stale axle-grease in equal quantities are boiled together, and the patients treated with this preparation, the sores having been previously

ea compositione curantur, ita ut prius scabies ferro 3 erasa perluatur urina. Saepe etiam scalpello usque ad vivum resecare et amputare scabiem profuit, atque ita factis ulceribus mederi liquida pice atque oleo, quae expurgant et replent vulnera. Quae[1] cum expleta sunt, ut celerius cicatricem et pilum ducant,[2] maxime proderit fuligo ex aeno ulceri infricata.

XXXIII. Muscas quoque vulnera infestantes summovebimus pice et oleo vel unguine infusis. Cetera ervi farina recte curantur. Cicatrices oculorum ieiuna saliva et sale defricatae[3] extenuantur: vel cum fossili[4] sale trita sepiae testa, vel semine agrestis[5] pastinacae pinsito et per linteum super 2 oculos expresso. Omnisque dolor oculorum inunctione succi plantaginis cum melle acapno,[6] vel si id non est, utique thymino celeriter levatur. Nonnunquam etiam per nares profluvium sanguinis periculum attulit, idque repressum est infuso naribus viridis coriandri succo.

XXXIV. Interdum et fastidio ciborum languescit pecus. Eius remedium est genus seminis quod git[7] appellatur, cuius duo cyathi triti diluuntur olei cyathis tribus et vini sextario, atque ita faucibus infunduntur. Sed[8] nausea discutitur etiam, si caput alii tritum cum vini hemina saepius potandum praebeas. Suppuratio melius ignea lamina quam frigido ferramento reseratur, et expressa postea linamentis

[1] aquae cum *S* : aeque quae cum *AR*.
[2] ducat *SAR* : ducant *a*.
[3] defricata *AR* : defricta *S*.
[4] fossili *S* : fusili *AR*.
[5] agrestibus *SAR*.
[6] acaprio *SAR*.
[7] git *S* : gis *AR*.
[8] sed *a* : det *S*¹*A*.

[a] Roman coriander (*Nigella sativa*).

scraped with a knife and thoroughly washed with urine. Often, too, it has been found beneficial to 3 cut the scab to the quick with a lancet and remove it and to treat the resulting sores with liquid pitch and oil, which both cleanse the wounds and cause them to fill up; when they have filled, soot from a brazen vessel rubbed into the sore will be found most beneficial in causing the wounds to scar over and grow hair.

Remedies against flies and for pain in the eyes.

XXXIII. We shall get rid of the flies which infest wounds by pouring on them pitch and oil or fat. The other kinds of sores are correctly treated with the flour of bitter-vetch. Scars on the eyes are reduced by rubbing with fasting spittle and salt or with the shell of a cuttle-fish pounded up with mineral salt or with the seed of the wild parsnip crushed and squeezed through linen over the eyes. Any kind of 2 pain in the eyes is quickly alleviated by anointing them with the juice of the plantain mixed with honey obtained without smoking out the bees, or, if this is not available, at any rate with thyme-honey. Sometimes bleeding at the nose has proved dangerous and has been stopped by pouring the juice of green coriander into the nostrils.

Remedies for nausea and emaciation.

XXXIV. A horse sometimes languishes through distaste for food. The remedy for this is a kind of seed called *git*,[a] two *cyathi* of which are crushed and dissolved in three *cyathi* of oil and one *sextarius* of wine and poured down the throat. Nausea can also be stopped by frequently giving the animal a bruised head of garlic in a *hemina* of wine to drink. It is better to open up an abscess with a red-hot metal plate than with a cold iron instrument, and when the pus has been squeezed out, it is dressed with lint.

2 curatur. Est etiam illa pestifera labes, ut intra paucos dies equae subita macie et deinde morte corripiantur: quod cum accidit, quarternos sextarios gari singulis per nares infundere utile est, si minoris formae sunt: nam si maioris, etiam congios. Ea res omnem pituitam per nares elicit, et pecudem expurgat.

XXXV. Rara quidem, sed et haec est equarum nota [1] rabies, ut cum in aqua imaginem suam viderint, amore [2] inani capiantur, et per hunc oblitae pabuli, tabe cupidinis intereant. Eius vesaniae [3] signa sunt, cum per pascua veluti extimulatae concursant, subinde ut circumspicientes requirere [4] ac desiderare 2 aliquid videantur. Mentis error discutitur, si decidas inaequaliter comas equae et eam [5] deducas ad aquam. Tum demum speculata [6] deformitatem suam, pristinae imaginis abolet [7] memoriam. Haec de universo equarum genere satis dicta sunt. Illa proprie praecipienda sunt iis [8] quibus mularum greges curae est submittere.

XXXVI. In educando genere mularum antiquissimum est diligenter exquirere atque explorare parentem futurae prolis feminam et marem: quorum si alter alteri [9] non est idoneus, labat etiam quod ex 2 duobus fingitur. Equam convenit quadrimam [10] usque in annos decem amplissimae atque pulcher-

[1] nota *S* : non *AR*.
[2] amore *S* : more *AR*.
[3] vasae sapiae *A* : vase sapie *SR*.
[4] requirĩt *S* : requirit *AR*.
[5] decidas—eam *emend. Lundström praeeunte Svennungio.*
[6] speculatae *ed. pr.* : speculata *codd.*
[7] abolent *ed. pr.* : abolet *codd.*
[8] his *SA*.
[9] alter alteri *Schneider* : alteri *SAR*.

There is also a pestilential malady the effect of which is that mares are attacked with sudden emaciation and carried off by death in the course of a few days. When this comes on, it is beneficial to pour four *sextarii* of fish-pickle into the nostrils of each victim if it be of small stature, one *congius* if it be of larger size. This remedy draws away all the phlegm through the nostrils and purges the animal. 2

XXXV. There is a form of madness which comes over mares and is rare but remarkable, namely, that, if they have seen their reflexion in the water, they are seized with a vain passion and consequently forget to eat and die from a wasting disease due to love. It is a sign of this form of insanity when they rush about over their pastures as though they were goaded on and at times seem to be looking about them and seeking and missing something. This delusion is dispelled if you cut off her mane unevenly and lead her down to the water; then beholding at length her own ugliness, she loses the recollection of the picture which was formerly before her eyes. What I have now remarked with regard to mares in general must suffice; special instructions must now be given for those who devote themselves to breeding droves of mules. 2

Madness in mares.

XXXVI. For the rearing of mules it is of the utmost importance to seek out and examine the male and female which are to be the parents of the future offspring; for if one of them is not suitable to the other, the result of their union is a failure. A mare should be chosen which is between four and ten years of age, physically very big and handsome, with stout 2

Mules and their breeding.

[10] quadrimam *Schneider*: quamam *S*: quam am *A*: quoniam *R*.

rimae formae, membris fortibus, patientissimam laboris eligere, ut discordantem utero suo generis alieni stirpem insitam facile recipiat ac perferat, et ad fetum [1] non solum corporis bona, sed et ingenium conferat. Nam cum difficulter iniecta genitalibus locis animentur semina, tum etiam concepta diutius in partum adolescunt, atque [2] peracto anno mense tertiodecimo vix eduntur, natisque inhaeret plus socordiae paternae quam vigoris materni. 3 Verumtamen equae dictos ut in usus minore cura reperiuntur,[3] maior est labor eligendi maris [4]: quoniam saepe iudicium probantis frustratur experimentum. Multi admissarii specie tenus mirabiles pessimam [5] sobolem forma [6] vel sexu [7] progenerant. Nam sive parvi corporis feminas fingunt, sive etiam speciosi plures mares quam feminas, reditum patrisfamiliae minuunt. At quidam contempti ab aspectu pretiosissimorum seminum feraces sunt. Nonnunquam aliquis generositatem suam natis exhibet, sed hebes in voluptate rarissime [8] solicitatur ad venerem. 4 Huiusce sensum [9] magistri lacessunt [10] admota [11] generis eiusdem femina, quoniam similia similibus familiariora fecit natura. Itaque obiectu asinae cum

[1] ad fetum *S* : adfectum *A*.
[2] atque *edd.* : utque *SA*.
[3] curereperiuntur *SA* : reperiuntur *ac*.
[4] magis *SAR*.
[5] mirabiles pessimam *Ursinus* : mirabilissimam *SAR*.
[6] formam *SAR*.
[7] sex *SAR*.
[8] rarissime *S* : rarissimi *AR*.
[9] sensum *R* : sensium *SA*.
[10] lacessunt *S* : *om*. *AR*.
[11] admota *S* : subadmota *A*.

limbs and well able to endure toil, that she may receive and bear in her womb an alien offspring of another race planted within her and confer on her progeny not only her good physical qualities but also her natural disposition. For not only are the seeds, which are injected into the genital parts, with difficulty quickened into life but also after conception they take longer to mature into the creature which is to be born, and it is only after the completion of a year that in the thirteenth month the offspring is brought forth with difficulty, and more of the sluggishness of the father is inherent in the offspring than the vigour of the mother. Nevertheless, while mares for breeding mules are less trouble to find, the task of selecting the male parent is greater, for often experience disappoints the judgment of the man who has to choose it. Many stallions which are admirable as far as appearance goes procreate offspring which are very inferior either in physique or sexual qualities—for if they produce she-mules of small size or more males than females of fine physique, they diminish the income of the proprietor of the estate—while some stallions which have been despised on account of their appearance are productive of the most valuable progeny. It sometimes happens that a stallion displays his high quality in his offspring but is sluggish in taking his pleasure and can be only very seldom induced to have intercourse. Owners of studs stimulate the senses of such a stallion by bringing up to him a female of the same race as himself,[a] since nature has made like more at home with like; then, when by

3

4

[a] *I.e.* an ass and not a mare.

superiectum[1] eblanditi sunt, velut incensum et obcaecatum cupidine, subtracta quam petierat, fastiditae imponunt equae.

XXXVII. Est et[2] alterum genus admissarii furentis in libidinem, quod nisi astu inhibeatur, affert gregi perniciem. Nam et saepe vinculis abruptis gravidas inquietat et, cum admittitur, cervicibus dorsisque feminarum imprimit morsus. Quod ne faciat, paulisper ad molam vinctus amoris saevitiam labore[3] temperat, et sic veneri modestior admittitur.
2 Nec tamen aliter admittendus est etiam clementioris libidinis, quoniam multum refert naturaliter sopitum pecudis ingenium modica exercitatione[4] concuti atque excitari, vegetioremque factum marem[5] feminae iniungi, ut tacita quadam[6] vi semina ipsa principiis[7] agilioribus figurentur.

3 Mula[8] autem non solum ex equa et asino, sed ex asina et equo, itemque onagro et equa generatur. Quidam vero non dissimulandi auctores, ut Marcus Varro, et ante eum Dionysius ac Mago prodiderunt mularum fetus regionibus Africae adeo non prodigiosos haberi, ut tam familiares sint incolis partus
4 earum, quam sunt nobis equarum. Neque tamen ullum est in hoc pecore aut animo aut forma[9] prae-

[1] superiectum *Lundström* : -u *SAR*.
[2] est et *AR* : et est *S*.
[3] labore *ed. pr.* : laborare *SAR*.
[4] exercitatione *ed. pr.* : excitatione *SAR*.
[5] marem *a* : mare *SA*[1]*R*.
[6] quadam *ed. pr.* : quadram *SA*.
[7] principis *SA*.
[8] mula *S* : multa *AR*.
[9] formam *SA* : forma *a*.

[a] In the translation of this part of Columella, *ass* is the female donkey.
[b] *R.R.*, II. 1. 27.
[c] See Book I. 1. 10.

putting the ass [a] in his way, they have lured on the stallion which has thrown himself upon her, while he is as it were inflamed and blinded by desire, they take away the ass, which he had wanted, and put him to the mare which he had scorned.

XXXVII. There is another type of stallion which is mad to gratify his lust and brings ruin on the stud unless cunning is used to restrain him, for he often breaks his bonds and disturbs the pregnant mares and, when he covers them, inflicts bites on their necks and backs. To prevent this he is harnessed for a time to a mill and tempers the fierceness of his passion with hard work and is only put to the mare when he has moderated his desires. Nor indeed should a stallion of milder passions be allowed to cover a mare under any other conditions, since it is very important that the naturally slumbering temperament of the animal should be stirred up and excited by moderate exercise and that the male should be put to the female when he has become more animated, in order that the seed itself, in virtue of some secret force, may be fashioned by more active elements.

The breeding of mules (continued).

2

A mule can be bred not only from a mare and a donkey, but also from an ass and a horse, and further from a wild ass and a mare. Indeed some authors, who ought not to be passed over in silence, such as Marcus Varro [b] and, before him, Dionysius [c] and Mago, have related that in some regions of Africa the production of offspring by mules is so far from being considered a prodigy that their offspring is as familiar to the inhabitants as those born from mares are to us. There is, however, nothing in the way of a mule superior either in disposition or in form to

3

4

stantius, quam quod seminavit asinus[1] quamvis[2] possit huic aliquatenus comparari[3] quod progenerat onager, nisi et indomitum, et servitio[4] contumax silvestris mores, strigosumque[5] patris praefert[6] habitum. Itaque eiusmodi admissarius nepotibus magis quam filiis utilior est. Nam ubi asina et onagro[7] natus admittitur equae, per gradus infracta[8] feritate quicquid ex eo provenit, paternam[9] formam et modestiam, fortitudinem celeritatemque avitam
5 refert. Qui ex equo et asina concepti generantur, quamvis a patre nomen traxerint, quod hinni vocantur, matri per omnia magis similes sunt. Itaque commodissimum est asinum destinare mularum generi seminando, cuius, ut dixi, species experimento
6 est speciosior. Verumtamen ab aspectu non aliter probari debet, quam ut sit amplissimi corporis, cervice valida, robustis ac latis costis, pectore musculoso et vasto, feminibus lacertosis, cruribus compactis, coloris nigri vel maculosi.[10] Nam murinus cum sit in asino vulgaris, tum etiam non optime respondet in
7 mula. Neque nos universa quadrupedis species decipiat, si qualem probamus conspicimus. Nam quemadmodum arietum quae sunt in linguis et palatis maculae, plerumque in velleribus agnorum deprehenduntur: ita si discolores pilos asinus in

[1] seminabituinsinus *S*: seminabitu in sinus *A*.
[2] quamvis *add. Lundström.*
[3] conpari *S*: comparari *a*.
[4] servili *SA*.
[5] mores trigo sunt quam *SAa*.
[6] praefert *Lundström*: praeferret *SAR*.
[7] onagro *S*: onager *AR*.
[8] infracta *ed. pr.*: infra et a *SAR*.
[9] paternam *S*: -a *AR*.
[10] macilis *AR*: magilis *S*.

that begotten by a male ass, though up to a certain point the progeny of a wild ass can be compared to it, except that, being both difficult to train and rebellious against servitude, it exhibits the wild character and lean condition of its sire. A stallion, therefore, of this kind is more useful for the production of descendants in the second than in the first generation; for, when the offspring of a she-ass and a wild ass is put to a mare, the ferocity of the wild animal has been broken down, and any offspring of this union reproduces the form and mild temper of its sire and the strength and quickness of its grandsire. The progeny conceived and procreated from 5 a horse and an ass, though they have derived their name of "hinny" from their sire,[a] show in every respect a greater resemblance to their dam; it is, therefore, most advantageous to choose a donkey as sire for a race of mules whose appearance, as I have said, is proved by experience to be handsomer. However, 6 from the point of view of appearance, it ought not to be approved unless it has an ample stature, a strong neck, robust and broad flanks, a vast and muscular chest, brawny thighs, solid legs and a black or spotted coat; for a mouse-colour, as it is commonplace in a donkey, is not very suitable in a mule either. We 7 must not let the general appearance of this quadruped deceive us if we see that it is such as we approve of; for just as the spots on the tongue and palates of rams are generally found repeated on the fleeces of the lambs which they sire, so if a donkey has different coloured hairs on its eyelids or ears, it often sires an offspring of diverse colouring also; and this

[a] Because their neighing (*hinnitus*) resembles that of a horse.

palpebris aut auribus gerit, sobolem[1] quoque frequenter facit diversi coloris, qui et ipse, etiam si diligentissime in admissario exploratus est, saepe tamen domini spem decipit. Nam interdum etiam citra praedicta signa dissimiles sui mulas fingit. Quod accidere non aliter reor, quam ut avitus color primordiis seminum mixtus[2] reddatur nepotibus.

8 Igitur qualem descripsi asellum, cum est protinus[3] genitus, oportet matri statim subtrahi, et ignoranti equae subici. Ea[4] optime tenebris fallitur. Nam obscuro loco partu eius amoto, praedictus quasi ex ea natus alitur. Cui deinde cum decem diebus insuevit equa, semper postea desideranti[5] praebet ubera. Sic nutritus[6] admissarius equas diligere condiscit. Interdum etiam, quamvis materno lacte sit educatus, potest a tenero[7] conversatus[8] equis familiariter earum consuetudinem appetere. Sed ei 9 non oportet minori quam trimo inire permitti.[9] Atque id ipsum si concedatur,[10] vere fieri conveniet, cum et desecto viridi pabulo et largo hordeo firmandus, nonnunquam etiam salivandus erit. Nec tamen tenerae feminae committetur. Nam nisi prius ea marem cognovit,[11] adsilientem admissarium calcibus proturbat, et iniuria depulsum etiam ceteris equis reddit inimicum. Id ne fiat, degener ac vulgaris

[1] sobolem *R* : subole *S*¹*A*¹ : sobole *S*².
[2] mixtus *S* : mixtu *A*.
[3] protinus *Lundström* : ptri *A* : ptris *S* : patri *c*.
[4] ae *SA*¹ : ac *A*²*R*.
[5] desideranti *S* : destinanti *AR*.
[6] nutritus *S* : nutritur *AR*.
[7] potest a tenero *S* : potestate vero *AR*.
[8] conversatus *S* : -ur *AR*.
[9] inire permitti *Lundström* : inaremitti *S* : in are mitti *A*.
[10] concidatur *SAR*.
[11] cognovit *ed. pr.* : concivit *SAR*.

colouring, although the stallion was most carefully examined to see if it was present, is often a cause of disappointment to the owner. For sometimes also a stallion shapes mules very different from himself in respects other than the signs mentioned above. This, I think, occurs for no other reason than that the colour of the grandsire is transmitted to the second generation mixed with the elements which form the seed.

8 As soon as the foal of the ass, such as I have described, is brought to birth, it should be taken away from its mother and put under a mare who has no knowledge of it. This deception is best carried out in dark conditions; for if her offspring has been taken away from her in a dark place and the aforesaid foal is put under her it is nourished by her as if it were her own offspring; and then, when she has become accustomed to it for ten days, she henceforward always gives it her dugs whenever it wants to feed. The future stallion fed in this manner learns to have an affection for mares. Sometimes also, although it has been reared on its own mother's milk, if it has lived familiarly amongst mares from its tender years, it may well seek their company. 9 It must not, however, be allowed to cover them when it is less than three years old, and when it is permitted to do so, it will be well that intercourse should take place in the spring, since it will have to be fortified with chopped green fodder and an abundance of barley and sometimes also given a drench. It ought not, however, to be put to a young mare; for unless she has already had experience of a male, she repulses the donkey with her hoofs when he leaps upon her, and the affront which he has received inspires him furthermore with an aversion for all other mares. To prevent this, a

asellus admovetur, qui solicitet obsequia feminae: neque is tamen inire sinitur. Sed, si iam est equa veneris patiens, confestim abacto viliore, pretioso
10 mari[1] subigitur.[2] Locus est ad hos usus extructus, machinam vocant rustici, duos parietes adverso clivulo inaedificatos qui angusto intervallo sic inter se distant, ne femina conluctari aut admissario ascendenti avertere se possit. Aditus est ex utraque parte, sed ab inferiore clatris[3] munitus: ad quae[4] capistrata in imo clivo constituitur equa, ut et prona[5] melius ineuntis semina recipiat, et facilem sui tergoris ascensum ab editiore parte minori quadrupedi praebeat. Quae cum ex asino conceptum edidit, partum sequenti anno vacua nutrit. Id enim utilius est, quam quod quidam faciunt, ut et fetam nihilominus
11 admisso equo impleant. Annicula mula recte a matre repellitur, et amota montibus aut feris[6] locis pascitur, ut ungulas duret, sitque[7] postmodum longis itineribus habilis. Nam clitellis aptior mulus. Illa quidem[8] agilior: sed uterque sexus et viam recte graditur, et terram commode proscindit, nisi si pretium quadrupedis rationem rustici onerat,[9] aut campus gravi gleba[10] robora boum deposcit.

[1] maris *SAR*.
[2] subigitur *S* : iniungitur *AR*.
[3] claris *SAR*.
[4] quod *SAR*.
[5] pronam *SAR*.
[6] seris *SAR*.
[7] sique *SA* : sitque *ac*.
[8] quod *SAR*.
[9] onerant *SAR*.
[10] gleba *S* : graebra *A*[1] : craebra *A*[2] : crebra *R*.

[a] Compare Chapter XIX above.

badly-bred, ordinary donkey is brought to seek her compliance; he should not, however, be allowed to cover her, but if the mare is inclined to submit to his desires, the more ignoble donkey is promptly driven away and the mare is covered by the valuable stallion. A special place is constructed for these purposes— 10 the countryfolk call it a "machine"[a]—it consists of two lateral walls built into gently-rising ground, having a narrow space between them, so that the mare cannot struggle or turn away from the donkey when he tries to mount her. There is an entrance at each end, that on the lower level being provided with cross-bars, to which the mare is fastened with a halter and stands with her forefeet at the bottom of the slope, so that, leaning forward she may the better receive the insemination of the donkey and make it easier for a quadruped smaller than herself to mount upon her back from the higher ground. When the mare has given birth to a foal of which the donkey is the sire, she rears it during the following year without being with foal again. This method is better than that which some people follow, who cause her to be covered again by the stallion and to be with foal, although she has only just foaled. When a she-mule 11 is a year old, it is right to take it away from its dam and put it to feed far away in the mountains or in wild places, so that it may harden its hoofs and presently be fit for long journeys. Now the male is better than the female mule for carrying a pack-saddle, whereas the latter is more nimble; but both sexes step out well on a journey and are useful for breaking up the soil, unless the price of the animal is too burdensome an expense for the farmer, or a soil, being of heavy sod, demands the strength of oxen.

XXXVIII. Medicinas huius pecoris plerumque iam in aliis generibus edocui: propria tamen quaedam vitia non omittam, quorum remedia subscripsi. Equienti mulae cruda brassica datur. Suspiriosae sanguis detrahitur, et cum sextario vini atque olei thuris semuncia marrubii succus instar heminae 2 mixtus infunditur. Suffraginosae hordeacea farina imponitur, mox suppuratio ferro reclusa linamentis curatur, vel gari [1] optimi sextarius cum libra olei per narem sinistram demittitur, admisceturque huic medicamini trium vel quattuor ovorum albus liquor sepa- 3 ratis vitellis. Flemina [2] secari, et interdum inuri solent. Sanguis demissus [3] in pedes, ita ut in equis emittitur: vel si est herba, quam veratrum vocant rustici, pro pabulo cedit. Est et ὑοσκύαμος, cuius semen detritum et cum vino datum praedicto vitio medetur.

Macies et languor submovetur saepius data potione, quae recipit semunciam sulphuris ovumque crudum, et myrrhae pondus denarii. Haec trita vino admiscentur,[4] atque ita faucibus infunduntur. 4 Sed et tussi dolorique ventris eadem ista aeque medentur. Ad maciem nulla res tantum quantum medica potest. Ea herba viridis celerius [5] nec tarde tamen arida faeni vice saginat iumenta: verum

[1] gari *S*: cari *AR*.
[2] flemina *Lundström*: femina *SAR*.
[3] demissus *S*: di- *AR*.
[4] admiscetur *SA*.
[5] celerius *S*: *om. AR*.

---

[a] A kind of hellebore. [b] *Medicago sativa.*

Remedies for the diseases of mules.

XXXVIII. Though, in dealing with other classes of animals, I have already described most of the medicines which mules require, I will not omit to mention certain maladies which are peculiar to these animals, the remedies for which I have here subjoined. If a mule is in heat, raw cabbage is administered; if it is asthmatic, blood is drawn off and about a *hemina* of the juice of horehound mixed with a *sextarius* of wine and half an ounce of oil of frankincense is poured down its throat. 2 If it is suffering from spavin, barley-flour is applied, and then the suppuration is opened with a lancet and dressed with lint, or else a *sextarius* of the best fish-pickle in a pound of oil is poured through the left nostril; the whites of three or four eggs from which the yolks have been separated are mixed with this medicament. 3 Blood-blisters round the ankles are usually cut and sometimes cauterized. When blood flows down into the feet, it is drawn off by the same method as is applied to horses, or, if the herb which the country-folk call *veratrum* [a] is available, it is given as fodder. Another remedy is henbane, the seed of which, crushed and administered with wine, cures this malady.

Emaciation and languor are dispelled by frequent potions containing half an ounce of sulphur, a raw egg and a *denarius* weight of myrrh; these are beaten up and mixed in wine and then poured down the animal's throat. 4 The same ingredients serve equally well as a remedy for a cough and for pain in the stomach. For emaciation nothing is as efficacious as lucerne [b]; this herb, when it is green, quickly fattens beasts of burden, and is not slow in doing so even when it is dry and used instead of hay, but it must be

modice danda, ne nimio sanguine stranguletur pecus. Lassae et aestuanti mulae adeps [1] in fauces demittitur,[2] vinumque [3] in os suffunditur. Cetera exequemur in mulis sic, ut prioribus huius voluminis partibus tradidimus, quae curam boum equarumque continent.

[1] ad eos *SA*¹*R*: adeps *A*²*a*.
[2] demittitur *S*: di- *AR*.
[3] virumque *SA*¹: vinumque *a*.

given in moderation, lest the animal be choked by an excess of blood. When a mule is exhausted and feeling the heat, fat is thrust down its throat and wine poured into its mouth. In all other respects in dealing with mules we shall follow the method which we have prescribed in the earlier parts of this book which deal with the care of oxen and horses.

BOOK VII

LIBER VII

I. De minore pecore dicturis, P. Silvine, principium tenebit minor in ora[1] Arcadiae vilis hic vulgarisque asellus, cuius plerique rusticarum rerum auctores in emendis tuendisque iumentis praecipuam rationem volunt esse; nec iniuria. Nam etiam eo rure,[2] quod pascuo caret, contineri potest, exiguo et qualicunque pabulo contentus. Quippe vel foliis spinisque vepraticis[3] alitur, vel obiecto fasce sarmentorum. Paleis vero, quae paene omnibus regionibus, abundant, etiam gliscit.

2 Tum imprudentis custodis negligentiam fortissime sustinet: plagarum et penuriae tolerantissimus: propter quae tardius deficit, quam ullum aliud armentum. Nam laboris et famis maxime patiens raro morbis afficitur. Huius animalis tam[4] exiguae tutelae plurima et necessaria opera supra portionem respondent, cum et facilem terram qualis in Baetica totaque Libye sit[5] levibus aratris[6] proscindat, et

[1] minor in ora *Lundström*: minor minora *SAac*.

[2] eorum re *SAc*: eo re *a*.

[3] vepratici salitur *S*: vel pratici salitur *A*: vel prati his alitur *a*.

[4] tam *S*: tamen *Aac*.

[5] sit *Ald.*: si *SAac*.

[6] aratis *SA*[1].

BOOK VII

I. Since, Publius Silvinus, we are now about to deal with the lesser farm-animals, our first subject shall be that cheap and common animal the lesser[a] ass from the region of Arcadia, to which the majority of writers on agriculture consider that particular attention should be paid when it is a question of buying and tending beasts of burden; and they are quite right, for it can be kept even in a country which lacks pasturage, since it is content with very little fodder of any sort of quality, feeding on leaves and the thorns of brier-bushes, or a bundle of twigs which is offered to it; indeed it actually thrives on chaff, which is abundant in almost every region.

The ass.

2 Further, it endures most bravely the neglect of a careless master and tolerates blows and want most patiently; for which reasons it is slower in breaking down than any other animal used for ploughing, for, since it shows the utmost endurance of toil and hunger, it is rarely affected by disease. The performance by this animal of very many essential tasks beyond its share is as remarkable as the very little care which it requires, since it can both break up with a light plough easily worked soil, such as is found in Baetica and all over Libya, and can draw on vehicles

[a] *I.e.* the ass as compared with the mule.

3 non minima pondera vehiculo trahat. Saepe etiam, ut celeberrimus poeta memorat:

> . . . tardi costas agitator aselli
> Vilibus aut onerat pomis, lapidemque revertens
> Incusum aut atrae massam picis urbe reportat.

Iam vero molarum et conficiendi frumenti paene solemnis est huius pecoris [1] labor. Quare omne [2] rus tanquam maxime necessarium instrumentum desiderat asellum, qui, ut dixi, pleraque utensilia et vehere [3] in urbem et reportare collo vel dorso commode potest. Qualis autem species eius vel cura probatissima sit, superiore libro, cum de pretioso praeciperetur, satis dictum est.

II. Post huius [4] quadrupedis ovilli pecoris secunda ratio est, quae prima fit, si ad utilitatis magnitudinem referas. Nam id praecipue nos contra frigoris violentiam protegit, corporibusque nostris liberaliora praebet velamina. Tum etiam casei lactisque abundantia non solum agrestes saturat, sed etiam elegantium mensas iucundis et numerosis dapibus

2 exornat. Quibusdam vero nationibus frumenti expertibus victum commodat, ex quo Nomadum Getarumque plurimi γαλακτοπόται dicuntur. Igitur id pecus, quamvis mollissimum sit, ut ait prudentissime Celsus, valetudinis tutissimae est, minimeque pestilentia laborat. Verum tamen eligendum est ad

[1] pecorū *SA*[1]. [2] omnem *SA*[1].
[3] vehere *ed. pr.*: e vere *S*: vere *A*. [4] huius *S*: *om. AR.*

---

[a] Vergil, *Georg.* I. 273 ff.
[b] *I.e.* the mule, treated in Book VI. Chapters XXXVI–XXXVIII.
[c] A tribe living north of the lower course of the Danube.

loads which are far from being small. Often too as 3 the most famous of poets says:

> The tardy donkey's driver loads its sides
> With cheap fruits and returning brings from town
> A hammered millstone or black lump of pitch.[a]

This animal's almost invariable task at the present day consists in turning a mill and grinding corn. Every estate, therefore, requires a donkey as that might be called a necessary instrument, since, as I have said, it can conveniently convey to town and bring back most things that are required for use either with load on its neck or on its back. What kind of ass and what method of looking after it is most approved, has been sufficiently described in a previous book, where instructions have been given about the valuable type of animal.[b]

On the purchase and care of sheep.

II. The importance of the sheep is secondary to that of the ass, though the sheep is of primary account if one has regard to the extent of its usefulness. For it is our principal protection against the violence of the cold and supplies us with a generous provision of coverings for our bodies. Then, too, it is the sheep which not only satisfies the hunger of the country folk with cheese and milk in abundance but also embellishes the tables of people of taste with a variety of agreeable dishes. Indeed to some 2 tribes, who have no corn, the sheep provides their diet; hence most of the nomadic tribes and the Getae[c] are called the "Milk-Drinkers." Though the sheep, as Celsus most wisely remarks, is a very delicate creature, it enjoys sound health and suffers very little from contagious disease. Nevertheless a breed of sheep must be chosen to suit local conditions, a prin-

naturam loci: quod semper observari non solum in hoc, sed etiam in tota ruris disciplina Vergilius praecipit, cum ait:

Nec vero terrae ferre omnes omnia possunt.

3 Pinguis et campestris situs proceras oves tolerat; gracilis et collinus quadratas; silvestris et montosus exiguas; pratis planisque novalibus tectum pecus commodissime pascitur. Idque non solum generibus, sed etiam coloribus plurimum refert. Generis eximii Calabras, Apulasque et Milesias[1] nostri existimabant, earumque optimas Tarentinas. Nunc Gallicae pretiosiores habentur, earumque praecipue Altinates. Item quae circa Parmam et Mutinam macris stabulantur campis. Color albus cum sit 4 optimus, tum etiam est utilissimus, quod ex eo plurimi fiunt, neque hic ex alio. Sunt etiam suapte natura pretio commendabiles[2] pullus atque fuscus, quos praebent in Italia Pollentia, in Baetica Corduba. Nec minus Asia rutilos,[3] quos vocant ἐρυθραίους. Sed et alias varietates in hoc pecoris genere docuit usus exprimere. Nam cum in municipium Gaditanum ex vicino Africae miri coloris silvestres ac feri arietes, sicut aliae bestiae, munerariis deportarentur, M. Columella patruus meus acris vir ingenii, atque illustris

[1] miles *SAac*. [2] commendabilis *SAac*.
[3] lutilos *SA*[1].

[a] *Georg*. II. 109.
[b] *I.e.* those which on account of the excellence of their wool are covered with skins to preserve their fleeces (Varro, *R.R.*, II. 2. 19: Horace, *Od*. II. 6. 10).
[c] A town near Venice.
[d] Both these towns were in Cisalpine Gaul, Mutina being the modern Modena.

ciple which ought always to be observed not only with regard to sheep but in every department of agriculture, as Vergil warns us, when he says:

Nor can all kinds of land all things produce.[a]

A rich, flat country supports tall sheep, a lean and hilly region those of square build, while a wooded, mountainous land produces small sheep. "Coated"[b] sheep are best pastured in meadows and flat fallow ground. Not only the question of the kinds of sheep but also that of their colour are matters of great importance. Our farmers used to regard the Calabrian, Apulian and Milesian as breeds of outstanding excellence, and the Tarentine as the best of all; now Gaulish sheep are considered more valuable, especially that of Altinum,[c] also those which have their folds in the lean plains round Parma and Mutina.[d] While white is the best colour, it is also the most useful, because very many colours can be made from it; but it cannot be produced from any other colour. By their very nature black and dark brown sheep also, which Pollentia[e] in Italy and Corduba[f] in Baetica produce, are esteemed for the price which they command; Asia likewise provides the red colour which they call "erythraean." Experience has also taught the way to produce other variations of colour in this kind of animal. For when fierce wild rams of a marvellous colour were brought across amongst other wild beasts from a neighbouring district of Africa to the municipal town of Gades for those who were giving public shows, my uncle Marcus Columella, a man of keen intelligence and a dis-

3

4

[e] A city of Liguria (the Italian Riviera).
[f] Cordova in Spain.

agricola, quosdam mercatus, in agros transtulit, et 5 mansuefactos tectis ovibus admisit. Eae primum hirtos,[1] sed paterni coloris agnos ediderunt, qui deinde et ipsi Tarentinis ovibus impositi, tenuioris velleris arietes progeneraverunt. Ex his rursus quicquid conceptum est, maternam mollitiem, paternum et avitum retulit colorem. Hoc modo Columella dicebat, qualemcunque speciem, quae fuerit[2] in bestiis, per nepotum gradus mitigata feritate reddi. Sed[3] ad propositum revertar.

Ergo duo genera sunt ovilli pecoris, molle et hirsutum. Sed in utroque vel emendo vel tuendo[4] plura communia, quaedam tamen sunt propria generosi, quae observari conveniat. Communia in emendis gregibus fere illa: si candor lanae maxime placet, nunquam nisi[5] candidissimos mares legeris: quoniam ex albo saepe fuscus editur partus; ex erythraeo vel pullo nunquam generatur albus.

III. Itaque non solum ea ratio est probandi arietis, si vellere candido vestitur, sed etiam si palatum atque lingua concolor lanae est. Nam cum hae corporis partes nigrae aut maculosae sunt, pulla vel etiam

[1] hirtus *Ald.*: ortos *SAR.*
[2] fuerint *SAa*: fuerunt *c.*
[3] reddi sed *S*²: reddis et *S*¹: reddisset *Ac.*
[4] vel tuendo *add. Lundström*: *om. SAR.*

tinguished agriculturist, bought some of them and transferred them to his estate, and, when he had tamed them, mated them with "coated" ewes. These produced in the first generation lambs with coarse wool but of the same colour as their sires. 5 When these in their turn were coupled with Tarentine ewes, they produced rams with a finer fleece. All the descendants of these latter in their turn reproduced the soft wool of their dams and the colours of their sires and grandsires. Columella used to claim that in this way whatever outward appearance the wild animals possessed was reproduced in the second and later generations of their descendants, while their savage nature was tamed. But I must return to my subject.

There are then two kinds of sheep, the soft-fleeced and the shaggy-coated; but, while there are several points common to both kinds when you are buying or looking after them, there are certain special characteristics of the well-bred sheep which it is well to observe. The following are generally the common points to be looked for when you are buying flocks: if whiteness of fleece is what pleases you most, you should never choose any but the whitest rams, for a dark lamb is often the offspring of a white ram, while a white lamb is never bred from a red or brown sire.

On the choosing of rams.

III. And so, if a ram has a white fleece, this is not itself a reason for approving of it, but only if its palate and tongue are also of the same colour as its wool; for if these parts of the body are black or spotted, the offspring is either dark or even parti-coloured. The same poet as I quoted above, amongst many other

[5] nisi *S* : *om. AR.*

varia nascitur proles; idque inter cetera eximie talibus numeris[1] significavit idem qui supra:

Illum autem, quamvis aries sit candidus ipse,[2]
Nigra subest udo tantum cui lingua palato,
Reice, ne maculis infuscet vellera pullis
Nascentum.[3]

2 Una eademque ratio est in erythraeis et nigris arietibus, quorum similiter, ut iam dixi, neutra pars esse debet discolor lanae, multoque minus ipsa universitas tergoris maculis variet. Ideo nisi lanatas oves emi non oportet, quo melius unitas coloris appareat: quae nisi praecipua est in arietibus, paternae notae plerumque natis inhaerent.[4]

3 Habitus autem maxime probatur, cum est altus atque procerus, ventre promisso atque lanato, cauda longissima, densique velleris, fronte lata, testibus amplis, intortis cornibus: non quia magis hic sit utilis, (nam est melior mutilus aries) sed quia[5] minime nocent intorta potius, quam surrecta et patula cornua. Quibusdam[6] tamen regionibus, ubi caeli status uvidus ventosusque est, capros et arietes optaverimus vel amplissimis cornibus, quod ea porrecta[7] altaque[8] maximam partem capitis a tempestate defendant.[9]

4 Itaque si plerumque[10] est atrocior hiems,[11] hoc genus eligemus: si clementior, mutilum probabimus marem: quoniam est illud incommodum in cornuto,

[1] numeris *ed. pr.*: numeri *SAac*.
[2] ipse *S*: ipsa *Aac*.
[3] nascentium *SAac*.
[4] inheret *SAac*.
[5] quia *S*: qui *Ac*.
[6] quibusdam *S*: quibus *Aac*.
[7] profecto *SAac*.
[8] altoque *SAac*.
[9] defendant *ed. pr.*: defendat *SAac*.
[10] plerum *SAa*.
[11] hiems *a*, *ed. pr.*: hiemis *SA*.

points, has expressed the same thing excellently in the following lines:

> But though the father-ram itself is white,
> If under his wet palate a black tongue
> Lurks, then reject it, lest with dusky spots
> It stain the fleeces of the future race.[a]

The same reasoning applies both to red and to black rams, in whom, likewise, as I said just now, neither the tongue nor the palate ought to be different in colour from the wool, still less should the whole skin be variegated with spots. Sheep, therefore, should never be bought unless they still have their wool on their backs, so that it may be easier to see that they are of one colour only, because, unless this is a prominent feature of the rams, the marks on the father generally persist in the offspring. 2

The points which are most highly esteemed in a ram are breadth and height of stature, a belly which hangs down and is woolly, a very long tail, a thick fleece, a broad forehead, large testicles and curling horns—not because such a ram is more useful (for it is better without horns), but because horns do much less harm if they are curling than if they are up-standing and spreading. In some localities, however, where the climate is damp and windy, we should prefer that both he-goats and rams should have very large horns, because, being thus wide-spreading and lofty, they protect most of the head from the storm. So, if the winter generally tends to be severe, we shall choose rams of this type; if it is milder, we shall prefer a ram which is hornless; for there is this disadvantage about a sheep with horns, that, being 3 4

[a] Vergil, *Georg.* III. 387 ff.

quod cum sentiat se velut quodam naturali telo[1] capitis armatum, frequenter in pugnam procurrit, et fit in feminas quoque procacior. Nam rivalem (quamvis solus admissurae non sufficit) violentissime persequitur, nec ab alio tempestive patitur iniri[2] 5 gregem, nisi cum est fatigatus libidine. Mutilus autem, cum se tanquam exarmatum intelligat, nec ad rixam promptus est, et in venere mitior. Itaque capri vel arietis[3] petulci saevitiam pastores hac astutia repellunt. Mensurae pedalis robustam tabulam configunt aculeis, et adversam fronti cornibus religant. Ea res ferum prohibet[4] a rixa, quoniam 6 stimulatum suo ictu ipsum se sauciat. Epicharmus autem Syracusanus, qui pecudum medicinas diligentissime conscripsit, affirmat pugnacem arietem mitigari terebra secundum auriculas foratis cornibus, qua curvantur in flexum. Eius quadrupedis aetas ad progenerandum optima est trima: nec tamen inhabilis usque in annos octo. Femina post bimatum maritari debet, iuvenisque habetur quin-7 quennis: fatiscit post annum septimum. Igitur, ut dixi, mercaberis[5] oves intonsas:[6] variam et canam[7] improbabis, quod sit incerti coloris. Maiorem trima dente[8] minacem sterilem repudiabis. Eliges bimam vasti corporis, cervice[9] prolixi villi, nec asperi,

[1] naturali telo *S* : naturate loco *A*.
[2] inire *ac* : ini *SA*. [3] arietis *a* : arietes *SA*.
[4] prohibet *a* : prohibita *SA*.
[5] mercaberis *ed. pr.* : mercaveris *S* : mercaris *Aac*.
[6] intonsas *S* : intonsis *Aac*.
[7] calvamque *Richter* : et canam *prior. edd.*
[8] trima dente *ed. pr.* : trime dentem *SAac*.
[9] cervi et *SAac*.

[a] See note on Book I. 1. 8.

conscious that its head is armed, as it were, with a natural weapon, it often rushes into the fray and also becomes too wanton towards the females. For (although it does not itself suffice to mate with the whole flock) it pursues its rival in the most violent manner and does not allow the flock to be covered at the proper time by any other ram, except when it is worn out by lust. On the other hand the hornless 5 ram, since it realizes that it is, as it were, disarmed, is not prompt to quarrel and is milder in its amours. Shepherds, therefore, use the following ruse to check the brutality of a butting he-goat or ram: they fix spikes in a strong board a foot in length and tie it to the horns with the spikes facing the forehead. This prevents the animal, fierce though he may be, from quarrelling, because by his blow he pricks and wounds himself. Epicharmus,[a] the Syracusan, who has written a very careful treatise on remedies for 6 cattle, declares that a pugnacious ram can be tamed by piercing its horns with a gimlet near the ears at the point where the horns bend into a curve. The best time for breeding from this animal is when it is three years old; but it continues to be suitable up to eight years of age. The female ought to be mated after its second year and is still regarded as young at five years; after its seventh year it becomes exhausted. You will, therefore, as I have said, buy 7 ewes before they have been sheared and you will reject those which are parti-coloured or bald, because its colour can not be determined. You will refuse a sterile ewe which has passed its third year and has projecting teeth: you will select a two-year-old with a large frame, a neck covered with shaggy hair which is abundant but not coarse, and a woolly and

lanosi et ampli uteri. Nam vitandus est glaber et exiguus.

8 Atque haec fere communia sunt in comparandis ovibus. Illa etiam tuendis: humilia facere stabula, sed in longitudinem potius quam in latitudinem porrecta,[1] ut simul et hieme calida sint, nec angustiae fetus oblidant.[2] Ea poni debent contra medium diem: namque id pecus, quamvis ex omnibus animalibus sit[3] vestitissimum, frigoris tamen impatientissimum est, nec minus aestivi vaporis. Itaque cohors clausa sublimi macerie praeponi vestibulo debet, ut sit in eam tutus exitus aestuanti;[4] deturque opera, ne quis humor consistat, ut semper quam aridissimis filicibus[5] vel culmis stabula constrata sint, quo purius[6] et mollius incubent foetae, sintque illa[7] mundissima, neque earum valetudo, quae praecipue custodienda est, infestetur uligine.

9 Omnia autem pecudi larga praebenda sunt alimenta. Nam vel exiguus numerus, cum pabulo satiatur, plus domino reddit, quam maximus grex, si senserit penuriam. Sequeris autem novalia non solum herbida, sed quae plerumque vidua sunt spinis; utamur[8] enim saepius auctoritate divini carminis:

> Si tibi lanitium curae est, primum aspera silva
> Lappaeque tribulique absint;

[1] porrectā *S*: profectam *A*.
[2] obligant *SAac*. [3] sit *S*: *om. AR*.
[4] aestuanti *Richter*: aestivandi *prior. edd.*
[5] filicibus *Sc*: felicibus *Aa*.
[6] plurius *SAR*.
[7] sint quala *Gesner*: sint quola *SAac*.
[8] utamur *ac*: utantur *c*: utam *S*¹*A*: utar *S*².

ample belly; for a small and hairless ewe must be avoided.

These are, roughly speaking, the general points which must be observed when you are buying sheep; the following points must be observed in their management. Their folds should be built low and extended in length rather than in breadth, so that they may be warm in winter and also that lack of space may not cause the ewes to cast their young. They should be placed so as to face the mid-day sun; for sheep, though naturally the best clothed of animals, can least endure cold, or summer heat either. For this reason a closed court with a high wall ought to be constructed in front of the entrance, so that there may be a safe way out for the animal when it is affected by the heat; and care must be taken to prevent there being any standing water by always keeping their folds strewn with the driest possible fern or straw, so that the ewes after lambing may have something clean and soft on which to lie, and that the folds may be very clean, and that the health of the ewes, which must be specially guarded, may not be impaired by dampness. Sheep must be supplied with an abundance of every kind of food; for even a small flock, if it is given its fill of fodder, brings its owner a bigger return than a very large one which has suffered from want. You must look for fallow land which is not only grassy but also for the most part free from thorns; for, to make our repeated appeal to the authority of inspired poesy,[a] 8 9

If wool is your desire, above all else
Avoid the prickly woods and burs and caltropses.

[a] Vergil, *Georg.* III. 384 f.

10 quoniam ea res, ut ait idem, scabras oves reddit,

cum tonsis illotus[1] adhaesit
Sudor, et hirsuti secuerunt corpora vepres:

tum etiam quotidie minuitur lanae fructus,[2] quae quanto prolixior in pecore concrescit, tanto magis obnoxia[3] est rubis, quibus velut hamis inuncata pascentium tergoribus avellitur. Molle vero pecus etiam velamen, quo protegitur, amittit,[4] atque id non parvo sumptu reparatur.

11 Inter auctores fere constat, primum esse admissurae tempus vernum Parilibus,[5] si sit ovis matura,[6] sin vero[7] feta, circa Iulium mensem. Prius tamen haud dubie probabilius,[8] ut messem vindemia,[9] fructum deinde vineaticum fetura pecoris excipiat, et totius autumni pabulo satiatus agnus ante maestitiam frigorum atque hiemis ieiunium confirmetur. Nam melior est autumnalis verno, sicut ait verissime Celsus; quia[10] magis ad rem pertinet, ut ante aestivum quam hibernum solstitium convalescat: solus-12 que ex omnibus bruma commode nascitur. Ac si res exigit, ut plurimi mares progenerandi sint, Aristoteles vir callidissimus rerum naturae praecipit

[1] inlutus S^1A : inlotus S^2.
[2] fructus *S* : *om. Aac.*
[3] obnoxia *S* : obnoxium *Aac.*
[4] admittit *SAc* : amittit *a.*
[5] Parilibus *S* : paribus *Aac.*
[6] matura *S* : mature *Aac.*
[7] sin vero *ed. pr.* : sincera *SAac.*
[8] probabilis *SAac.*
[9] mensem vindemiam *SAa* : messem vindemiam *c.*
[10] quia *S* : qui *Aac.*

For, as the same poet says,[a] it causes scab in sheep, 10

> When after shearing sweat unwashen clings
> And prickly briers tear away their flesh.

Moreover, the yield of wool is daily reduced, for the more abundantly it grows upon the animal, the more exposed it is to brambles, by which it is caught, as if by hooks, and torn from their backs as they feed. The sheep also loses the soft covering with which it is protected, and this can only be replaced at considerable expense.

The authorities are in general agreement that the 11 earliest time of the year at which the ewes should be mated is the spring, when the Parilia[b] is celebrated, if the ewe has just reached maturity, but, if she has already produced a lamb, about the month of July. The earlier date is, however, undoubtedly preferable, so that, just as the vintage follows the harvest, so the birth of the lamb may succeed to the gathering in of the grapes, and the lamb, having enjoyed its fill of food during the whole autumn, may gain strength before the gloomy cold season and the short rations of winter come on. For an autumn lamb is superior to a spring lamb, as Celsus very truly remarks, because it is more important that it should grow strong before the summer solstice than before the winter solstice, and it alone of all animals can be born without risk in mid-winter. If circumstances require 12 that more males than females should be produced, Aristotle,[c] that shrewd researcher into natural

[a] Vergil, *Georg.* III. 443.

[b] The feast of Pales, tutelary goddess of sheep and shepherds, which was celebrated on April 18th.

[c] *De Gen. Anim.*, 766, 35 ff.

admissurae tempore observare siccis diebus halitus septentrionales, ut [1] contra ventum gregem pascamus, et eum spectans admittatur pecus: at si feminae generandae sunt, austrinos flatus captare, ut eadem ratione [2] matrices ineantur. Nam illud, quod priore libro docuimus, ut admissarii dexter vel etiam sinister vinculo testiculos obligetur, in magnis gregibus operosum est.

13 Post feturam deinde longinquae regionis opilio villicus fere omnem sobolem pastioni [3] reservat; suburbanae, teneros agnos, dum adhuc herbae sunt expertes, lanio tradit, quoniam et parvo sumptu devehuntur, et iis submotis, fructus lactis ex matribus non minor [4] percipitur. Submitti tamen etiam in vicinia urbis quintum quemque [5] oportebit. Nam 14 vernaculum pecus peregrino longe est utilius: nec committi debet, ut [6] totus grex effetus [7] senectute dominum destituat: cum praesertim boni pastoris vel prima cura sit annis omnibus in demortuarum vitiosarumque ovium locum totidem vel etiam plura capita substituere: quoniam saepe frigorum atque hiemis saevitia pastorem decipit, et eas oves [8] interimit, quas ille tempore autumni ratus adhuc esse [9] 15 tolerabiles, non submoverat. Quo magis etiam propter hos casus, nisi quae validissima [10] non [11] com-

[1] et *SAac*.
[2] ratione *a*: -em *SA*.
[3] pastioni *S*: -e *Aac*.
[4] minor *ac*: mino *A*: *om. S*.
[5] quicumq; *A*.
[6] ut *S*: *om. Aac*.
[7] effectus *Aac*: effectu *S*.
[8] eas oves *Ald.*: exe ovis *S*: ex eo vix *Aac*.
[9] esse *S*²: essem *S*¹*A*.
[10] validissimo *SAac*.
[11] non *Sa*: anon *A*: anno *c*.

phenomena, advises that in the breeding season we should look out for breezes from the north on dry days, so as to pasture the flock facing this wind, and that the male should cover the female looking in that direction; if, on the other hand, female births are desired, we should seek for southern breezes, so that the ewes may be covered in the same manner. The device, which was described in the preceding book,[a] of tying up the right or left testicle of the ram with a band, is difficult to carry out in large flocks.

13 After the lambing season the bailiff in charge of the sheep on an outlying estate reserves almost all the young offspring for pasture; and in a section near town hands over the tender lambs, before they have begun to graze, to the butcher, since it costs only a little to convey them to the town and also, when they have been taken away, no slighter profit is made out of the milk from their mothers. Even in the neighbourhood of a town, however, one lamb in five will have to be left with its mother, for an animal born on the spot is much more profitable than one brought from a distance, nor ought the mistake be made of letting 14 the whole flock become exhausted by age and leave the owner without any stock, especially as it is the first duty of a good shepherd every year to substitute the same number of sheep, or even more, in place of those which have died or are diseased, since the severity of the cold and winter often surprises the shepherd and causes the death of those ewes which he had failed to remove from the flock in the autumn because he thought them still able to stand the cold. These 15 mishaps are also further reason why no ewe, unless it is very strong, should be caught unprepared by winter

[a] Book VI. 28.

prehendatur hieme, novaque progenie repleatur numerus. Quod qui faciet, servare debebit, ne minori quadrimae, neve ei, quae excessit annos octo, prolem submittat. Neutra enim aetas ad educandum est idonea: tum etiam quod ex vetere materia nascitur, plerumque congeneratum parentis senium[1] 16 refert. Nam vel sterile vel imbecillum est. Partus vero incientis[2] pecoris non secus quam obstetricum more custodiri debet. Neque enim aliter hoc animal quam muliebris sexus enititur, saepiusque etiam, quando[3] est omnis rationis ignarum, laborat in partu. Quare veterinariae medicinae prudens esse debet pecoris magister, ut, si res exigat, vel integrum conceptum, cum transversus haeret locis genitalibus, extrahat, vel ferro divisum citra matris perniciem partibus educat,[4] quod Graeci vocant ἐμβρυουλκεῖν.
17 Agnus autem, cum est editus, erigi debet, atque uberibus admoveri, tum[5] etiam eius diductum[6] os pressis humectare papillis, ut condiscat maternum trahere alimentum. Sed prius quam hoc fiat, exiguum lactis emulgendum[7] est, quod pastores colostrum vocant: ea nisi aliquatenus emittitur, nocet agno qui primo biduo[8] quo natus est, cum matre claudatur, ut et ea partum suum foveat, et ille ma-18 trem agnoscere condiscat. Mox deinde quamdiu non lascivit, obscuro et calido septo[9] custodiatur; postea

[1] senium S^2: se nimiũ S^1*Ac*.
[2] incientis *Ursinus*: incipientis *codd.*
[3] quando *Richter*: quanto *prior. edd.*
[4] educat *S*: ducat *Aa*: duca *c*.
[5] tum *S*: cum *Aac*.
[6] diductum *S*: de- *Aac*.
[7] emulgendus *SAac*.
[8] primo biduo *Heinsius*: moviduo S^1*Ac*.
[9] septo *Sa*: septe *Ac*.

and why the number should be made up with young stock. Whoever is going to follow this system will have to take care not to put a lamb under a ewe which is less than four years or more than eight years old, for a ewe of neither of these ages is fit to bring up its young; moreover, the offspring of aged stock generally reproduces the qualities of old age inherited from its parents, being either sterile or weakly. The delivery of a pregnant ewe 16 should be watched over with as much care as midwives exercise; for this animal produces its offspring just in the same way as a woman, and its labour is often even more painful since it is devoid of all reasoning. Hence the owner of a flock ought to have some knowledge of veterinary medicine, so that, if circumstances require it, when the foetus becomes stuck crosswise in the genital organs, he may either extract it whole, or be able to remove it from the womb, after dividing it with a knife without causing the mother's death—an operation which the Greeks call *embryulkein*.[a] The lamb, when it has been 17 brought forth, ought to be set upon its feet and put near its mother's udder; then its mouth should be opened and moistened by pressing the mother's teats, so that it may learn to derive its nourishment from her. But, before this is done, a little milk should be drawn off, which shepherds call "biestings," for, if this is not to some extent extracted, it does harm to the lamb, which for the first two days after its birth should be shut up with its mother, so that she may cherish her offspring, and that it may learn to know her. Then, as long as it has not begun to frisk 18 about, it should be kept in a dark and warm en-

[a] *I.e.* extracting the embryo.

luxuriantem virgea cum comparibus hara[1] claudi oportebit, ne velut puerili nimia exultatione macescat: cavendumque est, ut tenerior separetur a validioribus, quia robustus angit imbecillum. Satisque[2] est mane prius quam grex procedat in pascua; deinde etiam crepusculo redeuntibus saturis ovibus admiscere agnos. Qui cum firmi esse coeperint,[3] pascendi sunt intra stabulum cytiso, vel medica,[4] tum etiam furfuribus, aut, si permittat annona, farina hordei vel ervi: deinde, ubi convaluerint, circa meridiem pratis aut novalibus[5] villae continuis matres admovendae sunt, et septo emittendi agni, ut condiscant[6] foris pasci.

19

20 De genere pabuli iam et ante diximus, et nunc eorum, quae omissa sunt, meminerimus, iucundissimas herbas esse, quae aratro proscissis arvis nascantur; deinde quae pratis uligine carentibus; palustres silvestresque minime idoneas haberi. Nec tamen ulla sunt tam blanda pabula, aut[7] etiam pascua, quorum gratia[8] non exolescat usu continuo, nisi pecudum fastidio pastor occurrerit praebito sale, quod velut[9] aquae ac pabuli condimentum per aestatem canalibus ligneis impositum, cum e pastu redierint oves, lambunt, atque eo sapore cupidinem bibendi pascendique concipiunt. At contra penuriae

21

[1] area *SAac*.
[2] satisque *S* : statimque *A*[1]*ac*.
[3] coeperint *A* : ceperunt *S*.
[4] medimedicatum *S* : medicatum *A*.
[5] navalibus *SA*.
[6] condiscant *c* : -at *SAa*.
[7] aut *S* : ut *Aac*.
[8] gratio *S*[1]*A* : ratio *S*[2].
[9] velut aquae ac *Lundström* : vel atquae ac *S* : vel atq; *ARac*.

closure; afterwards, when it begins to be sportive, it will have to be shut up with the lambs of its own age in a pen fenced with osiers, so that it may not become thin from what we may call too much youthful frolicking, and care must be taken to separate a more tender lamb from the stronger ones, because the robust torments the feeble. It is enough to make this separation in the morning before the flock goes out to pasture, and then at dusk to let the lambs mingle with the ewes when they return home after eating their fill. When the lambs begin to get strong, they should be fed in the folds with shrub-trefoil or lucerne, and also with bran, or, if the price permits, with flour of barley or of bitter-vetch. Afterwards, when they have reached their full strength, their mothers should be brought about mid-day to the meadows or fallow lands adjoining the farm and the lambs released from their pen, so that they may learn to feed outside. 19

Concerning the nature of their food we have already spoken before and now call to mind what was not mentioned, namely, that the vegetation which is most acceptable is that which comes up when the fields have received their first ploughing; the next best is that which grows in meadows which are free from marsh; boggy and wooded lands are considered least suitable. There is, however, no fodder or even pasturage so agreeable that the pleasure which it gives does not grow stale with continuous use, unless the shepherd counteracts this aversion of his sheep by providing salt. This is placed in wooden troughs during the summer to serve as a kind of seasoning in their water and fodder and the sheep lick it up when they return from the pasture, and the taste of it makes them conceive a desire to eat and drink. But 20 21

hiemis succurritur obiectis intra tectum per praesepia cibis. Aluntur autem commodissime repositis ulmeis vel ex fraxino[1] frondibus, vel autumnali faeno, quod cordum vocatur. Nam id mollius et ob hoc
22 iucundius est, quam maturum. Cytiso quoque et sativa vicia[2] pulcherrime pascuntur. Necessariae tamen, ubi cetera defecerunt, etiam ex leguminibus paleae. Nam per se hordeum, vel fresa cum faba cicercula sumptuosior est, quam ut suburbanis regionibus salubri pretio[3] possit praeberi: sed sicubi vilitas[4] permittit,[5] haud dubie[6] sunt optima.
23 De temporibus autem pascendi, et ad[7] aquam ducendi per aestatem non aliter sentio, quam ut prodidit Maro:

Luciferi primo cum sidere frigida rura
Carpamus, dum mane novum, dum gramina canent,
Et ros in tenera pecori gratissimus herba,
Inde, ubi quarta sitim caeli collegerit hora,
Ad puteos, aut alta greges ad stagna . . .

perducamus, medioque die, ut idem, ad vallem,

Sicubi magna Iovis antiquo robore quercus
Ingentes tendit ramos, aut sicubi nigrum,
Ilicibus crebris sacra nemus accubat umbra.

24 Rursus deinde iam mitigato vapore compellamus

[1] fragino *SA*.
[2] vicia *A* : vitia *Sac*.
[3] potio *SAac*.
[4] si cubilitas *SA*.
[5] permittit *S* : mittit *AR*.
[6] dubie *S* : dubium *Aac*.
[7] ad *om*. *SAR*.

on the other hand the lack of food in winter is relieved by putting food for them under cover in their folds. They can be most conveniently fed on leaves of elm or ash which have been kept in store or on autumn hay, which is called the "after-crop"; for it is softer and therefore pleasanter than the early crop. Shrub-trefoil and cultivated vetch also make excellent fodder; but, when all else has failed, chaff of dried pulse must be used as a last resort, for barley by itself or chickling-vetch crushed with beans is too expensive to be provided at a reasonable price in districts near towns; but, wherever their cheapness allows, they are undoubtedly the best food. As for the times at which sheep ought to be fed and taken to water during the summer, my opinion is the same as that delivered by Maro: 22 23

At Lucifer's first rising let us haste
To the cool fields, while yet the dawn is new,
And turf still hoary, and on tender grass
The dew is sweetest to the feeding herd.
Then, when the sky's fourth hour brings thirst to all,
Let's lead the flocks to wells and deep-dug pools,[a]

and in the middle of the day, as the same poet says, let us conduct them to a valley,

Where haply Jove's great oak with hardwood old
Stretches its giant branches or a grove
Black with thick holm-oaks broods with holy shade.[b]

Then, when the heat is abated, let us again conduct 24

[a] Vergil, *Georg.* III. 324 ff. [b] *Ib.*, 332–334.

ad[1] aquam—etiam per aestatem id faciendum est—et iterum ad pabula[2] producamus.

> Solis ad occasum, cum frigidus aera vesper
> Temperat, et saltus reficit iam roscida luna.[a]

Sed observandum est sidus aestatis per emersum Caniculae ut ante meridiem grex in occidentem spectans agatur et in eam partem progrediatur, post meridiem in orientem. Siquidem plurimum refert, ne pascentium capita sint adversa soli, qui plerumque nocet animalibus oriente praedicto sidere.

25 Hieme et vere matutinis temporibus intra septa contineantur, dum dies arvis gelicidia detrahat. Nam pruinosa herba pecudi gravedinem[3] creat, ventremque[4] proluit. Quare etiam frigidis humidisque temporibus anni semel die[5] potestas aquae facienda est.

26 Tum qui sequitur gregem circumspectus ac vigilans (id quod omnibus et omnium quadrupedum custodibus praecipitur) magna clementia moderetur; idemque[6] propior[7] quia[8] silent, et in agendis recipiendisque ovibus adclamatione ac baculo minetur: nec unquam telum emittat in eas: neque ab his longius recedat: nec aut recubet,[9] aut considat. Nam nisi procedit, stare debet, quandoquidem

[1] ad *om. SAac.* [2] in pabula *S* : pabulo *AR.*
[3] glaudigine *S* : glaudinem *A.*
[4] ventem quae S^1A^1: ventrem S^2.
[5] die *Ursinus* : ei *SAR.*
[6] -que *add. ed. pr.* [7] proprior *SAR.*
[8] -que *SAR.* [9] recavet *SAR.*

---

[a] Vergil, *Georg.* III, 336 f.
[b] The text here gives no satisfactory sense and is certainly corrupt. The MS. reading *proprior* is meaningless and *propior*

them to the water (and this must be done even in the summer) and again drive them back to the pasture,

> Till sun-set, when chill evening cools the air
> And Luna's dews the thirsty glades refresh.[a]

But about the time when the Dogstar shows itself, we must carefully observe the position of the sun in summer, so that before mid-day the flock may be driven facing the west and may advance in that direction, but that after mid-day it may be driven towards the east, since it is of great importance that their heads, as they graze, should not face the sun, which is generally harmful to animals at the rising of the aforesaid constellation. 25 In winter and spring the sheep should be kept in their pens during the morning hours until the sun removes the rime from the fields, for grass with hoar-frost upon it causes catarrh in cattle and loosens the bowels; wherefore also in cold and damp seasons of the year they must be given the opportunity of drinking only once a day.

26 He who follows the flock should be observant and vigilant—a precept which applies to every guardian of every kind of four-footed animal—and should be gentle in his management of them and also keep close to them, because they are silent,[b] and when driving them out or bringing them home, he should threaten them by shouting or with his staff but never cast any missile at them, nor should he withdraw too far from them nor should he lie or sit down; for unless he is advancing he should stand upright, because the duty of a guardian calls for a lofty and com-

is scarcely better. A somewhat different line of thought is contained in the emendation *idemque pronior quam silens* suggested by Richter (*Hermes* LXXX. 213).

custodis officium sublimem celsissimamque oculorum veluti speculam desiderat, ut neque tardiores et gravidas, dum cunctantur,[1] neque agiles et fetas, dum procurrunt, separari [2] a ceteris sinat; ne fur, aut bestia [3] hallucinantem pastorem decipiat. Sed haec communia fere sunt in omni pecore ovillo. Nunc quae sunt generosi propria dicemus.

IV. Graecum pecus, quod plerique Tarentinum vocant, nisi cum domini praesentia est, vix expedit haberi: siquidem et curam et cibum maiorem desiderat. Nam cum sit universum genus lanigerum ceteris pecudibus mollius, tum ex omnibus Tarentinum est mollissimum, quod nullam domini aut magistrorum ineptiam sustinet, multoque minus 2 avaritiam; nec aestus, nec frigoris patiens. Raro foris, plerumque domi alitur, et est avidissimum cibi; cui si quid [4] detrahitur fraude villici,[5] clades sequitur gregem. Singula capita per hiemem recte pascuntur ad praesepia tribus hordei vel fresae cum suis valvulis fabae, aut cicerculae quattuor sextariis, ita ut et aridam frondem praebeas,[6] aut siccam vel viridem medicam cytisumve, tum etiam cordi faeni septena 3 pondo, aut leguminum paleas adfatim. Minimus agnis vendundis in hac pecude, nec ullus lactis reditus haberi potest. Nam et qui submoveri debent,[7] paucissimos post dies quam editi sunt, immaturi fere

[1] cunctantur *a* : cunctatur *SAc*.
[2] separare *SAR*.
[3] bestias *A* : bestius *S*.
[4] si quid *S* : *om. AR*.
[5] vilici *Sa* : vilicis *Ac*.
[6] praebeat *SAR*.
[7] debet *SAac*.

manding elevation from which the eyes can see as from a watch-tower, so that he may prevent the slower, pregnant ewes, through delaying, and those which are active and have already borne their young, through hurrying forward, from becoming separated from the rest, lest a thief or a wild beast cheat the shepherd while he is day-dreaming. These precepts are of general application and apply to sheep of all kinds; we will now deal with some points which are peculiar to the best breeds.

"Coated" sheep.

IV. It is scarcely advantageous to keep the Greek breed, which most people call the Tarentine, unless the owner is constantly on the spot, since it requires more care and food than other kinds. For, while all the sheep which are kept for their wool are more delicate than the others, the Tarentine breed is particularly so, for it does not tolerate any carelessness on the part of the owner or shepherd, much less niggardliness, nor can it stand heat or cold. 2 It is seldom fed out of doors but generally at home, and is most greedy of fodder and, if the bailiff fraudulently abstracts any of the food, disaster overtakes the flock. During the winter, when the sheep are fed in their pens, a satisfactory diet per head is three *sextarii* of barley or of beans crushed with their pods, or four *sextarii* of chickling-vetch provided you also supply them with dried leaves or lucerne, dry or fresh, or shrub-trefoil; also seven pounds of hay of the second crop is to their liking or plenty of pulse-chaff. 3 Only a very small profit can be made by selling the lambs of this kind of sheep and no return from the ewes' milk; for the lambs which ought to be taken away from their mother a very few days after birth, are generally slaughtered before they reach maturity, and

mactantur; orbaeque natis[1] suis matres alienae soboli praebent ubera: quippe singuli agni binis nutricibus submittuntur, nec quicquam subtrahi submissis expedit, quo saturior lactis agnus celeriter confirmetur, et parta nutrici consociata minus laboret in educatione fetus sui. Quam ob causam diligenti cura servandum est, ut et suis quotidie matribus et alienis non amantibus agni subrumentur.

4 Plures autem in eiusmodi gregibus quam in hirtis masculos enutrire oportet. Nam prius quam feminas inire possint mares castrati, cum bimatum expleverunt,[2] enecantur, et pelles eorum propter pulchritudinem lanae maiore pretio quam alia vellera mercantibus traduntur. Liberis autem campis et omni[3] surculo ruboque vacantibus ovem Graecam pascere meminerimus, ne, ut supra dixi, et lana

5 carpatur et tegumen. Nec tamen ea minus sedulam curam foris, quia[4] non quotidie procedit in pascua, sed maiorem[5] domesticam postulat. Nam saepius detegenda et refrigeranda est: saepius eius lana diducenda, vinoque et oleo insuccanda, nonnunquam etiam tota est eluenda, si diei permittit apricitas: idque ter anno fieri sat est. Stabula vero frequenter everrenda et purganda, humorque omnis urinae deverrendus est, qui commodissime siccatur perforatis tabulis, quibus ovilia consternuntur, ut grex

6 supercubet. Nec tantum caeno aut stercore, sed

[1] nates *SA*: nate *ac*.
[2] expleverunt A^2a: expleverint *c*: expluerunt SA^1.
[3] omnis *SA*.
[4] quia *addidi*.

their dams, deprived of their own lambs, are given the offspring of others to suckle; for each single lamb is put under two nurses and it is inexpedient that it should be deprived of any of their milk, that so, receiving a more satisfying quantity of milk, it may quickly grow strong, and that the ewe which has borne a lamb, having a nurse to share her duties, may have less difficulty in bringing her offspring up. Therefore you must be very careful to see that the lambs are daily put to the udders of their own mothers and also of strange ewes who have no maternal affection for them. 4 But in flocks of this kind more males must be brought up than in those of coarse-woolled sheep; for the males are castrated before they can be mated, when they have completed two years, and are killed, and their skins sold to dealers at a much higher price than other fleeces because of the beauty of their wool. We shall remember to feed a Greek sheep on open fields free from all shoots and brambles, lest, as I have already said, its wool and its covering be torn away. Nor, because it does not go out to pasture every day, 5 does it require less but more diligent care at home than out of doors; for it must frequently be uncovered and allowed to cool and its wool pulled apart and soaked with wine and oil. Sometimes too the whole animal must be washed, if sunny weather allows it, but it is enough to do this three times a year. The fold must be frequently swept and cleansed and all moisture due to urine must be brushed away, the best method of keeping it dry being the use of boards with holes in them with which the sheep-folds are paved, so that the flock may lie down on them. The shelters must be free 6

[5] maiorem *ed. pr.*: maioris *SAac.*

exitiosis quoque serpentibus tecta liberentur: quod ut fiat,

> Disce et odoratam stabulis incendere cedrum,
> Galbaneoque agitare graves nidore chelydros.
> Saepe sub immotis praesepibus, aut mala tactu,
> Vipera delituit, caelumque exterrita fugit:
> Aut tecto assuetus coluber.

Quare, ut idem iubet:

> cape robora, pastor,
> Tollentemque minas, et sibila colla tumentem
> Deice.

Vel ne istud cum periculo facere necesse sit, muliebres capillos,[1] aut cervina saepius ure cornua, quorum odor maxime non patitur stabulis praedictam pestem consistere.

7 Tonsurae certum tempus anni per omnes regiones servari non potest, quoniam nec ubique tarde, nec celeriter aestas ingruit: et est modus optimus considerare tempestates, quibus ovis neque frigus, si lanam detraxeris, neque aestum, si nondum detonderis,[2] sentiat. Verum ea quandoque detonsa fuerit, ungi debet tali medicamine: succus excocti [3] lupini, veterisque vini faex, et amurca pari mensura 8 miscentur, eoque liquamine tonsa ovis imbuitur,[4] at que ubi per triduum dilibuto tergore medicamina [5]

[1] capillos *S* : capillus *Aac*.
[2] detonseris *c* : detoderis *S*[1] : detonderis *S*[2] : detodoris *A*.
[3] excocti *S*[2] : excoleti *AR*.
[4] imbuitur *A*[2]*R* : inbitur *S*[1] : imbitur *A*[1].
[5] medicamina *Ald.* : media *SAR*.

---

[a] Galbanum was the resinous sap of an umbelliferous plant (*Bubon galbanum*) growing in Syria.

not only from mud and ordure but also from deadly snakes; with this end in view,

> Learn too to burn the fragrant cedar-wood
> And from the stalls to drive dread water-snakes
> With fumes of Syrian gum;[a] a viper oft,
> Dangerous to the touch, 'neath unmoved pens
> Has lurked and, frightened, shunned the light of heaven,
> Or else a grass-snake wont to haunt the shed.[b]

Therefore, at the bidding of the same poet,

> Seize, shepherd,
> A club of oak, and when it rears its head
> In threatening wise and swells its hissing neck,
> Then strike it down.[c]

Or, to avoid the necessity of this dangerous expedient, burn a woman's hair continually or a stag's horn, the odour of which is the best thing to prevent this pestilential creature from settling in the sheep-folds.

7 It is impossible to observe in all regions the same fixed time of year for shearing, because summer does not everywhere advance with the same speed or slowness. The best plan is to watch carefully for weather when the sheep will not feel the cold if you deprive them of their wool, nor the heat if you put off shearing them. But, whenever a sheep has been sheared, it must be anointed with the following preparation: the juice of boiled lupines, the dregs of old wine and the lees of olives are mixed in equal portions and the sheep is soaked with this liquid after it has been sheared, and when, after its skin 8 has been anointed during three days and it has

[b] Vergil, *Georg.* III. 414 ff. [c] *Ib.*, 419–421.

perbiberit, quarto die, si est vicinia maris, ad litus deducta mersatur: si minus, caelestis [1] aqua sub dio [2] salibus in hunc usum durata paulum decoquitur, eaque grex perluitur. Hoc modo curatum pecus toto anno scabrum fieri non posse Celsus affirmat: nec dubium est, quin etiam ob eam rem lana mollior atque prolixior renascatur.

V. Et quoniam censuimus cultum curamque recte valentium, nunc quemadmodum vitiis aut morbo laborantibus subveniundum sit, praecipiemus: quanquam pars haec exordii paene tota iam exhausta [3] est, cum de medicina maioris pecoris priore libro disputaremus. Quia [4] cum sit fere eadem corporis natura minorum maiorumque quadrupedum, paucae parvaeque morborum et remediorum differentiae possunt inveniri; quae tamen quantulaecunque sunt, non omittentur a nobis.

2 Si aegrotat universum pecus, ut et ante praecepimus, et nunc, quia remur esse [5] maxime salutare, iterum adseveramus, in hoc casu, quod est remedium praesentissimum, pabula mutemus et aquationes, totiusque regionis alium quaeramus statum caeli, curemusque, si ex calore et aestu concepta pestis invasit, ut opaca rura: si invasit frigore, ut eligantur

3 aprica. Sed modice ac sine festinatione persequi pecus oportebit, ne imbecillitas eius longis itineribus

[1] si minus celestis *S* : si minusca et gustis *A*.
[2] sub dio *ed. pr.* : subsidio *SAR*.
[3] exhausta *ed. pr.* : excausa *A*[1] : exhausa *S* : exausta *ac*.
[4] qua *SAR*.
[5] quia remur esse *S*[2] : qui aremus res se *S*[1] : quiremus resse *Aac*.

---

[a] Book VI. 6–19 and 30–35.

absorbed this preparation, on the fourth day, if the sea is near at hand, the sheep should be driven down to the shore and plunged in; but, if this is impossible, rain-water, after being hardened for this purpose with salt in the open air, is boiled for a short time and the flock thoroughly washed with it. Celsus declares that a sheep treated in this manner cannot possibly suffer from scab for a whole year, and there is no doubt that, as a result, its wool too will grow again more soft and luxuriant than before.

The diseases of sheep and their cure.

V. Since we have now considered the management and care which sheep require when in good health, we will now give directions how to come to the help of those which are suffering from ailments or diseases, although almost all this part of my treatise has already been entirely exhausted when we were discussing in the previous book [a] the medical treatment of the larger cattle; for since the physical nature of the smaller and of the larger quadrupeds is practically the same, only a few trifling differences are to be found in their diseases and the remedies of them; but, however unimportant they are, we will not omit them.

2 If the whole flock is sick, we again prescribe in this case also as the most efficacious remedy what we directed before, because we regard it as the most salutary, namely, to change both the fodder and the watering-places and to seek another climate for the grazing-ground as a whole, and to take care to choose densely shaded country, if the malady which has attacked the flock is the result of heat, but, if it is the result of cold, to choose a sunny district. 3 But it will be advisable to drive the flock at a moderate pace and not to hurry it for fear of aggravating its enfeebled

aggravetur: nec tamen in totum pigre ac segniter agere. Nam quemadmodum fessas morbo pecudes vehementer agitare et extendere non convenit, ita conducit mediocriter exercere, et quasi torpentes excitare, nec pati veterno consenescere atque extingui. Cum deinde grex ad locum fuerit perductus, 4 in lacinias colonis [1] distribuatur. Nam particulatim facilius quam universus convalescit, sive quia ipsius morbi halitus minor est in exiguo numero, seu quia expeditius cura maior adhibetur paucioribus. Haec ergo et reliqua, ne nunc eadem repetamus, quae superiore exordio percensuimus, observare debemus si universae laborabunt: illa, si [2] singulae.

5 Oves frequentius, quam ullum aliud animal infestantur [3] scabie: quae fere nascitur, sicut noster memorat poeta,

> Cum frigidus imber
> Altius ad vivum persedit, et horrida cano
> Bruma gelu,

vel post tonsuram, si remedium praediciti medicaminis non adhibeas, si aestivum sudorem mari vel flumine non abluas, si tonsum gregem patiaris silvestribus rubis ac spinis sauciari, si stabulo utaris, in quo mulae aut equi aut asini steterunt: praecipue

[1] coloniis *SAac*.
[2] illa si *S* : illas in *Aac*.
[3] infestantur *A*²*R* : -atur *S* : infertur *A*¹.

---

[a] Vergil, *Georg.* III. 441 f.

condition with long journeys; on the other hand it should not be driven at an absolutely slow and sluggish rate; for, while it is not expedient to urge sheep on forcibly when they are worn out by disease and put a strain upon them, yet it is good to give them moderate exercise and, as it were, to rouse them from their torpor and not allow them to lose strength through inactivity, and so perish. Next, when the flock has been conducted to its new station, it should be distributed in small groups amongst the farmers; for it recovers more easily when it is 4 divided up than when it is kept together, either because the infectiousness of the disease itself is less in a small number or because a more effective cure can be applied more expeditiously to fewer victims. These precepts, then, and the others which we laid down in the earlier part of our treatise (to avoid repeating here what we have already said) should be observed when the whole flock is sick; but if individual animals are affected, the following rules should be observed.

Sheep more often than any other animals are 5 attacked by the scab, which generally occurs, as our poet says,[a]

When the cold shower and shivering winter, chill
With hoary frost, have pierced them to the quick,

or else after they have been sheared, if you do not apply the remedy already described, or if you do not wash out the summer sweat in the sea or in a river, or if you allow the flock, after having been shorn, to suffer wounds from wild brambles or thorns, or if you are using a pen in which mules or horses or donkeys have stood; but, above all things, scantiness of

tamen exiguitas cibi maciem, macies autem scabiem
6 facit. Haec ubi coepit irrepere, sic intelligitur: vitiosum locum pecudes aut morsu scalpunt, aut cornu vel ungula tundunt, aut arbori adfricant, parietibusve[1] detergent: quod ubi aliquam facientem videris, comprehendere oportebit, et lanam[2] diducere: nam subest aspera cutis, et velut quaedam porrigo.[3] Cui primo quoque tempore occurrendum est, ne totam progeniem coinquinet, si quidem celeriter cum et alia pecora, tum praecipue oves
7 contagione vexentur. Sunt autem complura medicamina, quae idcirco enumerabimus, non quia cunctis uti necesse sit, sed quoniam nonnullis regionibus quaedam reperiri[4] nequeunt, ut[5] ex pluribus aliquod inventum remedio sit. Facit autem commode primum ea compositio, quam paulo ante demonstravimus, si ad faecem et amurcam succumque decocti lupini misceas portione aequa detritum
8 album elleborum.[6] Potest etiam scabritiem tollere succus viridis cicutae: quae verno tempore, cum iam caulem nec adhuc semina facit, decisa contunditur, atque expressus humor eius fictili vase reconditur, duabus urnis liquoris admixto salis torridi semodio. Quod ubi factum est, oblitum vas in stercilino[7] defoditur, ac toto anno fimi vapore concoctum mox promitur,[8] tepefactumque medicamentum illinitur scabrae parti, quae tamen prius aspera testa defricta

[1] parietibus vel *SAR*.
[2] lanam diducere *Ald.*: lana rudi deucere *S*: lana rudi ducere *Aac*.
[3] porrigo *S*[1]: prurigo *R*.
[4] repperiri *S*[2]: repperi *S*[1]*A*: reperiri *R*.
[5] ut *om. SAR*.
[6] eleborum *c, ed. pr.*: helleboreos *S*: -em *A*[1]: -um *A*[2].
[7] intercilino *S*[1]*A*[1]: instercilino *A*[2]: in sterquilino *S*[2].

fodder causes emaciation, and emaciation causes the scab. This disease can be diagnosed in the following 6 way when it begins to creep in: the sheep either gnaw the part affected, or strike it with horn or hoof, or rub it against a tree or wipe it upon the walls. When you see any sheep acting in these ways, it will be best to take hold of the animal and draw its wool apart, for there is a rough skin underneath it and a kind of crust. This must be treated at the first possible opportunity, lest it infect the whole flock, since, while other cattle are readily attacked by contagious disease, sheep are particularly so. There are, how- 7 ever, several remedies, which we will on this account enumerate, not because it is necessary to use them all at one time but in order that, since some of them are not to be met with in certain regions, one out of many may be found in order to effect a cure. First, the preparation which I explained just now can be used with advantage, namely, a mixture in equal portions of crushed white hellebore with lees of wine and dregs of oil and the juice of boiled lupine. The juice of green hemlock can also be used to remove 8 scabbiness; this plant is cut in spring-time, when it is already producing stalk but not seeds, and crushed, and the juice is pressed out and stored in an earthenware vessel, half a *modius* of dried salt being mixed with two *urnae* of the liquid. Next the vessel is sealed up and buried in a dung-pit and, after having matured for a whole year in the heat of the dung, it is taken out and the preparation is heated and smeared over the part affected by scab after it has been previously reduced to a state of soreness by being rubbed with a rough potsherd or a piece of

---

[8] promittitur S^1A^1.

9 vel pumice redulceratur. Eidem remedio est amurca duabus partibus decocta: item vetus hominis urina testis candentibus inusta. Quidam[1] tamen hanc ipsam[2] subiectis ignibus quinta parte minuunt, admiscentque pari mensura succum viridis cicutae: deinde singulis urnis eius[3] liquaminis[4] singulos 10 fricti salis sextarios[5] infundunt. Facit etiam sulfuris triti et picis liquidae modus aequalis igne lento[6] coctus. Sed Georgicum carmen affirmat nullam esse praestantiorem medicinam,

> Quam si quis ferro potuit rescindere summum
> Ulceris os: alitur vitium, vivitque tegendo.

Itaque reserandum est, et ut cetera vulnera, medicamentis curandum. Subicit deinde aeque prudenter, febricitantibus ovibus de talo vel inter duas ungulas sanguinem emitti[7] oportere:[8] nam plurimum, inquit[9]

> Profuit incensos aestus avertere, et inter
> Ima ferire pedis salientem sanguine venam.

11 Nos etiam sub oculis et de auribus sanguinem detrahimus. Clodigo[10] quoque dupliciter infestat[11] ovem, sive cum subluvies atque intertrigo in ipso discrimine ungulae nascitur: seu cum idem locus

[1] quadam *SAR*.
[2] hac ipsa *SA*: hanc ipsam *a*: hac ipsam *c*.
[3] singulis urnis eius *Lundström ex cit. Palladii*; singularis triti et picis eius *S*: singularis triti et picis *A*.
[4] liquaminis A^1.
[5] fricti sali sestarios S^1: frictis aliis extarios A^1: fruti salis sextarios *a*.
[6] lento *S*: lente *AR*.
[7] mitti *SA*.
[8] oportet *SAR*.
[9] id quid *S*: id quod A^1*ac*.
[10] clodigo *Svennung*: Clodi *S*: cludi A^1: cladi *ac*.
[11] infestato *SR*.

pumice-stone. The same disease is also treated with 9 oil-lees boiled down by two-thirds, and also with stale human urine in which red-hot tiles have been plunged. Some people, however, put the urine itself upon the fire and reduce its volume by one-fifth and mix with it an equal quantity of the juice of green hemlock and then pour into each urn of this liquid a *sextarius* of crushed salt.[a] An equal 10 quantity of ground sulphur and liquid pitch boiled over a slow fire has a good effect. A passage in the *Georgics*, however, declares that there is no more sovereign remedy,

> Than if with knife one cuts the ulcer's head;
> The scab, if covered, gains fresh food and life.[b]

That is why it must be opened and treated, like other wounds, with medicaments. The poet presently adds, with equal wisdom, that, when sheep are in a state of fever, they should be bled either from the pastern or between the two parts of the hoof; for, as he says,

> It oft has greatly helped to keep away
> The kindled flames of fever, if you strike
> The vein which throbs with blood beneath the foot.[c]

We also draw off blood beneath the eyes and from 11 the ears. Lameness also troubles sheep in two ways, either when fouling or galling occurs in the actual division of the hoof, or when the same place harbours

[a] The reading here is uncertain, but *triti et picis* has probably come in from the following sentence.
[b] *Ib.* 453 f.
[c] *Ib.* 459 f.

tuberculum habet, cuius media fere parte canino 12 similis extat pilus, eique [1] subest vermiculus. Subluvies et intertrigo pice per se liquida, vel alumine et sulfure atque aceto mixtis litae curentur, vel tenero [2] punico malo, prius quam grana faciat, cum alumine pinsito, superfusoque aceto vel aeris rubigine infriata,[3] vel combusta galla cum austero vino levigata 13 et inlita. Tuberculum, cui subest vermiculus, ferro quam cautissime circumsecari oportet, ne, dum amputatur, etiam, quod infra est, animal vulneremus: id enim cum sauciatur, venenatam saniem mittit, qua respersum [4] vulnus ita insanabile facit, ut totus pes amputandus sit: sed cum tuberculum diligenter circumcideris, candens sebum vulneri per ardentem tedam instillato.

14 Ovem pulmonariam similiter ut suem curari convenit, inserta per auriculam, quam veterinarii consiliginem vocant: de ea iam diximus, cum maioris pecoris medicinam traderemus.[5] Sed is [6] morbus aestate plerumque concipitur, si defuit aqua, propter quod vaporibus omni quadrupedi largius bibendi 15 potestas danda est. Celso placet, si est in pulmonibus vitium, acris aceti tantum dare, quantum ovis sustinere possit: vel humanae veteris urinae tepefactae trium heminarum instar per sinistram narem

[1] ei quae *SA*.
[2] tenero *ed. pr.*: tero S^1*Aac*: austero S^2.
[3] infriata *ed. pr.*: infrita S^1*AR*.
[4] repressum *AR*: res pressũ *S*.
[5] traderemus S^2: trademus S^1: tradimus *Aa*.
[6] is *ac*: his *SA*.

[a] *Pulmonaria officinalis*.
[b] Book VI. 5. 3; 14. 1.

a tubercule from about the middle of which a hair projects like that of a dog, which has a small worm beneath it. Fouling and galling are removed by being anointed with liquid pitch by itself or with alum and sulphur and vinegar mixed together, or young pomegranate, before it forms its seeds, crushed up with alum and with vinegar poured over it, or copper-rust sprinkled over it, or else burnt oak-apples pulverized and mixed with rough wine and smeared on the sore. A tubercule which has a worm inside it should be cut round with a knife with the greatest possible care, lest, in the course of cutting, we should also wound the part of the animal which is underneath it; for, if this is damaged, it discharges poisonous matter and, if this is sprinkled over the wound, it makes it so difficult to heal that the whole foot has to be amputated. But when you have carefully cut round the tubercule, burning fat should be made to drip over the wound by means of a lighted torch. 12 13

Any sheep which is suffering from a disease of the lungs should be treated in the same way as a pig is treated for the same disease, by the insertion through the ear of what the veterinary surgeons call lungwort.[a] We have already spoken [b] of this plant when we dealt with the treatment of the larger cattle. This disease is usually contracted in the summer if the water has been in short supply, and for this reason opportunity must be given to all quadrupeds of drinking more freely in hot weather. Celsus is of opinion that, if there is trouble in the lungs, one should give the sufferer as much sour vinegar as it can stand, or else pour down the left nostril through a small horn about three *heminae* of stale human urine which has been 14 15

corniculo infundere, atque axungiae sextantem faucibus inserere.

16 Est etiam insanabilis sacer ignis, quam pusulam [1] vocant pastores: ea nisi compescitur intra primam pecudem, quae tali malo correpta est, universum gregem contagione prosternit, si [2] quidem nec medicamentorum nec ferri remedia patitur. Nam paene ad omnem tactum excandescit: sola tamen fomenta non aspernatur lactis caprini, quod infusum tactu suo velut [3] eblanditur [4] igneam saevitiam, differens

17 magis occidionem gregis, quam prohibens. Sed Aegyptiae gentis auctor [5] memorabilis [6] Bolus [7] Mendesius, cuius commenta, quae appellantur Graece *χειρόκμητα*,[8] sub nomine Democriti falso produntur, censet [9] propter hanc pestem [10] saepius ac diligenter [11] ovium terga perspicere, ut si forte sit in aliqua tale vitium deprehensum, confestim scrobem [12] defodiamus in limine stabuli, et vivam pecudem, quae fuerit pusulosa,[13] resupinam obruamus, patiamurque super [14] obrutam meare totum gregem, quod eo facto morbus propulsetur.

18 Bilis aestivo tempore non minima [15] pernicies potione depellitur humanae veteris urinae, quae ipsa [16] remedio est etiam pecori arcuato. At si molesta

[1] pusulam *Ac* : pusillam *a* : pustulam S^2.
[2] si *S* : sic *AR*.
[3] tactu suo velut *Svennung* : tactus volet ut (et *a*) *SAc*.
[4] eblanditur *S* : et blanditur *AR*.
[5] auctor *S* : auctore *A* : auctorem *ac*.
[6] memorabilis S^2 : memorabis S^1AR.
[7] Bolus *Reinesius* : dolus *SAR*.
[8] Χειρόκμητα *Schneider* : Χειροκίμητα *S*.
[9] censet *S* : gens et *AR*.
[10] pestem *om. Aac*.
[11] ac diligenter S^2 : adliganter S^1A. [12] scribom S^1A.

heated, and put a *sextans* of axle-grease down its throat.

Erysipelas, which the shepherds call *pusula*, is incurable. Unless it is confined to the first sheep which is attacked by this kind of trouble, it infects and lays low the whole flock, if it does not yield to medical or surgical treatment; for it blazes forth at almost any touch. The only remedy which it does not reject is fomentation with goat's milk, which, when poured upon it, as it were, charms by its touch the fiery raging of the disease, postponing rather than preventing the destruction of the flock. The celebrated writer of Egyptian race, Bolus of Mendesium,[a] whose commentaries, which in Greek are called *Hand-wrought Products* and are published under the pseudonym of Democritus, is of opinion that as a precaution against this disease the hides of the sheep ought to be frequently and carefully examined, so that if any trace of disease is by chance discovered in any one of them, we may immediately dig a trench on the threshold of the sheep-fold and, laying it on its back, inter alive the animal which is suffering from erysipelas and allow the whole flock to pass over its buried body; for by doing this the disease is driven away. 16 17

Bile, not the least fatal disease in summer, is dispelled by making the victim drink stale human urine. The same remedy is also given to a sheep which is suffering from jaundice. If rheum is trouble- 18

[a] Pliny, *N.H.* XXIV. 102; Vitruvius IX. 3. His work was entitled συμπαθειῶν καὶ ἀντιπαθειῶν.

[13] pusillosa S^1AR. [14] super *S* : sub *AR*.
[15] minima *ed. pr.* : nimia *SAR*.
[16] ipsa *S* : ipse *AR*.

pituita est, cunelae bubulae, vel surculi nepetae silvestris lana involuti naribus inseruntur, versanturque donec sternuat ovis. Fracta pecudum non aliter quam hominum crura sanantur, involuta lanis oleo atque vino insuccatis, et mox circumdatis ferulis

19 conligata. Est etiam gravis pernicies herbae sanguinariae, qua si pasta est ovis, toto ventre distenditur, contrahiturque, et spumat et[1] quaedam tenuia[2] taetri odoris excernit.[3] Celeriter sanguinem mitti oportet sub cauda in ea parte quae proxima est clunibus, nec minus in labro superiore vena[4] solvenda est. Suspiriose[5] laborantibus auriculae ferro rescindendae, mutandaeque regiones; quod in omnibus morbis ac pestibus fieri debere censemus.

20 Agnis quoque succurrendum est vel febricitantibus, vel aegritudine alia affectis. Qui ubi morbo laborant, admitti ad matres non debent, ne in eas perniciem transferant. Itaque separatim mulgendae sunt oves, et caelestis aqua pari mensura lacti miscenda est, atque ea potio febricitantibus danda. Multi lacte caprino iisdem medentur, quod per

21 corniculum infunditur faucibus. Est etiam mentigo, quam pastores ostiginem vocant, mortifera lactentibus. Ea plerumque fit, si per imprudentiam[6] pastoris emissi agni vel etiam haedi roscidas herbas

[1] et *om. SAR.*
[2] tenui *SAR.*
[3] expernit *S*¹*Aac.*
[4] veno *S*¹*A*¹.
[5] suspiriose *ac* : suspiriore *SA.*
[6] per imprudentiam *ed. pr.* : prudentiam *SAR.*

some, stalks of ox-marjoram or wild mint, wrapped round with wool, are inserted in the nostrils and turned round and round until the sheep sneezes. The broken legs of sheep are treated in the same manner as those of human beings; they are wrapped in wool soaked in oil and wine and then bound up in splints which are placed round them. Knotgrass[a] 19 has also bad effects which are serious; for, if the sheep feeds on it, its whole belly becomes distended and then contracts, and the animal foams at the mouth and emits a thin kind of matter which has a foul odour. The victim must immediately be bled underneath the tail in the region nearest to the buttocks, and also a vein must be opened on the upper lip. Sheep whose breathing is asthmatical must have their ears cut with the knife and be transferred to other districts, a precaution which, in my opinion, ought to be taken in all diseases and plagues.

Succour must also be given to lambs when they are 20 suffering from fever or affected by any other sickness; those which are labouring under any disease ought not to be admitted to their dams, lest they pass on the malady to them. The ewes, then, must be milked separately, and rain-water must be mixed in equal measure with the milk and this potion given to the lambs which have fever. Many people use goats' milk as a remedy for these same lambs, pouring it down their throats through a small horn. There is 21 also an eruptive disease, called by the shepherds *ostigo* (lamb-scab), which is fatal to sucking lambs. This generally occurs, if, through the carelessness of the shepherd, the lambs or even kids have been let loose and have fed on grass which is covered with dew,

[a] See Book VI. 12. 5 and note.

depaverint, quod minime committi oportet. Sed cum id factum[1] est, velut ignis sacer os atque labra foedis ulceribus obsidet.[2] Remedio sunt hyssopus et sal aequis ponderibus contrita. Nam ea mixtura palatum atque lingua,[3] totumque os perfricatur. Mox ulcera lavantur aceto, et tunc pice liquida cum adipe suilla perlinuntur. Quibusdam placet rubiginis aeneae tertiam[4] partem duabus veteris axungiae portionibus commiscere, tepefactoque uti medicamine. Non nulli folia cupressi trita[5] miscent aquae, et ita perluunt ulcera atque palatum. Castrationis autem ratio iam tradita est. Neque enim alia in agnis, quam in maiore quadrupede servatur.

22

VI. Et quoniam de oviarico satis dictum est, ad caprinum pecus nunc revertar. Id autem genus dumeta potius, quam campestrem[6] situm desiderat: asperisque etiam locis ac silvestribus optime pascitur. Nam nec rubos aversatur, nec vepribus offenditur, et arbusculis frutectisque maxime gaudet. Ea sunt arbutus, atque alaternus[7] cytisusque agrestis, nec minus ilignei querneique frutices, qui in altitudinem non prosilierunt.

2 Caper, cui sub maxillis binae verruculae collo dependent, optimus habetur, amplissimi corporis, cruribus crassis,[8] plena et brevi cervice, flaccidis[9] et praegravantibus auribus, exiguo capite, nigro densoque et nitido atque longissimo pilo. Nam et ipse tondetur

Usum in castrorum ac miseris velamina nautis.

[1] id factum S^2 : infactum S^1*Aac*. [2] obsident *SAR*.
[3] lingua S^2A^1 : longua S^1 : linquam *AR*. [4] tertia SA^1.
[5] cupressi trita *S* : cum pressurita *AR*.
[6] campestre *R* : campreste S^1A.
[7] alaternus *S* : alternus *Aac*.
[8] crassis *S* : erasis *AR*.

which they certainly should not be allowed to do. But when this has happened, a kind of erysipelas surrounds their mouths and lips with filthy sores. 22 The cure consists of hyssop and salt crushed together in equal quantities, the palate, the tongue and the whole mouth being rubbed with this mixture. Next the sores are washed with vinegar and then thoroughly anointed with liquid pitch and lard. Some people prefer a mixture of one part of verdigris to two parts of stale axle-grease heated and used as a medicine; some make a mixture of crushed cypress-leaves and water and thoroughly wash the sores and the palate. The method of castration has already been described, for the operation is performed on lambs in the same manner as on the larger quadrupeds.

Goats.

VI. Now that enough has been said about sheep, I will next turn to goats. This species of animal prefers thickets to open country and is best pastured in rough and wooded districts; for it has no aversion to brambles and has no fault to find with briers and takes a particular pleasure in bushes and shrubs, such as the strawberry-tree, the buck-thorn, the wild trefoil and shrubs of holm-oak and oak which have not yet reached any great height.

2 The points of the best type of he-goat are two excrescences which project downwards from its throat below its jaws, a large frame, thick legs, a full, short neck, flabby and drooping ears, a small head, and black, thick, glossy and very long hair; for the he-goat is also shorn

For use in camps and hapless sailors' coats.[a]

[a] Vergil, *Georg.* III. 313.

[9] flaccidis *ac*: placcidis S^1A^1.

3 Est autem mensum[1] septem satis habilis ad progenerandum:[2] quoniam immodicus libidinis, dum adhuc uberibus[3] alitur, matrem stupro supervenit, et ideo ante sex annos celeriter consenescit, quod immatura veneris cupidine primis pueritiae temporibus exhaustus est. Itaque quinquennis 4 parum idoneus habetur feminis implendis. Capella praecipue probatur simillima hirco, quem descripsimus, si etiam est uberis maximi et lactis abundantissimi. Hanc pecudem mutilam[4] parabimus quieto caeli statu: nam procelloso atque imbrifero cornuta. Semper autem et omni regione maritos gregum mutilos esse oportebit: quia cornuti fere perniciosi 5 sunt propter petulantiam. Sed numerum huius generis maiorem, quam centum capitum sub uno clauso non expedit habere, cum lanigerae mille pariter commode stabulentur.[5] Atque ubi caprae primum comparantur, melius est unum gregem totum, quam ex pluribus particulatim mercari, ut nec in pastione separatim laciniae diducantur, et in caprili maiore concordia quietae consistant. Huic pecudi nocet aestus, sed magis frigus, et praecipue fetae, quia gelicidiosior[6] heims conceptum vitiat.[7] Nec tamen[8] ea sola creat[9] abortus,[10] sed etiam glans cum citra satietatem data est. Itaque nisi potest affatim praeberi, non est gregi permittenda.

[1] mensum *S*²: mensuum *S*¹*A*: mensium *R*.
[2] progenerandum *S*: procerandum *A*¹: procreandum *R*.
[3] uberius *SAR*.
[4] mutilam *ed. pr.*: milã *S*: mila *A*.
[5] stabuleanter *S*¹*A*¹: stabulantur *S*².
[6] gelicidiosior *Lundström*: geliciorior *S*¹: geliciodior *S*²: gelicior *Aac*.
[7] vitiat *S*: fecit *AR*.
[8] tamen *S*: tantum *AR*.
[9] creat *Aac.*: creant *Ald.*
[10] abortu *S*: abortat *AR*.

The he-goat is quite ready for breeding purposes 3 at the age of seven months; for it is immoderate in its desires and, while it is still being fed at its mother's udder, it leaps upon her and tries to do her violence. Hence, before it has reached six years of age, it is fast becoming old, because it has worn itself out in early youth by premature indulgence of its desires; and so, when it is only five years old, it is regarded as unfit for impregnating the female. A 4 she-goat is most highly approved which most closely resembles the he-goat which we have described, if it also has a very large udder and a great abundance of milk. If we live in a calm climate we shall acquire a she-goat without horns; for in a stormy and rainy climate we shall prefer one with horns; but always and in every district the fathers of the herd will have to be hornless, because those which have horns are generally dangerous because of their viciousness. One ought not to keep a larger number than a hundred 5 head of goats in one enclosure, though one can equally easily keep a thousand sheep in the folds. When one is acquiring she-goats for the first time, it is better to buy a whole herd at once than to purchase them one by one from a number of sources; this prevents them from splitting up into small groups while they are pasturing and makes them settle down quickly and in greater harmony in goat-stalls. The heat is harmful to this creature, but the cold is even more so, especially to pregnant she-goats, for an unusually frosty winter destroys the embryo. But not only the abnormally frosty winter causes abortion; it also occurs if less than a sufficiency of mast is given them; and so the herd should not be allowed to eat mast unless a plentiful supply can be provided.

6 Tempus admissurae per autumnum fere ante mensem Decembrem praecipimus, ut iam propinquante vere, gemmantibus frutectis, cum primum silvae nova germinant fronde, partus edatur. Ipsum vero caprile vel naturali saxo, vel manu constratum [1] eligi debet, quoniam huic pecori nihil substernitur. Diligensque pastor quotidie stabulum converrit, nec patitur stercus aut humorem consistere lutumve fieri,

7 quae cuncta sunt capris inimica. Parit autem, si est generosa proles, frequenter duos, nonnunquam trigeminos. Pessima est fetura cum matres binae ternos haedos efficiunt. Qui ubi editi sunt, eodem modo, quo [2] agni educantur, nisi quod magis haedorum [3] lascivia compescenda, et arctius cohibenda est. Tum super lactis abundantiam samera, vel cytisus, aut hedera praebenda, vel etiam cacumina lentisci, aliaeque tenues [4] frondes obiciendae sunt. Sed ex geminis singula capita, quae videntur esse robustiora, in supplementum gregis reservantur,

8 cetera mercantibus traduntur. Anniculae vel bimae capellae (nam utraque aetas partum edit) submitti haedum non oportet. Neque enim educare nisi trima debet. Sed anniculae confestim depellenda suboles. Bimae [5] tamdiu admittenda, dum possit esse vendibilis. Nec ultra octo annos matres servandae sunt, quod assiduo partu fatigatae steriles [6]

9 existant. Magister autem pecoris acer, durus,

[1] constratu *SAR*. [2] qui *SAR*. [3] edorum *SAc* : odorum *a*. [4] tenue *SAR*. [5] bimae *SA* : bime *ac*. [6] fatigatae steriles *S*[2] : fatigata steriles *S*[1] : fatigata est exsteriles *A*.

[a] Recently the reading *bimae*, instead of *binae* has been strongly urged on the basis of palaeography and the sense of the passage (Richter, *Hermes* LXXX. 215).

The time which we advise for covering the she-goats is during the autumn, some time before the month of December, so that the kids may be born when spring is already approaching and the shrubs are coming into bud and the woods just sprouting with new foliage. A site for the goats' stable should be chosen which has a natural or artificial stone floor, since no litter is provided for this animal. A careful goatherd sweeps out the stable every day and does not allow any ordure or moisture to remain or any mud to form, all of which things are prejudicial to goats. If a she-goat is of good stock, it frequently bears twins and sometimes triplets. It is a very poor increase when two mothers produce only three kids between them.[a] When the kids are born, they are reared in the same manner as lambs except that their wantonness must be more repressed and kept within stricter bounds. Besides an abundance of milk, elm-seed or shrub-trefoil or ivy must be provided, or else tops of mastic and other delicate foliage must be put before them. When there are sets of twins, from each pair one, whichever seems to be the more robust, is reserved to fill up the herd, while the rest are handed over to the dealers. A she-goat of only one or two years (for both ages are capable of bearing young) should not be given kids to rear; for it ought not to bring up a kid till it is three years old. And a mother of one year ought to be immediately deprived of its offspring, but a kid of a two-year-old mother ought to be left with it until it is ready to be sold. The mother-goats ought not to be kept beyond eight years, because, worn out by continual bearing, they end by becoming barren. The herd-master ought to be keen, hardy, energetic,

strenuus, laboris patientissimus, alacer atque audax esse debet, et qui per rupes, per solitudines, per vepres facile vadat, et non, ut alterius generis pastores, sequatur, sed plerumque ut antecedat gregem. Maxime strenuum pecus est capella[1] praecedens;[2] subinde quae incedit,[3] compesci debet, ne procurrat,[4] sed placide ac lente pabuletur, ut et largi sit uberis, et non strigosissimi corporis.

VII. Atque alia genera pecorum, cum pestilentia vexantur, prius morbo et languoribus macescunt, solae capellae quamvis opimae[5] atque hilares subito concidunt et velut aliqua ruina gregatim prosternantur.[6] Id autem accidere[7] maxime solet ubertate pabuli. Quamobrem cum adhuc paucas pestis perculit, omnibus sanguis detrahendus: nec tota die pascendae, sed mediis quattuor horis intra septa 2 claudendae. Sin alius languor infestat, poculo medicantur arundinis et albae spinae radicibus, quas cum ferreis pilis diligenter contudimus,[8] admiscemus aquam pluvialem, solamque potandam pecori praebemus. Quod si ea res aegritudinem non depellit, vendenda sunt pecora; vel, si neque id contingere potest, ferro necanda saliendaque.[9] Mox interposito spatio, conveniet olim gregem reparare. Nec tamen antequam pestilens tempus anni, sive id fuit

[1] capelle *SAR*.

[2] praecedens *Schneider*: capraecedent *S*: capre cedent *Aac*.

[3] quae incedit *Lundström*: quem cedit *SAR*.

[4] procurret *SAR*.

[5] optima *SA*¹.

[6] prosternatur *SAR*.

[7] id autem accidere *Lundström*: id actim cedere *S*¹: id actum cedere *Aa*: id accidere *S*².

[8] contudimus *S*: contundimus *AR*.

[9] saliendaque *S*: saltenda quae *A*¹.

well able to endure toil, active and bold—the sort of man who can make his way without difficulty over rocks and deserts and through briers; he ought not to follow the herd like keepers of the other kind of cattle,[a] but should usually precede it. The she-goat which leads the herd is a very energetic animal; the one which so advances ought from time to time to be restrained in order that it may not race out in front but may browse quietly and slowly, so that it may have a large udder and not be lean of body.

Diseases of goats and their cure.

VII. Other kinds of domestic animals, when they are afflicted with pestilence, begin by wasting away with disease and weakness, but she-goats are the only animals which, though they are plump and lively, are suddenly cut off and over-whelmed, as it were, with sudden ruin, the whole herd at a time. This usually occurs as a result of too rich a diet. Therefore, when the plague has still stricken only a few of the herd, the goats should all be bled and given no food for a whole day and be kept shut up in their pens for the four middle hours of the day. 2 If besides this, a languor attacks them, they are dosed with a beverage consisting of the roots of reeds and white thorn, with which, after we have carefully bruised them with an iron pestle, we mix rain-water and give this, and nothing else, to the goats to drink. If this does not dispel their sickness, the animals must be sold; or, if this cannot be managed, they should be slaughtered with the knife and their flesh salted. Then, after an interval, the fitting time will come to replace the flock, but not before the pestilential season, if it was winter,

[a] *I.e.* oxen and cows and sheep.

hiemis, vertatur[1] aestate, sive autumni,[2] vere[3] 3 mutetur. Cum vero[4] singulae morbo[5] laborabunt, eadem remedia, quae etiam ovibus, adhibebimus; nam cum distendetur aqua cutis, quod vitium Graeci vocant ὕδρωπα, sub armo pellis leviter incisa perniciosum transmittat humorem, tum factum vulnus 4 pice liquida curetur. Cum effetae[6] loca genitalia tumebunt, aut secundae non responderint, defruti sextarius, vel cum id defuerit, boni vini tantundem faucibus infundatur, et naturalia ceroto[7] liquido repleantur. Sed ne nunc singula persequar, sicut in ovillo pecore praedictum est, caprino medebimur.

VIII. Casei quoque faciendi non erit omittenda cura, utique longinquis regionibus, ubi mulctram[8] devehere non expedit. Is porro si tenui liquore conficitur, quam celerrime vendendus est, dum adhuc viridis succum retinet: si pingui et opimo, longiorem patitur[9] custodiam. Sed lacte fieri debet sincero et quam recentissimo. Nam requietum vel aqua mixtum[10] celeriter acorem concipit. Id plerumque cogi agni aut haedi coagulo; quamvis possit et agrestis[11] cardui[12] flore conduci, et seminibus cneci, nec minus ficulneo lacte, quod emittit arbor, si 2 eius virentem saucies corticem. Verum optimus

[1] vertatur *ed. pr.*: vertantur *SAR*.
[2] autumnum *SAR*.
[3] verumutetur *SA*: vere mutentur *a*: ver utetur *c*.
[4] vero *R*: vere *SA*.
[5] domo *SAR*.
[6] & faetae *S*: et facte *AR*.
[7] geratori *A*: geroctori *S*.
[8] mulcram *S*: mulcra *A*[1]: multra *R*.
[9] patitur *S*: patimur *AR*.
[10] aqua mixtum *Heinsius*: quã mixtum *S*: maximum *AR*.
[11] agrestius *SAR*.
[12] cardiu *S*: cardius *AR*.

has changed to summer, or, if it was autumn, has changed to spring. If only individual goats are suffering from the disease, we shall apply the same remedies as to sheep; for when the skin is distended with water—the malady which the Greeks call *hydrōps* (dropsy)—a slight incision should be made in the skin under the shoulder, causing the fatal liquid to flow away; then the wound thus caused should be treated with liquid pitch. If, after a she-goat has borne young, the genital parts swell up and the after-birth has not put in an appearance, a *sextarius* of boiled down must, or, if this is not available, the same quantity of good wine, should be poured down the throat and the sexual parts filled with a liquid solution of wax. But, not to enter into more detail now, we shall give goats the same remedies as we have prescribed for sheep. 3 4

Cheese-making.

VIII. It will be necessary too not to neglect the task of cheese-making, especially in distant parts of the country, where it is not convenient to take milk to the market in pails. Further, if the cheese is made of a thin consistency, it must be sold as quickly as possible while it is still fresh and retains its moisture; if, however, it is of a rich and thick consistency, it bears being kept for a longer period. Cheese should be made of pure milk which is as fresh as possible, for if it is left to stand or mixed with water, it quickly turns sour. It should usually be curdled with rennet obtained from a lamb or a kid, though it can also be coagulated with the flower of the wild thistle or the seeds of the safflower,[a] and equally well with the liquid which flows from a fig-tree if you make an incision in the bark while it is still green. The best cheese, how- 2

[a] *Carthamus tinctorius*.

caseus est, qui exiguum medicaminis habet. Minimum autem coagulum[1] recipit sinum lactis argentei pondus denarii.[2] Nec dubium quin fici ramulis 3 glaciatus caseus iucundissime sapiat. Sed mulctra,[3] cum est repleta lacte, non sine tepore aliquo debet esse. Nec tamen admovenda est flammis, ut quibusdam placet, sed haud procul igne constituenda, et confestim cum concrevit liquor, in fiscellas aut in calathos vel formas transferendus est. Nam maxime refert primo quoque tempore serum percolari et a 4 concreta materia separari. Quam ob causam rustici nec patiuntur quidem sua sponte pigro humore defluere, sed cum paulo solidior caseus factus est, pondera superponunt, quibus exprimatur serum: deinde ut formis aut calathis exemptus[4] est, opaco ac frigido loco, ne possit vitiari: quamvis mundissimis tabulis componitur, aspergitur tritis salibus, ut exudet acidum liquorem: atque ubi duratus est, vehementius premitur, ut conspissetur. Et rursus torrido sale contingitur, rursusque ponderibus con-5 densatur. Hoc cum per dies novem factum est, aqua dulci abluitur,[5] et sub umbra cratibus in hoc factis[6] ita ordinatur, ne alter alterum caseus contingat, et ut modice siccetur: deinde, quo tenerior permaneat, clauso neque ventis obnoxio loco stipatur per com-

[1] coagulo *SR* : coaculo *A*[1].
[2] argenteis . . . denariis *SAR*.
[3] mulctrat *SA*.
[4] exemptus *ac* : exemtus *S*[2] : exemtis *S*[1]*A*.
[5] dulci abluitur *S* : dulcia bibitur *A* : dulci ebibitur *R*.
[6] fatis *SA*.

ever, is that which contains only a very small quantity of any drug. The least amount of rennet that a pail of milk requires weighs a silver *denarius*; and there is no doubt that cheese which has been solidified by means of small shoots from a fig-tree has a very pleasant flavour. 3 A pail when it has been filled with milk should always be kept at some degree of heat; it should not, however, be brought into contact with the flames, as some people think it proper to do, but should be put to stand not far from the fire, and, when the liquid has thickened, it should immediately be transferred to wicker vessels or baskets or moulds; for it is of the utmost importance that the whey should percolate as quickly as possible and become separated from the solid matter. 4 For this reason the country-folk do not even allow the whey to drain away slowly of its own accord, but, as soon as the cheese has become somewhat more solid, they place weights on the top of it, so that the whey may be pressed out; then, when the cheese has been taken out of the moulds or baskets, it is placed in a cool, shady place, that it may not go bad, and, although it is placed on very clean boards, it is sprinkled with pounded salt, so that it may exude the acid liquid; and, when it has hardened, it is still more violently compressed, so that it may become more compact; and then it is again treated with parched salt and again compressed by means of weights. 5 When this has been done for nine days it is washed with fresh water. Then the cheeses are set in rows on wickerwork trays made for the purpose under the shade in such a manner that one does not touch another, and that they become moderately dry; then, that the cheese may remain the more tender, it is closely packed on several shelves in an enclosed

plura tabulata. Sic neque[1] fistulosus neque salsus neque aridus provenit. Quorum vitiorum primum solet accidere, si parum pressus; secundum, si nimio 6 sale imbutus: tertium,[2] si sole exustus est. Hoc genus casei potest etiam trans maria permitti. Nam is, qui recens intra paucos dies absumi debet, leviore cura conficitur. Quippe fiscellis exemptus in salem muriamque[3] demittitur, et mox in sole paulum siccatur. Nonnulli antequam pecus numellis induant,[4] virides pineas nuces in mulctram demittunt, et mox super eas emulgent, nec separant, nisi cum transmiserint[5] in formas coactam materiam. Ipsos quidam virides conterunt nucleos, et lacti permiscent, 7 atque ita congelant. Sunt qui thymum contritum cribroque colatum cum lacte cogant. Similiter qualiscunque velis saporis efficere possis, adiecto quod elegeris condimento. Illa vero notissima est ratio faciendi casei, quem dicimus manu[6] pressum.[7] Namque is paulum gelatus[8] in mulctra dum[9] est tepefacta,[10] rescinditur et fervente aqua perfusus vel manu figuratur,[11] vel buxeis formis exprimitur. Est etiam non ingrati saporis muria perduratus, atque ita malini ligni vel culmi fumo coloratus. Sed iam redeamus ad originem.

[1] sic neque *S* : -s igneas *A* : ligneas *c*.
[2] tertium *S* : tertio *AR*.
[3] muriamque *S* : murtamq; *A*.
[4] induant *Brouckhusius* : indurat *SAR*.
[5] transmiserint *ed. pr.* : transierunt *SAR*.
[6] vanu *SA*.
[7] pressum *a* : -us *c* : pressu *SA*.
[8] caelatus *S* : celatus *AR*.
[9] mulctra dum *S* : mulctrandum *AR*.
[10] tepefacta *ed. pr.* : neres phata *S* : neres fata *A*.
[11] figuratur *Ald.* : figuratus *SAR*.

place which is not exposed to the winds. Under these conditions it does not become full of holes or salty or dry, the first of these bad conditions being generally due to too little pressure, the second to its being over-salted, and the third to its being scorched by the sun. This kind of cheese can even be 6 exported beyond the sea. Cheese which is to be eaten within a few days while still fresh, is prepared with less trouble; for it is taken out of the wicker-baskets and dipped into salt and brine and then dried a little in the sun. Some people, before they put the shackles [a] on the she-goats, drop green pine-nuts into the pail and then milk the she-goats over them and only remove them when they have transferred the curdled milk into the moulds. Some crush the green pine-kernels by themselves and mix them with the milk and curdle it in this way. Others 7 allow thyme which has been crushed and pounded through a sieve to coagulate with the milk; similarly, you can give the cheese any flavour you like by adding any seasoning which you choose. The method of making what we call "hand-pressed" cheese is the best-known of all: when the milk is slightly congealed in the pail and still warm, it is broken up and hot water is poured over it, and then it is either shaped by hand or else pressed into box-wood moulds. Cheese also which is hardened in brine and then coloured with the smoke of apple-tree wood or stubble has a not unpleasant flavour. But let us now return to the point from which we digressed.[b]

[a] *I.e.* to restrain them while they are being milked.

[b] The author regards this chapter on cheese-making as a digression from his real subject, which is a description of the smaller domestic animals.

IX. In omni genere quadrupedum species maris diligenter eligitur, quoniam frequentius[1] patri similior est progenies, quam matri. Quare etiam in suillo pecore verres probandi sunt totius quidem corporis amplitudine[2] eximii,[3] sed qui quadrati potius quam longi aut rotundi sint, ventre promisso, clunibus vastis, nec proinde cruribus aut ungulis proceris, amplae et glandulosae cervicis, rostri[4] brevis[5] et resupini.[6] Maximeque ad rem pertinet, quam 2 salacissimos esse ineuntes.[7] Ab annicula aetate commode progenerant, dum quadrimatum agant: possunt tamen etiam semestres implere feminam. Scrofae probantur longissimi[8] status, et ut sint reliquis membris similes descriptis verribus. Si regio frigida et pruinosa est, quam durissimae densaeque et nigrae setae[9] grex eligendus est; si temperata atque aprica, glabrum pecus vel etiam 3 pistrinale album potest pasci. Femina sus[10] habetur ad partus edendos idonea[11] fere usque in annos septem, quae quanto fecundior est, celerius senescit. Annicula non improbe concipit, sed iniri[12] debet mense Februario. Quattuor quoque mensibus feta, quinto parere, cum iam herbae solidiores sunt, ut et firma lactis maturitas porcis contingat, et cum desie-

[1] frequentius *S* : frequenter *AR*.
[2] amplitudine *S* : -em *Aac*.
[3] eximii *S* : eximit *AR*.
[4] rostri *S* : rostribus *Aac*.
[5] brevis *SA* : brevibus *R*.
[6] resupina *SAR*.
[7] esse ineuntes *Lundström* : esseminant et *SAR*.
[8] longissimis *SAR*.
[9] nigrae sete *S*[1] : nigrae sedet *S*[2] : nigraes et egrex *A*.
[10] suus *SA*.
[11] edendo nea *SA*[1]
[12] iniri *Ald.* : inire *SAR*.

IX. In every kind of quadruped it is a male of the fine appearance which is the object of our careful choice, because the offspring is more often like its father than like its mother. So too, when it is a question of pigs, those boars must meet with our approval which are remarkable for their outstanding bodily size in general, provided that they are square rather than long or round, and which have a belly which hangs down, huge haunches, but not correspondingly long legs and hoofs, a long and glandulous neck, and a snout which is short and snub; also it is especially important that they should be as lustful as possible when they have sexual intercourse. They are fit for breeding purposes from a year old until they are four years old, though they can also impregnate a sow at six months old. Breeding sows are esteemed which are very long in shape, provided that in their other limbs they resemble the description which we have given of the boars. If the district is cold and frosty, a herd should be selected with very hard, dense, black bristles; if it is temperate and sunny, smooth pigs and even white ones such as are kept by bakers[a] may be pastured there. A sow is considered fit for breeding purposes until it is about seven years old, but the more prolific it is the more quickly it becomes old. It can quite well conceive at a year old, but ought to be covered by the boar in the month of February and, having been four months with young, it should farrow in the fifth month, when the grass is already of stronger growth, so that the porkers may find the milk at the perfection of its full strength and also, when they cease to be suckled at

Pigs.

2

3

[a] It was customary for bakers to keep pigs and feed them on the superfluous bran (Plaut., *Capt.*, 4. 2. 28).

rint uberibus ali,[1] stipula pascantur, ceterisque
4 leguminum caducis frugibus. Hoc autem fit longinquis regionibus, ubi nihil nisi submittere expedit. Nam suburbanis lactens [2] porcus aere [3] mutandus est: sic enim mater non educando labori subtrahitur, celeriusque iterum conceptum partum edet. Idque bis [4] anno faciet. Mares, vel cum primum ineunt semestres, aut cum saepius progeneraverunt, trimi [5] aut quadrimi castrantur, ut possint pinguescere.
5 Feminis quoque vulvae ferro exulcerantur, et cicatricibus clauduntur, ne sint genitales. Quod facere non intelligo quae ratio compellat,[6] nisi penuria cibi. Nam ubi est ubertas pabuli, submittere prolem semper expedit.

6 Omnem porro situm ruris pecus hoc usurpat. Nam et montibus et campis commode pascitur, melius tamen palustribus agris, quam sitientibus. Nemora sunt convenientissima, quae vestiuntur [7] quercu, subere, fago, cerris, ilicibus, oleastris, termitibus, corylis, pomiferisque silvestribus, ut sunt albae spinae, Graecae siliquae, iuniperus, lotus, pampinus, cornus, arbutus, prunus, et paliurus, atque achrades pyri. Haec enim diversis temporibus mitescunt, ac
7 paene toto anno gregem saturant. At ubi penuria est arborum, terrenum pabulum consectabimur, et

[1] ali *Sc* : alti *Aa*.
[2] lactis *a* : lactens *c* : lactes *SA*.
[3] aere *R* : ae ru *S* : eru *A*.
[4] vis *A* : quis *S*.
[5] primi *SA*.
[6] compellat *R* : -ant *SA*.
[7] vertuntur *SA*.

[a] Schneider is probably right in thinking that *termes* represents the Greek τέρμινθος.

the udder, they may feed on stubble and the fruits also which fall from leguminous plants. This is the 4 practice in out-of-the-way regions where raising stock is the only thing which pays; for in districts near towns the sucking pig must be turned into money, for then its mother is saved trouble by not having to rear it and will more quickly conceive and produce another offspring, and so bear twice in the same year. The males are castrated, so that they may be enabled to grow fat, either at six months, when they first begin to cover the sows, or else at three or four years of age, when they have been often used for breeding. An operation is also performed with the 5 knife on the wombs of the females to make them suppurate and close up as a result of scarring over, so that they cannot breed. I do not know the reason for doing this, unless it is lack of food; for where there is abundance of fodder, it always pays to rear stock.

Moreover, pigs can make shift in any sort of 6 country wherever situated. For they find suitable pasture both in the mountains and in the plains, though it is better on marshy ground than on dry. The most convenient feeding-grounds are woods covered with oaks, cork-trees, beeches, Turkey oaks, holm-oaks, wild olive trees, terebinth-trees,[a] hazels, wild fruit-trees like the whitethorn, carob-trees, junipers, nettle-trees, vine-tendrils, cornel-trees, strawberry-trees, plum-trees, Christ's thorn, and wild pear-trees. For these ripen at different times and provide plenty of food for the herd almost all the year round. But where there is a lack of trees, we 7 shall have recourse to fodder which grows near the ground and prefer muddy to dry ground, so that the

sicco limosum praeferemus, ut paludem rimentur, effodiantque [1] lumbricos, atque in luto volutentur, quod est huic pecudi gratissimum; quin etiam aquis abuti possint: namque id fecisse maxime per aestatem profuit, et dulces eruisse radiculas aquatilis silvae, tanquam scirpi [2] iuncique et degeneris arundinis, quam vulgus cannam vocat. Nam cultus quidem ager opimas reddit sues, cum est graminosus, et pluribus generibus [3] pomorum consitus, ut per anni diversa tempora mala, pruna, pyrum, multiformes nuces ac ficum praebeat. Nec tamen propter haec parcetur horreis. Nam saepe de manu dandum est, cum foris deficit pabulum. Proper quod plurima glans vel cisternis in aquam vel fumo tabulatis recondenda [4] est. Fabae quoque et similium leguminum, cum vilitas permittit, facienda est potestas, et utique vere, dum adhuc lactent [5] viridia pabula, quae suibus plerumque nocent.[6] Itaque mane priusquam procedant in pascua, conditivis cibis sustinenda [7] sunt, ne immaturis herbis citetur alvus, eoque vitio pecus emacietur. Nec ut ceteri greges universi claudi debent, sed per [8] porticus harae [9] faciendae sunt, quibus aut a partu [10] aut etiam praegnates includantur. Nam praecipue sues cater-

8 (margin, at "dinis"); 9 (margin, at "condenda")

[1] effodiantque *R* : et fodiantque *SA*.
[2] stirpi *SAac*.
[3] generibus *om. SA*.
[4] recondenda *S*² : reconda *S*¹*A* : recondita *ac*.
[5] lactent *ex cit. Palladii* (III. 26. 3) : lantiunt *SA* : lanciunt *R*.
[6] nocet *SA*.
[7] sustinenda *SAac*.
[8] per *om. SA*.
[9] harae *om. SA*.
[10] parte *SA*.

pigs may root about in the marsh and turn up worms and wallow in the mud, which pigs love to do; and may they also be able to use water freely; for it has proved a great benefit for them to do this in the summer and to tear up the sweet-flavoured rootlets of under-water growths, such as the reed-mace, the rush, and the bastard reed, which the vulgar call the "cane." Sows indeed 8 grow fat on cultivated ground when it is grassy and planted with fruit-trees of several kinds, so as to provide at different seasons of the year apples, plums, pears, nuts of many kinds and figs. You should not, however, on the strength of these fruits be sparing of the contents of the granary, which should often be handed out when out-door food fails. For this purpose plenty of mast should be stored either in cisterns of water or in lofts exposed to the smoke.[a] They should also be given the opportunity of feeding 9 on beans and similar leguminous vegetables, when their cheapness makes this possible, especially in the spring when green fodder is still in a juicy condition, which is generally harmful to pigs. Early in the morning, therefore, before they go out to pasture, they should be given a nourishing meal of food from the store, that the bowel may not be irritated by grass which is immature and that the herd may not waste away by the trouble which it causes. Pigs ought not to be shut up all together, like all other herds, but sties ought to be constructed after the manner of colonnades, in which the sows can be shut up after farrowing and even during pregnancy; for sows more than any other animals, when they are

[a] *Cisternis—tabulatis*, these words are possibly corrupt but the general meaning is clear. Pontedera suggests *cisternis sine aqua vel fumosis tabulatis*.

vatim atque incondite cum sunt pariter inclusae, 10 super alias aliae cubant et fetus elidunt. Quare, ut dixi, iunctae parietibus harae construendae sunt in altitudinem pedum quattuor, ne sus transilire septa queat. Nam contegi non debet, ut a superiore parte custos numerum porcorum recenseat, et si quem decumbens mater oppresserit, cubanti subtrahat. Sit autem vigilax, impiger, industrius, navus. Omnium, quas pascit, et matricum et iuniorum meminisse debet, ut uniuscuiusque partum consideret. Semper observet enitentem, claudatque ut in [1] hara 11 fetum edat. Tum denotet [2] protinus quot et quales sint [3] nati, et curet maxime ne quis [4] sub nutrice aliena educetur [5]: nam facillime porci, si evaserint haram, miscent se, et scrofa cum decubuit, aeque 12 alieno ac suo praebet ubera. Itaque porculatoris maximum officium est, ut unamquamque [6] cum sua prole claudat. Qui si memoria deficitur, quo minus agnoscat cuiusque progeniem, pice liquida eandem [7] notam scrofae et porcis imponat, et sive per literas sive per alias formas unumquemque fetum cum matre distinguat. Nam in maiore numero diversis notis opus est, ne confundatur memoria custodis. 13 Attamen quia id facere gregibus amplis videtur

[1] claudatq; ut in *R*: claudat in *SA*.
[2] dinotet *SA*.
[3] sunt *SA*.
[4] ne quis S^2: nutrix equis S^1A.
[5] alienę (-e *A*) ducetur *SA*: aliena educatur *R*.
[6] unamque *SA*.
[7] eandem *R*: eadem *AS*.

penned together in a crowd and pell-mell, lie one on top of another and abortions are thus caused. Therefore, 10 as I have said, sties should be built joined by party walls each to the other and four feet in height, so that the sow may not be able to jump over the these barriers. They ought not to be roofed over, so that the man in charge may be able to look in from above and count the number of piglings, and that if any mother is lying on top of its litter and squeezes one of them, he may extract it from under her. The swineherd must be watchful, energetic, painstaking and active: he ought to be able to remember all the sows under his charge, both those which have produced offspring and the younger sows, so that he may identify the offspring of each separately. He must be on the watch for sows which are farrowing and shut them up, so that they may produce their litter in a sty; 11 he must then take note immediately of the number and quality of the piglings which are born and take special care that none of them is brought up by a sow which is not its mother; for the sucking-pigs, if they have escaped from the sty, very easily become mixed up, and the sow, when it lies down, offers its dugs as freely to the offspring of other sows as to her own. Thus the most important duty of the swine breeder is 12 to keep each sow shut up with its own litter. If he has not a good memory and so cannot recognize the offspring of each sow, he should put the same mark on the sow and its piglings with liquid pitch, so that he may distinguish the different litters and their mothers by means of letters or some other device; for where a large number is involved, it is necessary to employ distinctive marks, so that the swineherd's memory may not be confused. Since, however, it seems a 13

operosum, commodissimum est haras ita fabricare, ut limen earum[1] in tantam altitudinem consurgat, quantam[2] possit nutrix evadere; lactens autem supergredi[3] non possit. Sic nec alienus irrepit, et in cubili suam quisque matrem nidus[4] expectat, qui tamen non debet octo capitum numerum excedere: non quia ignorem fecunditatem scrofarum maioris esse numeri, sed quia celerrime fatiscit, quae plures educat. Atque eae quibus partus submittitur, cocto sunt hordeo sustinendae, ne ad maciem summam perducantur,[5] et ex ea ad aliquam perniciem. 14 Diligens autem porculator frequenter suile converrit, et saepius haras. Nam quamvis praedictum animal in pabulatione spurce versetur, mundissimum tamen cubile desiderat. Hic fere cultus est pecoris suilli recte valentis. Sequitur ut dicamus, quae sit cura vitiosi.

X. Febricitantium signa sunt, cum obstipae sues transversa capita ferunt, ac per pascua subito, cum paulum procurrerunt, consistunt, et vertigine correptae concidunt. Earum notanda sunt capita, 2 quam in partem proclinent,[6] ut ex diversa parte de auricula sanguinem mittamus. Item sub cauda duobus digitis a clunibus intermissis venam feriamus, quae est in eo loco satis ampla, eamque sarmento prius oportet verberari, deinde ab ictu[7] virgae tu-

[1] limen earum *R*: minearum *SA*.
[2] quantã *S*: quantum *Aac*.
[3] supergredi *R*: ut pergredi *SA*.
[4] nidus *S*: -os *AR*. [5] perducatur *SA*.
[6] proclinent *S*: proclinentur *AR*.
[7] avictu *S*: abiectu *a*.

laborious task to carry out this plan in large herds, the most convenient method is to construct the sties in such a way that their thresholds are low enough for the sow to be able to get out but too high for the sucking pig to climb over; thus no strange porker can creep in, and each litter awaits its own mother in the place where they sleep. A litter ought not to number more than eight, not that I am ignorant that the fecundity of breeding-sows can produce more than this number, but because a sow which rears more than eight quickly becomes worn out. Those sows which are given a litter to rear, must be sustained with cooked barley, so that they may not be reduced to a state of extreme emaciation and from that to some fatal sickness. 14 The careful swineherd will frequently sweep out the piggery and the sties still more often; for, though the animal in question behaves in a filthy manner when it is at pasture, it likes its sleeping-place to be very clean. Such, more or less, is the manner in which pigs should be kept when they are in good health; our next task is to deal with the care of the pig in disease.

Diseases of pigs and their cures.

X. The signs of fever in pigs are when they lean over and hold their heads awry, and, after running forward a little way over their feeding-ground, suddenly halt and are seized with giddiness and fall down. 2 Notice must be taken in which direction they lean their heads forward, so that we may let blood from the ear on the opposite side; we shall also smite under the tail, at two fingers' distance from the haunches, the vein which at this point is fairly big, but it ought first to be beaten with a vine-twig, and then, as it swells up from the stroke of the rod, it should be opened with a knife, and, after the blood

mentem ferro rescindi, detractoque sanguine colli-
3 gari saligneo libro vel etiam ulmeo. Quod cum fecerimus, uno aut altero die sub tecto pecudem continebimus, et aquam modice calidam quantam volent, farinaeque hordeaceae singulos sextarios praebebimus. Strumosis sub lingua sanguis mittendus est, qui cum profluxerit, sale trito cum farina triticea confricari totum os conveniet. Quidam praesentius putant esse remedium cum per[1] cornu singulis ternos cyathos gari[2] demittunt. Deinde fissas taleas ferularum lineo funiculo religant: et ita collo suspendunt, ut strumae ferulis contingantur.
4 Nauseantibus quoque salutaris habetur eburnea scobis sali[3] fricto et fabae minute fresae commixta, ieiunisque prius quam in pascua prodeant obiecta.[4] Solet etiam universum[5] pecus aegrotare ita, ut emacietur, nec cibos capiat, productumque[6] in pascua medio campo procumbat, et quodam veterno pressum
5 somnos aestivo sub sole captet. Quod cum facit, totus grex tecto clauditur stabulo, atque uno die abstinetur potione et pabulo:[7] postridie radix anguinei[8] cucumeris trita et commixta cum aqua datur sitientibus: quam cum pecudes biberunt, nausea correptae vomitant, atque expurgantur, omnique bile depulsa, cicercula vel faba dura muria con-

[1] cum per *R*: compea *S*: cumpea *A*.
[2] gari *R*: cari *S*¹.
[3] sale *SAR*.
[4] abiecta *AR*: obiecta *S*.
[5] universam *SA*¹: -um *A*²*R*.
[6] productusque *SA*.
[7] paulo *SA*.

has been drawn off, the vein ought to be bound up with bark of a willow or even of an elm-tree. After 3 this we shall keep the animals under cover for a day or two and give them as much moderately warm water as they shall desire and a *sextarius* each of barley-flour. If pigs are scrofulous, they must be bled under the tongue and, when the blood has flowed, it will be well to rub the whole mouth with powdered salt mixed with wheaten flour. Some people think that a more efficacious remedy is to make them swallow three *cyathi* each of fish-pickle through a horn; they then tie together split sticks of fennel with a linen cord and hang them round their necks in such a way that the scrofulous tumours are in contact with the fennel-stalks. For pigs suffer- 4 ing from vomiting, ivory-dust is regarded as a good remedy mixed with powdered salt and beans ground very small and given to them on an empty stomach before they go out to pasture. Sometimes also the whole herd suffers at the same time, which causes them to become thin and to refuse their food and to lie down in the middle of the field when they are driven out to pasture and to want to go to sleep in the summer sunshine overcome by a kind of drowsiness. When this happens, the whole herd is shut up 5 in a covered stable and deprived of drink and food for one day; then on the following day the root of the snake-like cucumber, crushed and mixed with water, is given to quench their thirst, and when the animals have drunk it they are seized with nausea and vomit and so are purged; when all the bile has been discharged, they are given chick-pea or beans sprinkled with hard brine, after which they are allowed to drink

[8] anguinei *R*: sanguinei *SA*.

spersa, deinde, sicut hominibus, aqua calida potanda permittitur.

6 Sed cum omni quadrupedi per aestatem sitis sit infesta, tum suillo maxime est inimica. Quare non ut capellam vel ovem, sic et hoc animal bis [1] ad aquam duci praecipimus: [2] sed si fieri potest, iuxta flumen aut stagnum per ortum Caniculae detineri: quia cum sit aestuosissimum, non est contentum potione aquae, nisi obesam ingluviem atque distentam pabulis alvum demerserit ac refrigeraverit: nec ulla re magis gaudet, quam rivis atque caenoso lacu volu-

7 tari. Quod si locorum situs repugnat, ne ita fieri possit, puteis extracta et large canalibus immissa praebenda sunt pocula, quibus nisi affatim satientur, pulmonariae fiunt. Isque [3] morbus optime sanatur auriculis inserta consiligine: de qua radicula dili-

8 genter ac saepius iam locuti sumus. Solet etiam vitiosi splenis dolor eas infestare, quod accidit, cum siccitas [4] magna provenit, et, ut Bucolicum loquitur poëma,

Strata iacent passim sua quaeque sub arbore poma.

Nam pecus inexsatiabile [5] sues, dum dulcedinem pabuli consectantur supra modum, aestate splenis [6] incremento laborant. Cui succurritur, si fabricentur canales tamaricis [7] et rusco, repleanturque aqua, et deinde sitientibus admoveantur; quippe

[1] bis *S*²: vis *S*¹*A*: *om. R.*
[2] praecipimus *R*: precepimus *SA*.
[3] isque *ed. pr.*: quiq; *R*: quisq; *S*¹*A*.
[4] ficitas *A*: sicitas *S*.
[5] inexitiabiles bis *SA*²: inexitiaviles bis *A*¹.
[6] esbatae splenis *SAac*.
[7] tramaricis *R*: tramaricus *SA*.

warm water, as men are allowed to do in similar circumstances.

While thirst in the summer is pernicious to all quadrupeds, it is specially hurtful to pigs. We, therefore, advise that they should not be taken to water twice a day, like goats and sheep, but that, if possible, they should be kept in the neighbourhood of a river or pool at the time of the rising of the Dogstar; for, when a pig is feeling the intense heat, it is not content with drinking the water, if it cannot also plunge into it and so cool its fat maw and its belly distended with fodder, and there is nothing in which it takes so much pleasure as wallowing in streams and muddy lakes. But if the nature of the district makes this impossible, drinking water should be drawn from wells and poured into troughs in generous supply; for, unless they are abundantly satisfied, their lungs become affected. This disease is best treated by inserting lungwort into the ears, a small root of which we have already more than once spoken about and in detail. Pain from a diseased spleen also often attacks them; this happens when a serious drought occurs and when, as the Bucolic poem says,[a] 6 7 8

Fruits lie on all sides, each strewn 'neath its tree.

For pigs, being insatiable animals, make for sweetness in their food beyond measure and suffer exceedingly in the summer from swelling of the spleen. This can be relieved if troughs made of tamarisk wood and butcher's broom are constructed and filled with water and put in their way when they are thirsty; for the juice of the wood has a

[a] Vergil, *Ecl.* VII. 54.

ligni succus medicabilis epotus intestinum tumorem compescit.

XI. Castrationis autem in[1] hoc pecore duo tempora servantur, veris et autumni: et eius administrandae duplex ratio. Prima illa, quam iam tradidimus, cum duobus vulneribus impressis per unamquamque plagam singuli exprimuntur testiculi. Altera est speciosior, sed magis periculosa, quam 2 tamen non omittam. Cum virilem partem unam ferro reseratam[2] detraxeris, per impressum vulnus scalpellum inserito, et mediam quasi cutem, quae intervenit duobus membris genitalibus, rescindito, atque uncis digitis alterum quoque testiculum educito: sic fiet una cicatrix adhibitis ceteris remediis, quae prius docuimus. Illud autem, quod pertinet ad religionem[3] patrisfamilias, non reticen- 3 dum putavi. Sunt quaedam scrofae, quae mandunt fetus suos: quod cum fit, non habetur prodigium. Nam sues ex omnibus pecudibus[4] impatientissimae famis aliquando sic indigent pabuli, ut non tantum alienam, si liceat, sobolem, sed etiam suam consumant.[5]

XII. De armentis ceterisque pecudibus et magistris, per quos quadrupedum greges humana solertia[6] domi forisque curantur atque observantur, nisi fallor, satis accurate disserui. Nunc ut exordio priore sum

[1] in *om. SAR.*
[2] reseratam *S* : resecatam *Aac.*
[3] regionem *SA.*
[4] pecudibus *R* : *om. SA.*
[5] consummant *a* : consumat *SAc.*
[6] solertia *R* : sollerti *SA*[1].

---

[a] *I.e.* one testicle.
[b] *I.e.* which may suggest superstitious fancies to his mind.

medicinal effect and, being swallowed, stops intestinal swelling.

XI. Two seasons are observed for castrating the pig, spring and autumn. There are two methods of carrying out this operation. The first, which we have already described, consists of making two incisions and squeezing out a testicle through each of them. The other is more spectacular but more dangerous; but I will not pass it over in silence. 2 When you have opened up with the knife and drawn out one of the male organs,[a] insert a lancet through the wound that has been made; then cut the middle skin, as it were, which intervenes between the two genital members, and with your bent fingers draw out the other testicle also; the result will be that there will be only one scar after the application of the other remedies which we have described earlier. 3 But there is one point, which concerns the religious scruples of the head of the family,[b] and which I have thought that I ought not to pass over in silence, namely, that there are some breeding-sows which devour their young. When this happens, it is not regarded as a prodigy; for pigs, of all farm-animals, are the least able to endure hunger, and sometimes feel such need of food that they consume not only the offspring of other sows, if they are allowed to do so, but also their own young.

Dogs.

XII. I have now, unless I am mistaken, dealt in sufficient detail with animals used for ploughing and other cattle and with the herdsmen who are employed to look after and watch over flocks of four-footed animals at home and out of doors with all the resources of human intelligence. Now, as I promised in the earlier part of my treatise, I will speak of the

pollicitus, de mutis custodibus loquar; quamquam canis falso dicitur mutus custos. Nam quis hominum clarius aut tanta vociferatione bestiam vel furem praedicat, quam iste latratu? quis famulus amantior domini? quis fidelior comes? quis custos incorruptior? quis excubitor inveniri potest vigilantior? quis denique ultor aut vindex constantior? Quare vel in primis hoc animal mercari tuerique debet agricola, quod et villam et fructus familiamque et pecora custodit. Eius autem parandi tuendique 2 triplex ratio est. Namque unum genus adversus hominum[1] insidias eligitur, et id villam quaeque iuncta sunt villae custodit. At alterum[2] propellendis iniuriis hominum ac ferarum; et id observat domi stabulum, foris pecora pascentia. Tertium venandi gratia comparatur; idque non solum nihil agricolam iuvat, sed et avocat desidemque ab opere suo reddit. 3 De villatico[3] igitur et pastorali dicendum est: nam venaticus nihil pertinet ad nostram professionem.

Villae custos eligendus est amplissimi corporis, vasti latratus canorique, ut prius auditu maleficum, deinde etiam conspectu terreat, et tamen nonnunquam ne visus quidem horribili fremitu suo fuget insidiantem. Sit autem coloris unius; isque magis eligatur albus

[1] *post* hominum *add.* et ferarum *R*.
[2] laterum S^1A^1.
[3] villatigo *S* : vit latigo A^1.

dumb guardians of the flocks, though it is wrong to speak of the dog as a dumb guardian; for what human being more clearly or so vociferously gives warning of the presence of a wild beast or of a thief as does the dog by its barking? What servant is more attached to his master than is a dog? What companion more faithful? What guardian more incorruptible? What more wakeful night-watchman can be found? Lastly, what more steadfast avenger or defender? To buy and keep a dog ought, therefore, to be among the first things which a farmer does, because it is the guardian of the farm, its produce, the household and the cattle. There are three different reasons for procuring and keeping a dog. One type of dog is chosen to oppose the plots of human beings and watches over the farm and all its appurtenances; a second kind for repelling the attacks of men and wild beasts and keeping an eye at home on the stables and abroad on the flocks as they feed; the third kind is acquired for the purposes of the chase, and not only does not help the farmer but actually lures him away from his work and makes him lazy about it. We must, therefore, speak of the farm-yard dog and the sheep-dog; for the sporting hound has nothing to do with the art which we profess.

2

3

As guardian of the farm a dog should be chosen which is of ample bulk with a loud and sonorous bark in order that it may terrify the malefactor, first because he hears it and then because he sees it; indeed, sometimes without being even seen it puts to flight the crafty plotter merely by the terror which its growling inspires. It should be the same colour all over, white being the colour which should rather be chosen for a

in pastorali, niger in villatico: nam varius in neutro est laudabilis. Pastor album probat, quoniam est ferae dissimilis, magnoque opus interdum discrimine est in propulsandis lupis sub obscuro mane vel etiam 4 crepusculo, ne pro bestia[1] canem feriat. Villaticus, qui hominum maleficiis opponitur, sive luce clara fur advenit,[2] terribilior niger conspicitur: sive noctu,[3] ne conspicitur quidem propter umbrae similitudinem, quamobrem tectus tenebris canis tutiorem accessum habet ad insidiantem. Probatur quadratus potius quam longus aut brevis, capite tam magno, ut corporis videatur pars maxima, deiectis et propendentibus auribus, nigris vel glaucis oculis acri lumine radiantibus, amplo villosoque pectore, latis armis, cruribus crassis et hirtis, cauda brevi, vestigiorum articulis[4] et unguibus amplissimis, qui Graece δράκες appellantur. Hic erit villatici canis status 5 praecipue laudandus. Mores autem neque mitissimi, neque rursus truces atque crudeles; quod illi furem quoque adulantur, hi etiam domesticos invadunt. Satis est severos esse nec blandos, ut nonnunquam etiam conservos iratius intueantur, semper excandescant in exteros. Maxime autem debent in custodia vigilantes conspici, nec erronei,[5] sed assidui

[1] vestio *S* : bestico *A*¹.
[2] advenit *S* : -erit *AR*.
[3] noctu ne *S* : nocte ne *A* : nocte nec *ac*.
[4] auriculis *SA*.
[5] errore ne *S*¹ : errore *A* : erronei *a* : arronei *c*.

sheep-dog and black for a farm-yard dog; for a dog of varied colouring is not to be recommended for either purpose. The shepherd prefers a white dog because it is unlike a wild beast, and sometimes a plain means of distinction is required in the dogs when one is driving off wolves in the obscurity of early morning or even at dusk, lest one strike a dog instead of a wild beast. 4 The farmyard dog, which is pitted against the wicked wiles of men, if the thief approaches in the clear light of day, has a more alarming appearance if it is black, whereas at night it is not even seen because it resembles the shadow and so, under the cover of darkness, the dog can approach the crafty thief in greater security. A squarely built dog is preferred to one which is long or short, and it should have a head so large as to appear to form the largest part of it; it should have ears which droop and hang down, eyes black or grey, sparkling with rays of bright light, a broad and shaggy chest, wide shoulders, thick, rough legs and a short tail; the joints of its feet and its claws, which the Greeks call *drakes*, should be very large. Such are the points which will meet with most approval in all farm-yard dogs. 5 In character they should neither be very mild nor, on the other hand, savage and cruel; if they are mild, they fawn on everyone, including the thief; if they are fierce they attack even the people of the house. It is enough that they should be stern but not fawning, so that they sometimes look even upon their companions in servitude with a somewhat wrathful eye, while they always blaze with anger against strangers. Above all they should be seen to be vigilant in their watch and not given to wandering, but diligent and cautious rather than rash; for the

et circumspecti magis quam temerarii.[1] Nam illi nisi[2] quod certum compererunt, non indicant: hi
6 vano strepitu et falsa suspicione concitantur. Haec idcirco memoranda credidi, quia non natura tantum, sed etiam disciplina[3] mores facit, ut et cum emendi potestas fuerit, eiusmodi probemus, et cum educabi-
7 mus domi natos, talibus institutis[4] formemus. Nec multum refert an[5] villatici corporibus graves et parum veloces sint: plus enim cominus et in gradu, quam eminus et in spatioso cursu facere debent. Nam semper circa septa et intra aedificium consistere, immo ne longius quidem recedere debent, satisque pulchre funguntur officio, si et advenientem sagaciter odorantur,[6] et latratu conterrent, nec patiuntur propius[7] accedere,[8] vel constantius appropinquantem violenter invadunt. Primum est enim non adtentari, secundum est lacessitum fortiter et perseveranter vindicari. Atque haec de domesticis custodibus; illa de pastoralibus.[9]

8 Pecuarius canis neque tam strigosus aut pernix debet esse, quam qui damas cervosque et velocissima sectatur animalia, nec tam obesus aut gravis, quam villae horreique custos: sed et robustus nihilominus,
9 et aliquatenus promptus ac strenuus, quoniam et ad rixam et ad pugnam, nec minus[10] ad cursum comparatur, cum et lupi[11] repellere insidias, et raptorem

[1] temeri *SA*.
[2] nam illi nisi *ac* : quam inlinisi *SA*.
[3] disciplinā *SA*.
[4] institutis *S* : -i *A*.
[5] refert an *R* : refertam an *A*² : refertam *SA*¹.
[6] adoriantur *a* : odorantur *c* : oderantur *SA*¹.
[7] proprius *SAac*.
[8] accidere *SA*¹.
[9] pastoribus *SAa* : pastoralibus *c*.
[10] tamen *SA*.
[11] rupi *SA*.

cautious do not give the alarm unless they have discovered something for certain, whereas the rash are aroused by any vain noise and groundless suspicion. I have thought it necessary to mention 6 these points, because it is not nature alone but education as well which forms character, so that, when there is an opportunity of buying a dog, we may choose one with these qualities and that when we are going to train dogs which have been born at home, we may bring them up on such principles as these. It does 7 not matter much if farm-yard dogs are heavily built and lack speed, since they have to function rather at close quarters and where they are posted than at a distance and over a wide area; for they should always remain round the enclosures and within the buildings, indeed they ought never go out farther from home and can perfectly well carry out their duties by cleverly scenting out anyone who approaches and frightening him by barking and not allowing him to come any nearer, or, if he insists on approaching, they violently attack him. Their first duty is not to allow themselves to be attacked, their second duty to defend themselves with courage and pertinacity if they are provoked. So much for the dogs which guard the house; our next subject is sheep-dogs.

A dog which is to guard cattle ought not to be as 8 lean and swift of foot as one which pursues deer and stags and the swiftest animals, nor so fat and heavily built as the dog which guards the farm and granary, but he must, nevertheless, be strong and to a certain extent prompt to act and vigorous, since the purpose 9 for which he is acquired is to pick quarrels and to fight and also to move quickly, since he has to repel the stealthy lurking of the wolf and to follow the

ferum consequi fugientem praedam excutere atque auferre debeat. Quare status[1] eius longior productiorque ad hos casus magis habilis est quam brevis aut etiam quadratus: quoniam, ut dixi, nonnunquam necessitas exigit celeritate bestiam[2] consectandi. Ceteri[3] artus similes membris villatici canis aeque probantur.

10 Cibaria fere eadem sunt utrique generi praebenda. Nam si tam laxa rura sunt, ut sustineant[4] pecorum greges, omnes sine discrimine canes hordeacea farina cum sero commode pascit. Sin autem surculo consitus ager sine pascuo est,[5] farreo vel triticeo pane satiandi sunt, admixto tamen liquore coctae fabae, sed tepido: nam fervens[6] rabiem creat.

11 Huic quadrupedi neque feminae neque mari nisi post annum permittenda venus est: quae si teneris conceditur, carpit et corpus et vires[7] animosque degenerat. Primus effetae partus amovendus est, quoniam tiruncula nec recte nutrit, et educatio totius habitus aufert incrementum. Mares iuveniliter usque in annos decem progenerant: post id tempus ineundis feminis non videntur habiles, quoniam seniorum pigra soboles existit. Feminae concipiunt usque in annos novem, nec[8] sunt utiles

12 post decimum. Catulos sex mensibus primis, dum corroborentur,[9] emitti non oportet, nisi ad matrem lusus ac lasciviae causa. Postea catenis per diem

[1] debeat quare status *om. SA.*

[2] celeritate bestiam *R*: celeriteratem bestii *A*[1]: celeriter autē bestii *S*.

[3] consectandi ceteri *R*: consectam dicere *SA*.

[4] sustineat *SA*.

[5] est *R*: et *SA*.

[6] fervens *R*: nam ferventi *S*: non aferventi *A*.

[7] veteres *SA*.

[8] nec *om. SA.*

[9] corroboretur *SA*.

wild beast as he escapes with his prey and make him drop it and to bring it back again. Therefore a dog of a rather long, slim build is better able to deal with these emergencies than one which is short or even squarely built, since, as I have said, sometimes the necessity of pursuing a wild beast with speed demands this. The other joints in sheep-dogs if they resemble the limbs of farm-yard dogs meet with equal approval.

Practically the same food should be given to both types of dog. If the farm is extensive enough to support herds of cattle, barley-flour with whey is a suitable food for all dogs without distinction; but if the land is closely planted with young shoots and affords no pasture, they must be given their fill of bread made from emmer or wheaten flour, mixed, however, with the liquid of boiled beans, which must be lukewarm, for, if it is boiling, it causes madness. 10

Neither dogs nor bitches must be allowed to have sexual intercourse until they are a year old; for if they are allowed to do so when they are quite young, it enfeebles their bodies and their strength, and causes them to degenerate mentally. The first puppies which a bitch produces must be taken from her, because at the first attempt she does not nourish them properly and the rearing of them hinders her general bodily growth. Dogs procreate vigorously up to ten years of age, but beyond that they do not seem suitable for covering bitches, for the offspring of an elderly dog turns out to be slow and lazy. Bitches conceive up to nine years of age, but are not serviceable after the tenth year. Puppies should not be allowed to run loose during the first six months, until they are grown strong, except to join their mother in sport and play; later they should be kept 11 12

continendi, et noctibus solvendi. Nec unquam eos, quorum generosam [1] volumus indolem conservare, patiemur alienae nutricis uberibus educari: quoniam semper et lac et spiritus maternus longe magis ingenii 13 atque incrementa corporis auget.[2] Quod si effeta [3] lacte deficitur, caprinum maxime conveniet praeberi catulis,[4] dum fiant [5] mensum quattuor.

Nominibus autem non longissimis appellandi sunt, quo celerius quisque vocatus exaudiat: nec tamen brevioribus quam quae duabus syllabis enuntientur, sicuti Graecum est σκύλαξ, Latinum *ferox*, Graecum λάκων, Latinum *celer*: vel femina, ut sunt Graeca σπουδή, ἀλκή, ῥώμη: Latina, *lupa*, *cerva*, *tigris*.

14 Catulorum caudas post diem quadragesimum, quam sint editi, sic castrare [6] conveniet. Nervus est, qui per articulos spinae prorepit usque ad ultimam partem caudae: is mordicus [7] comprehensus [8] et aliquatenus eductus abrumpitur: quo facto neque in longitudinem cauda foedum capit incrementum, et, ut plurimi pastores affirmant, rabies arcetur letifer morbus huic generi.

XIII. Fere autem per aestatem sic muscis aures canum exulcerantur, saepe ut totas amittant: quod ne fiat, amaris nucibus contritis linendae sunt. Quod si ulceribus iam praeoccupatae fuerint, coctam picem liquidam suillae adipi mixtam [9] vulneribus

[1] generosa *SA*. [2] aget *SA*¹.
[3] et fata *SA*. [4] catulus *SA*.
[5] fiat *SAR*. [6] siccatrare *S*¹*A*¹.
[7] modice *SAR*.
[8] comprehensus *R* : compressus *SA*.
[9] mixtam *add. Aldus*.

[a] Xenophon, *Cyneg.*, VII. 5, gives a list of some fifty names of dogs. They all are words of two syllables.

on the chain during the day and let loose at night. We should never allow those whose noble qualities we wish to preserve, to be brought up at the dugs of any strange bitch, since its mother's milk and spirit always does much more to foster the growth of their minds and bodies. But if a bitch which has a litter is 13 deficient in milk, it will be best to provide goats' milk for the puppies until they are four months old.

Dogs should be called by names which are not very long, so that each may obey more quickly when he is called, but they should not have shorter names than those which are pronounced in two syllables,[a] such as the Greek Σκύλαξ (puppy) and the Latin *Ferox* (savage), the Greek Λάκων (Spartan) and the Latin *Celer* (speedy) or, for a bitch, the Greek Σπουδή (zeal), Ἀλκή (Valour), Ῥωμή (strength) or the Latin *Lupa* (she-wolf), *Cerva* (hind) and *Tigris* (tigress). 14 It will be found best to cut the tails of puppies forty days after birth in the following manner: there is a nerve, which passes along through the joints of the spine down to the extremity of the tail; this is taken between the teeth and drawn out a little way and then broken. As a result, the tail never grows to an ugly length and (so many shepherds declare) rabies, a disease which is fatal to this animal, is prevented.[b]

XIII. It commonly happens that in the summer the ears of dogs are so full of sores caused by flies, that they often lose their ears altogether. To prevent this, the ears should be rubbed with crushed bitter almonds. If, however, the ears are already covered with sores, it will be found a good plan to drip boiled liquid pitch mixed with lard on the wounds. Ticks

Remedies for the diseases of dogs.

[b] This is quoted by Pliny, *N.H.* VIII. § 153.

stillari conveniet. Hoc eodem[1] medicamine contacti ricini decidunt. Nam manu non sunt vellendi, 2 ne, ut et ante praedixeram, faciant[2] ulcera.[3] Pulicosae cani remedia sunt sive cyminum tritum pari pondere cum veratro, aquaque mixtum et inlitum; seu cucumeris anguinei[4] succus: vel si haec non sunt, vetus amurca per totum corpus infusa. Si scabies infestabit, gypsi et sesami tantundem conterito, et cum pice liquida permisceto, vitiosamque partem linito: quod medicamentum putatur etiam hominibus esse conveniens. Eadem pestis si fuerit vehementior, cedrino[5] liquore aboletur. Reliqua vitia sicut in ceteris animalibus praecepimus, curanda erunt.

3 Hactenus de minore pecore. Mox de villaticis pastionibus, quae continent volucrum pisciumque et silvestrium quadrupedum curam, sequenti volumine praecipiemus.

[1] edem *S* : eadem *AR*.
[2] faciant *R* : faciunt *SA*.
[3] ulcerã *S* : ultra *A* : vulnera *R*.
[4] anguinei *R* : sanguinei *SA*.
[5] vehementior cedrino *R* : cedrino vehementer *SA*.

also fall off if they are touched with this same preparation; for they ought not to be plucked off by hand, lest, as we have remarked also before, they cause sores. A dog which is infested with fleas should be treated either with crushed cumin mixed in water with the same quantity of hellebore and smeared on, or else with the juice of the snake-like cucumber, or if these are unobtainable, with stale oil-lees poured over the whole body. If a dog is attacked by the scab, gypsum and sesame should be ground together in equal quantities and mixed with liquid pitch and smeared on the part affected; this remedy is reported to be suitable also for human beings. If this plague has become rather violent, it is got rid of by the juice of the cedar-tree. The other diseases of dogs will have to be treated according to the instructions which we have given for the other animals. 2

So much for the lesser domestic animals. In the next book we will give instructions about the keeping of live stock at the farm-house, which includes the care of fowls, fish and four-footed wild creatures. 3

BOOK VIII

LIBER VIII

I. Quae fere consummabant, Publi Silvine, ruris experiendi[1] scientiam, quaeque pecuariae negotiationis exigebat ratio, septem memoravimus libris. Hic nunc sequentis numeri titulum possidebit: nec[2] quia proximam propriamque rustici curam desiderent ea, quae dicturi sumus, sed quia non alio loco, quam in agris aut villis debeant administrari, et tamen 2 agrestibus magis, quam urbanis prosint. Quippe villaticae pastiones, sicut pecuariae, non minimam colono stipem conferunt, cum et avium stercore macerrimis vineis et omni surculo atque arvo medeantur; et eisdem familiarem focum[3] mensamque pretiosis[4] dapibus opulentent;[5] postremo venditorum animalium pretio villae reditum augeant. Quare[6] de hoc quoque genere pastionis dicendum 3 censui. Est autem id fere[7] vel in villa, vel circa villam.

In villa est, quod appellant Graeci ὀρνιθῶνας, καὶ περιστερεῶνας; atque etiam cum datur liquoris[8] facultas ἰχθυοτροφεῖα sedula cura exercentur. Ea

[1] experiendi *SA*: exercendis experiendique *a*: et exercendique *c*.
[2] nec *om. SA*.
[3] focum *Aac*: locum *S*.
[4] pretiosis *Sac*: pretioribus *A*.
[5] opulentent *A*: -ant *c*: -et *S*: -em *a*.
[6] quare *ac*: que *SA*.
[7] fere *Sac*: ferre *A*.
[8] liquoris *Aac*: litoris *S*.

BOOK VIII

I. We have now, Publius Silvinus, dealt in seven books with what practically constituted a complete account of the science of gaining knowledge of the land and all that was required for the business of raising cattle. Our present book shall bear the next number, eight, for its title, not that the subject of which we are going to speak demands the close and particular attention of the farmer, but because it ought not to be undertaken except in the country and on the farm, and brings benefit to country-folk rather than to town-dwellers. For the keeping of animals at the farm, as of cattle on the pasture, brings no small profit to farmers, since they use the dung of fowls to doctor the leanest vines and every kind of young tree and every kind of soil, and with the fowls themselves they enrich the family kitchen and table by providing rich fare; and, lastly, with the price which they obtain by selling animals they increase the revenue of the farm. Therefore I have thought it fitting that I should speak also of the keeping of this kind of animal. But it is generally carried on either at the farm or in its neighbourhood.

Of the keeping of birds and fishes on the farm.

2

At the farm there are what the Greeks call ὀρνιθῶνες and περιστερεῶνες (poultry-houses and dovecotes), and also, where a supply of water is available, ἰχθυοτροφεῖα (fish-ponds), the management of which requires unremitting care. All these, to use by

3

sunt omnia, ut Latine potius loquamur, sicut avium cohortalium stabula, nec minus earum, quae conclavibus saeptae saginantur, vel aquatilium animalium 4 receptacula. Rursus circa villam ponuntur μελισσῶνες καὶ χηνοτροφεῖα, quin etiam λαγοτροφεῖα studiose administrantur, quae nos similiter appellamus apum cubilia, apiaria, vel nantium volucrum, quae stagnis piscinisque laetantur, aviaria, vel etiam pecudum silvestrium, quae nemoribus clausis custodiuntur, vivaria.

II. Prius igitur de his praecipiam, quae intra saepta villae pascuntur.[1] Ac de aliis quidem forsitan ambigatur,[2] an sint agrestibus possidenda: gallinarum vero plerumque agricolae cura solennis est. Earum genera sunt vel cohortalium,[3] vel rusticarum 2 vel Africanarum. Cohortalis est avis, quae vulgo per omnes fere villas conspicitur: rustica, quae non dissimilis villaticae per aucupem decipitur, eaque plurima est in insula, quam nautae in Ligustico mari sitam producto nomine alitis Gallinariam vocitaverunt: Africana est, quam plerique Numidicam dicunt, Meleagridi similis, nisi quod rutilam galeam et cristam capite[4] gerit, quae utraque sunt in 3 Meleagride caerulea. Sed ex his tribus generibus cohortales feminae proprie appellantur gallinae, mares autem galli, semimares capi, qui hoc nomine

[1] *post* pascuntur *add.* quod sint genera gallinarum *ac*.
[2] ambigatur *c* : ambigantur *SAa*.
[3] cohortalium *A* : chortalium *S*. [4] capite *om. S*.

[a] Variously identified as hazel-hen, heath-hen, field-hen and red-legged partridge.
[b] This island is still called by this name and lies off Albengo, three miles E. of Alassio on the Italian Riviera.
[c] Probably the guinea-fowl.

preference the terms employed in our own language, are enclosures for farm-yard fowls and likewise for birds which are fattened in coops, or else for aquatic animals. On the other hand, in the neighbourhood 4 of the farm μελισσῶνες and χηνοτροφεῖα (bee-hives and goose-pens) find their place, and there are also carefully managed λαγοτροφεῖα (feeding-places for hares). To these we give a set of similar names, speaking of *apiaries*, where bees are lodged, *aviaries* for swimming birds which take their pleasure in pools and fish-ponds, and *vivaria* for wild creatures which are confined in enclosed woodlands.

Of the various kinds of farm-yard poultry.

II. First then I will give instructions about the creatures which are fed within the precincts of the farm. With regards to other animals it may perhaps be doubted whether country people should possess them; but the keeping of hens by farmers is quite a general practice. They fall into three classes, the farm-yard fowl, the "rustic"-hen[a] and the African fowl. The farm-yard fowl is the bird 2 commonly to be seen on almost every farm. The "rustic"-cock which is not very different from the farm-yard bird and is caught by the wiles of the fowler, is found in the greatest number in the island in the Ligurian sea to which sailors have given the name Gallinaria,[b] a lengthened form of the Latin word for hen. The African fowl,[c] which most people call Numidian, resembles the *meleagris*,[d] except that it has on its head a red helmet and crest, both of which are blue on the *meleagris*. Of these three kinds 3 the female farm-yard fowls alone are properly called hens, its males being called cocks and the half-males

[d] Our term for the turkey family, Meleagridae, is derived from this word.

vocantur, cum sunt castrati libidinis abolendae causa. Nec tamen id patiuntur amissis genitalibus, sed ferro candente calcaribus inustis, quae cum ignea vi consumpta sunt, facta ulcera dum consanescant, figulari creta linuntur.

4 Huius igitur villatici generis non spernendus est reditus, si adhibeatur educandi scientia, quam plerique Graecorum et praecipue celebravere Deliaci. Sed et hi, quoniam procera corpora et animos [1] ad proelia pertinaces requirebant, praecipue Tanagricum genus et Rhodium probabant, nec minus Chalcidicum [2] et Medicum, quod ab imperito vulgo litera 5 mutata Melicum appellatur. Nobis nostrum vernaculum maxime placet: omisso tamen illo studio Graecorum, qui ferocissimum quemque alitem certaminibus et pugnae praeparabant. Nos enim censemus instituere vectigal industrii patrisfamilias, non rixosarum [3] avium lanistae, cuius plerumque totum patrimonium, pignus aleae, victor gallinaceus pyctes abstulit.

6 Igitur cui placebit sequi nostra praecepta, consideret oportet primum quam multas, et cuiusmodi parare debeat matrices, deinde qualiter eas tutari [4] et pascere; mox quibus anni temporibus earum partus excipere; tum demum ut incubent et excludant efficere; postremo ut commode pulli educentur

[1] animos *ac*: animos-a (a *erasa*) *A*: animosa *S*.
[2] calchidicum *Sc*: calcidicum *Aa*.
[3] rixiosarum *Sa*.
[4] tutari *SAc*: tueri *a*.

[a] From Tanagra in Boeotia.
[b] From Chalcis in the island of Euboea.
[c] *I.e.* Persian.

capons; they are given this name because they have been castrated to rid them of sexual desire. They do not, however, suffer castration by the loss of their genital organs but by having their spurs burnt with a red-hot iron; when these have been consumed by the force of the fire, they are smeared with potter's clay until the sores which have been caused heal up.

4 The profit from keeping the farm-yard type of fowl is not to be despised if a scientific method of rearing them is put into operation, which most of the Greeks and in particular the people of Delos have made famous. The Greeks, however, since they desired height of body and determined courage in the fray, esteemed most highly the Tanagran[a] and Rhodian breeds and likewise the Chalcidian[b] and Median[c] (called by the ignorant vulgar Melian,[d] by the change of one letter). 5 We take most pleasure in our own native breed; however, we lack the zeal displayed by the Greeks who prepared the fiercest birds they could find for contests and fighting. Our aim is to establish a source of income for an industrious master of a house, not for a trainer of quarrelsome birds, whose whole patrimony, pledged in a gamble, generally is snatched from him by a victorious fighting[e]-cock.

6 He, therefore, who shall be minded to follow our instructions, should consider first with how many and what kind of breeding-hens he ought to provide himself, and then how he ought to look after and feed them; next, at what seasons of the year he ought to reserve the eggs which they produce; then he should arrange for their setting and hatching, and finally take thought for the proper rearing of the

[d] *I.e.* from the island of Melos, one of the Cyclades.
[e] A 'boxer.'

operam dare. His enim curis et ministeriis exercetur ratio cohortalis, quam Graeci vocant ὀρνιθοτροφίαν.

7 Parandi autem modus est ducentorum capitum, quae pastoris unius curam dispendant[1]: dum tamen anus sedula vel puer adhibeatur custos vagantium, ne obsidiis hominum, aut insidiatorum[2] animalium diripiantur. Mercari porro nisi fecundissimas aves non expedit. Eae sint rubicundae vel infuscae[3] plumae, nigrisque pinnis[4]: ac si fieri poterit, omnes huius, et ab hoc proximi coloris eligantur. Sin aliter, vitentur[5] albae; quae fere cum sint molles ac minus vivaces, tum ne fecundae quidem facile reperiuntur:[6] atque etiam conspicuae propter insigne candoris ab accipitribus et aquilis saepius abripi-

8 untur. Sint ergo matrices robii coloris[7] quadratae, pectorosae, magnis capitibus, rectis rutilisque cristulis,[8] albis auribus, et sub hac specie quam amplissimae, nec paribus ungulis:[9] generosissimaeque[10] creduntur, quae quinos habent digitos, sed ita ne cruribus emineant transversa calcaria. Nam quae hoc virile gerit insigne, contumax ad concubitum dedignatur[11] admittere marem, raroque fecunda, etiam cum incubat, calcis aculeis ova perfringit.

[1] dispendeat *c*: distendant *SAa*.
[2] insidiatorum *SA*: insidiosorum *ac*.
[3] infuscae *SAc*: fuscae *a*.
[4] pinnis *SAc*: pennis *a*.
[5] vitentur *a*: evitentur *c*: viterbitentur *SA*.
[6] reperiantur *codd.*
[7] robii coloris *S*: robusta coloris *A*: probi coloris *ac*.
[8] rectis rutulisque cristulis *c*: rectis rutilis *SA*: rectilis (rectis a^2) rutulisque cristulis a^1.
[9] ungulis *ac*: unguibus *SA*.
[10] generosissimeque *ac*: generosis eque *S*: generosis seque *A*.

chickens. For it is by attention to these points and management that the business of poultry-keeping, which the Greeks call ὀρνιθοτροφία (bird-rearing), is carried out.

7 Two hundred head are the limit which should be acquired fully to employ the care of one person to feed them, provided, however, that an industrious old woman or a boy be set to watch over the fowls which go astray, so that they may not be carried off by the wiles of men or of animals which lie in wait for them. Further only the most prolific fowls should be bought. They should have red or darkish plumage and black wings; and, if this is possible, they should be chosen of the latter colour all over and of the nearest colour to it. Failing these colours, white hens should be avoided; for, while they are delicate and not very long-lived, it is also not easy to find white fowls which are prolific: also, being conspicuous owing to their remarkably light colour they are rather often carried off by hawks and eagles. 8 Let your brood-hens, therefore, be of a red colour, square-built, big-breasted, with large heads, straight, red crests, white ears; they should be the largest obtainable which present this appearance and should not have an even number of claws. Those are reckoned the best-bred which have five toes [a] but without any cross-spurs projecting from their legs; for a hen which has this masculine characteristic is refractory and disdains to admit the male to intercourse and is rarely prolific, and, when she does sit, breaks the eggs with the sharp points of her spurs.

[a] *I.e.* four claws and one spur on each leg.

---

[11] dedignatur *Sac*: dedignatam *A*.

9 Gallinaceos mares nisi salacissimos habere non expedit. Atque in his quoque sicut in feminis, idem color, idemque numerus unguium, status altior quaeritur: sublimes, sanguineaeque, nec obliquae cristae: ravidi,[1] vel nigrantes oculi: brevia et adunca rostra: maximae candidissimaeque aures: paleae [2] ex rutilo albicantes, quae velut incanae barbae dependent: iubae [3] deinde variae, vel ex auro flavae, per colla 10 cervicesque in humeros diffusae: tum lata et musculosa pectora, lacertosaeque similes bracchiis alae, tum procerissimae caudae, duplici ordine, singulis utrinque prominentibus pinnis inflexae: quinetiam vasta femina [4] et frequenter horrentibus plumis hirta: robusta crura, nec longa, sed infestis velut 11 sudibus nocenter armata. Mares [5] autem, quamvis non ad pugnam neque ad victoriae laudem praeparentur, maxime tamen generosi probantur, ut sint elati, alacres, vigilaces, et ad saepius canendum prompti, nec qui facile terreantur: nam interdum resistere debent, et protegere coniugalem gregem: quin et attollentem minas serpentem, vel aliud noxium animal interficere.

12 Talibus autem maribus quinae singulis feminae comparantur.[6] Nam Rhodii generis aut Medici propter gravitatem neque patres nimis salaces, nec fecundae matres: quae tamen ternae singulis maritantur. Et cum pauca ova posuerunt, inertes ad incubandum, multoque magis ad excludendum,

[1] ravidi *edd.*: rabidi *c*: rubidi *SA*: rubicundi *a*.
[2] paleae *Aac*: galeae *S*.
[3] iubae *om. A*.
[4] femina *Aac*: femini *S*.
[5] mares *SAac*.
[6] comparantur *Aac*: comparant *S*.

It is advisable not to keep any but the most salacious cock-birds and the same colour as in hens, and the same number of claws is looked for in them, but a loftier stature. Their crest should be high, blood-red and not crooked, their eyes darkish or tending towards black, their beaks short and hooked, their ears very large and white, their wattles bright-red tending towards white and hanging down like grey beards, their head-feathers of different colours or gold shading into yellow and extending over their throats and necks on to their shoulders. Their chests should be broad and muscular, their wings brawny and like arms, and their tails very prominent and divided into two halves, bending over with a single projecting feather on each side. They should also have huge thighs, thickly covered with bristling feathers; their legs should be robust but not long, and armed for offence with what may be described as stakes ready for the attack. These male birds, though they are not being trained for fighting and the glory of winning prizes, are, nevertheless, esteemed as well-bred if they are proud, lively, watchful and ready to crow frequently and not easily to be frightened; for on occasion they have to act on the defensive and protect their flock of wives, nay, even to slay a snake which rears its threatening head or some other hurtful animal. 9 10 11

For such male birds as these five hens each are provided. Of the Rhodian and Median breeds the father-birds are not very salacious on account of their heavy build, nor are the mother-birds very prolific: however, three hens are mated with each cock-bird. And when they have laid a few eggs, they are lazy about sitting on them and much more so about hatching 12

raro fetus suos educant. Itaque quibus cordi est ea genera propter corporum speciem possidere, cum exceperunt ova generosarum, vulgaribus gallinis
13 subiciunt, ut ab his exclusi[1] pulli nutriantur. Tanagrici plerumque Rhodiis[2] et Medicis amplitudine pares, non multum moribus a vernaculis distant, sicut et Chalcidici. Omnium tamen horum generum nothi[3] sunt optimi[4] pulli, quos conceptos ex peregrinis maribus nostrates ediderunt. Nam et paternam speciem gerunt, et salacitatem fecunditatemque
14 vernaculam retinent. Pumiles[5] aves, nisi quem humilitas earum delectat, nec propter fecunditatem, nec propter alium reditum nimium probo, tam[6] hercule, quam nec pugnacem ac[7] rixosae[8] libidinis marem. Nam plerumque ceteros infestat, et non patitur inire feminas, cum ipse pluribus sufficere non
15 queat. Impedienda est itaque procacitas eius ampullaceo corio; quod cum in orbiculum formatum est, media pars eius rescinditur, et per excisam partem galli pes inseritur: eaque quasi compede cohibentur feri mores. Sed, ut proposui, iam de tutela generis universi praecipiam.

III. Gallinaria constitui debent parte villae, quae hibernum spectat orientem: iuncta sint ea furno vel culinae, ut ad avem perveniat fumus, qui est huic

[1] exclusi *ac* : excussi *S* : excusi *A*.
[2] Rhodiis *ac* : Hrodiis *S* : Hordiis *A*. [3] noti *ac*.
[4] *post* optimi *add.* sunt *SA*.
[5] pumiles *Ac* : pumileas *S* : humiles *a*.
[6] probo tam *c* : probatam *Sa* : -um *A*.
[7] ac *scripsi* : nec *codd.*
[8] rixose *a* : risose *c* : rixo *SA*.

them, and they rarely bring up their own offspring. Those, therefore, whose hearts are set on possessing these breeds on account of their fine appearance, when they have set aside the eggs of the well-bred hens, put them under ordinary hens, in order that the chickens when they are hatched may be brought up by the latter. Tanagran fowls, which are usually 13 equal in size to the Rhodian and Median, do not differ greatly from our native fowls in disposition, and the same is true of the Chalcidian. But of all these breeds the cross-bred chickens are the best, which our own hens have produced after conceiving them by foreign male birds; for they show the fine appearence of their fathers and their own native salaciousness and productivity. I do not highly commend 14 bantam-hens either for their fecundity or for any other return which they give—unless one takes a pleasure in their low stature—just as indeed I do not commend the bantam-cock either, which is given to fighting and whose lust makes him quarrelsome. For it generally attacks the other cock-birds and does not allow them to cover the hens, though it cannot itself suffice for a large number of hens. Its petulance, 15 therefore, must be checked by means of a piece of leather from an old flask, of which, after it has been formed into a round shape, the middle part is cut away and the cock's foot is inserted through this cut-out part, and by this kind of shackle its fierce disposition is restrained. But, as I proposed, I will now give directions for the care of poultry in general.

How to make a hen-house.

III. Hen-houses should be placed in the part of the farm which faces the rising sun in winter and should adjoin the oven or the kitchen, so that the smoke, which is particularly beneficial to this kind of animal,

generi praecipue salutaris. Totius autem officinae, id est ornithonis, tres continuae extruuntur cellae, quarum, sicuti dixi, perpetua frons orienti[1] sit 2 obversa. In ea deinde fronte exiguus detur unus omnino aditus mediae cellae; quae ipsa e tribus minima esse debet in altitudinem et quoquoversus pedes septem. In ea singuli[2] dextro laevoque pariete aditus ad utramque cellam faciendi sunt, iuncti parieti, qui est intrantibus adversus. Huic autem focus applicetur tam longe, ut nec impediat praedictos aditus, et ab eo fumus perveniat in utramque cellam: eaeque longitudinis et altitudinis duodenos pedes habeant, nec plus latitudinis quam media. 3 Sublimitas dividatur tabulatis, quae supra se quaternos, et infra septenos liberos pedes habeant, quoniam ipsa singulos occupant. Utraque tabulata gallinis servire debent, et ea parvis ab oriente singulis illuminari fenestellis, quae et ipsae matutinum exitum praebeant avibus[3] ad cohortem, nec minus vespertinum introitum. Sed curandum erit, ut semper noctibus claudantur, quo tutius aves maneant. Infra tabulata maiores fenestellae[4] aperiantur, et eae clatris muniantur, ne possint noxia irrepere animalia: sic tamen, ut illustria sint loca, quo commodius 4 aditet[5] aviarius, qui[6] subinde debet speculari aut incubantes aut parturientes fetus. Nam etiam in iis ipsis locis ita crassos parietes aedificare convenit,

[1] orienti *Schneider*: orientem *codd.*
[2] singuli *Sac*: singula *A.*
[3] avibus *ac*: animos *SA.*
[4] fenestellae *SAa*: fenestrae *c, Schneider.*
[5] aditet *Schneider*: habitet *codd.*
[6] qui *ac*: quia *S*: qua *A.*

may reach the fowls. Three adjacent cells are constructed to form the whole building or poultry-house and, as I have said, their continuous front should face the east. In this front there should be one small 2 entrance provided leading into the middle cell, which in itself should be the smallest of the three, being seven feet in height and in its other dimensions. In this cell entrances should be made in the right and left party walls, one leading to each of the other two cells and adjoining the wall which faces those who enter the central cell. To this wall a hearth should be fixed of such a length as not to block the entrances already mentioned and to allow the smoke from it to penetrate into each of the other two cells. These latter should have a length and height of twelve feet and no more breadth than the middle cell. The 3 height should be divided up by lofts with four unoccupied feet above them and seven below, since they themselves take up one foot. Both lofts ought to be used to accommodate the hens and should each be lighted by a small window on the east side, which may also provide the birds with a means of exit in the morning into the poultry-yard and a means of entrance in the evening; but care must be taken that they are always kept closed at night that the fowls may remain in greater safety. Below the lofts larger windows should be opened up and secured with lattice-work, that harmful animals may not be able to creep in, but at the same time so constructed that the interior may be well lighted, so that the poultry-keeper, who ought from time to time to keep an eye upon the hens when they are sitting and hatching their young, may more conveniently visit them. For in the hen-houses themselves too the walls should 4

ut excisa per ordinem gallinarum cubilia recipiant: in quibus aut ova edantur, aut excludantur pulli: hoc enim et salubrius et elegantius est, quam illud, quod quidam faciunt, ut, palis in parietes vehementer actis
5 vimineos qualos superimponant.[1] Sive autem parietibus ita, ut diximus, cavatis, sive qualis[2] vimineis[3] praeponenda erunt vestibula, per quae[4] matrices ad cubilia vel pariendi vel incubandi causa perveniant. Neque enim debent ipsis nidis involare,[5]
6 ne dum adsiliunt, pedibus ova confringant. Ascensus deinde avibus ad tabulata per utramque cellam datur iunctis parieti modicis asserculis, qui paulum formatis gradibus asperantur, ne sint advolantibus lubrici. Sed ab cohorte forinsecus praedictis[6] fenestellis scandulae similiter iniungantur, quibus irrepant aves ad requiem nocturnam. Maxime autem curabimus ut et haec aviaria et cetera, de quibus mox dicturi sumus, intrinsecus et extrinsecus[7] poliantur opere tectorio, ne[8] ad[9] aves feles habeant aut coluber accessum, et aeque noxiae prohibeantur pestes.
7 Tabulatis insistere dormientem avem non expedit, ne suo laedatur stercore; quod cum pedibus uncis adhaesit, podagram creat. Ea pernicies ut evitetur, perticae dolantur in quadrum, ne teres levitas earum supersilientem volucrem non recipiat. Conquadratae

[1] superponant *a* : -ent *SA* : -at *c*.
[2] qualis *c* : qualem *SA* : qualos *a*.
[3] vimineis *c* : -os *SAa*.
[4] que *SAac*.
[5] inbolare *SA*.
[6] praedictis *SAac*.
[7] et extrinsecus *ac* : *om*. *SA*.
[8] ne *a* : neque *SAc*.
[9] ad *om*. *A*.

be built so thick as to allow nesting-places for the hens to be cut out of them in a row, where either the eggs may be laid or the chickens hatched; for this is both healthier and neater than what some people do when they forcibly drive pegs into the walls and support wicker-work baskets on them. But in front of either 5 the walls which have been hollowed, as we have described, or of the wicker-work basket, porches must be placed through which the breeding-hens may reach their nests for the purpose of either laying eggs or sitting on them; for they ought not to fly into the nests themselves, lest, as they leap into them, they break the eggs with their feet. Next a means of 6 ascent for the hens to the lofts across each of the cells is provided by attaching to the wall moderately sized planks which are roughened a little by having steps made on them, so that the hens may not find them slippery when they fly on to them. Similarly little ladders should be attached on the outside leading from the poultry-yard to the little windows mentioned above, by which the birds may creep in for their nightly repose. But we shall take particular care that these poultry-houses and those about which we shall be speaking presently, are made smooth, within and without, with plaster-work, so that no cat or snake may have access to the fowls and that equally hurtful pests may be kept away.

It is not expedient that the hen should rest on a 7 loft's floor when it is asleep, lest it be harmed by its own dung, because this, if it has adhered to its crooked feet, causes gout. That this calamity may be avoided, perches should be hewn square lest their rounded smoothness should fail to give the bird a good hold when it springs up. After being squared

deinde foratis duobus adversis parietibus induuntur,[1] ita ut a tabulato pedalis altitudinis, et inter se bipedalis latitudinis spatio distent.

8 Haec erit cohortalis officinae dispositio. Ceterum cohors ipsa, per quam vagantur, non tam stercore, quam uligine careat. Nam plurimum refert aquam non esse in ea nisi uno loco, quam bibant, eamque mundissimam: nam stercorosa pituitam concitat. Puram tamen servare non possis, nisi clausam vasis[2] in hunc usum fabricatis. Sunt autem, qui aut aqua replentur aut cibo plumbei canales, quos magis utiles 9 esse ligneis aut fictilibus[3] compertum est. Hi superpositis operculis clauduntur, et a lateribus super mediam partem altitudinis per spatia palmaria modicis forantur cavis, ita ut avium capita possint admittere. Nam nisi operculis muniantur, quantulumcunque aquae[4] vel ciborum inest, pedibus everritur. Sunt qui a superiore parte foramina ipsis operculis imponant; quod fieri non oportet. Nam[5] supersiliens avis proluvie ventris cibos et aquam conspurcat.

IV. Cibaria gallinis praebentur optima pinsitum hordeum et vinacea[6] nec minus cicercula, tum etiam milium, aut panicum: sed haec ubi vilitas annonae permittit. Ubi vero ea est carior, excreta tritici minuta commode dantur. Nam per se id frumentum, etiam quibus locis vilissimum est, non utiliter praebetur, quia obest avibus. Potest etiam lolium

[1] induuntur *a*: induunt *SAc*.
[2] vasi *a*: vasis *c*: basis *SA*.
[3] ligneis et (aut *A*) fictilibus *S*: ligneos aut fictiles *ac*.
[4] aquae *om. A*.
[5] quam *SA*: nam *a*: tam *c*. [6] vinacia *SA*: vicia *ac*.

[a] *I.e.* chaff.

the poles should be fixed in holes in two walls which face one another, so that they may be a foot in height above the loft floor and two feet in breadth away from one another.

8 Such will be the arrangement of the hen-house in thepoultry-yard. But the poultry-yard itself, through which the hens wander, should be free not so much from dung as from moisture; for it is extremely important that there should be no water in it except in one place, namely, the water for them to drink and that water should be very clean (for water which has dung in it gives fowls the pip), yet you cannot keep it clean unless it is enclosed in vessels made for the purpose. But there are leaden troughs which are filled with either water or food, and it has been found that they are more useful than troughs of wood or pottery. 9 These are closed by having lids placed over them and are pierced with small holes above the middle of their height a palm's breadth apart from one another and large enough to admit the birds' heads. For if they are not provided with covers, any small quantities of water or food that is inside is swept out by the birds' feet. Some people make holes above in the top part of the covers themselves; this should not be done, for the bird leaping on the top befouls the food and water with its excrement.

How to feed hens.

IV. The best foods to be given to hens are bruised barley and grape-husks, likewise chick-pea and also millet and panic-grass, but these last two only when the low price of cereals permits. When cereals are dearer, small refuse [a] from wheat is a convenient food to give; for this grain by itself, even in places where it is very cheap, is not a suitable food because it is injurious to fowls. Boiled darnel can also be put

decoctum obici, nec minus furfures modice a farina excreti: qui si nihil habent farris, non sunt idonei, 2 nec tantum appetuntur ieiunis. Cytisi folia seminaque maxime probantur, et sunt huic generi gratissima: neque est ulla regio, in qua non possit[1] huius arbusculae copia esse vel maxima. Vinacea quamvis tolerabiliter pascant, dari non debent, nisi quibus anni temporibus avis fetum non edit: nam et partus 3 raros, et ova faciunt exigua. Sed cum plane post autumnum cessant a fetu, possunt[2] hoc cibo sustineri. Attamen quaecunque dabitur esca per cohortem vagantibus, die incipiente, et iam in vesperum declinato,[3] bis dividenda est, ut et mane non protinus a cubili latius evagentur, et ante crepusculum propter cibi spem temporius ad officinam redeant, possitque[4] numerus capitum saepius recognosci. Nam volatile pecus facile custodiam pastoris decipit.

4 Siccus etiam pulvis et cinis, ubicunque cohortem porticus vel tectum protegit, iuxta parietes reponendus est, ut sit quo aves se perfundant. Nam his rebus plumam pinnasque emundant: si modo credimus Ephesio Heraclito, qui ait sues caeno, 5 cohortales aves pulvere vel cinere[5] lavari. Gallina post primam emitti, et ante horam diei undecimam claudi debet: cuius vagae cultus hic, quem diximus,

[1] possit *ac*: possint *SA*.
[2] possunt *edd*: potest *SAac*.
[3] declinato *SAac*: declinante *edd*.
[4] possitque *ac*: possintque *SA*.
[5] cinere *om*. *SA*.

---

[a] The well-known Ionian philosopher of the late 6th century B.C.

before them and likewise bran if only partly separated from the meal; for if there is no meal with the food, it is not suitable nor have they much appetite for it, though they be hungry. The leaves 2 and seeds of the shrub-trefoil are very highly approved and are greatly appreciated by fowls, and there is no region in which it is not possible to find a very great abundance of this shrub. Grape-husks, although they tolerate them as food, should not be given to fowls except at times of year when they are not laying; for they cause them to lay seldom and only small eggs. But when they obviously stop laying after the autumn, they can be kept on this food. Whatever food is to be given 3 them when they are loose in the poultry-yard should be distributed in two parts, one when day is beginning and the other when it has already declined towards evening, so that in the morning they may not immediately wander too far away from their sleeping-quarters and that they may return before dusk to the poultry-house in better time in hopes of finding food there, and that the number of head may be verified more often. For winged creatures easily delude the watchfulness of the man who looks after them.

Dry dust and ashes should be placed near the party 4 walls wherever a porch or a roof shelters the poultry-yard, so that the birds may have the means to sprinkle themselves; for it is with these that they clean their feathers and wings, if we believe Heraclitus [a] the Ephesian who says that pigs wash themselves with mud, farm-yard fowls with dust or ashes. A hen 5 ought to be let out after the first hour of the day and be shut up again before the eleventh hour. Its manner of life when it is let loose will be as we have

erit: nec tamen alius clausae, nisi quod ea non emittitur sed intra ornithonem ter die pascitur maiore mensura. Nam singulis capitibus quaterni cyathi diurna cibaria sunt, cum vagis[1] bini praebeantur.
6 Habeat tamen etiam clausa oportet amplum vestibulum, quo prodeat, et ubi apricetur: idque sit retibus munitum, ne[2] aquila vel accipiter involet. Quas impensas et curas, nisi locis,[3] quibus harum rerum vigent pretia, non expedit adhiberi. Antiquissima est autem cum in omnibus pecoribus tum in hoc fides pastoris; qui nisi[4] eam domino servat, nullus ornithonis quaestus vincet[5] impensas. De tutela satis dictum est: nunc reliquum ordinem prosequemur.

V. Confecta bruma parere[6] fere id genus avium consuevit. Atque earum quae sunt fecundissimae, locis tepidioribus circa calendas Ianuarias ova edere incipiunt; frigidis autem regionibus eodem mense
2 post idus. Sed cibis idoneis fecunditas earum elicienda est, quo maturius partum edant. Optime praebetur ad satietatem hordeum semicoctum: nam et maius facit ovorum incrementum, et frequentiores partus. Sed is[7] cibus quasi condiendus est interlectis cytisi foliis ac semine eiusdem, quae[8] maxime putantur augere fecunditatem avium. Modus autem cibariorum sit, ut dixi, vagis binorum cyathorum hordei. Aliquid tamen admiscendum erit

[1] *post* vagis *add.* terni vel *c.*
[2] ne *om. A.*
[3] *post* locis *add.* et *SAa.*
[4] qui nisi *Aac* : quin si *S.*
[5] vincet *c* : vigit *A* : vingit *a* : vincit *S.*
[6] parare *c* : *om. SAa.*
[7] sed is *ac* : et his *SA.*
[8] *post* quae *add.* utraque *ac.*

described, and it will be no different when it is shut up except that it is not allowed to go out but is kept within the hen-house and fed three times a day with a larger quantity of food; for the daily ration is four *cyathi* per head, whereas that of the wandering bird is only two *cyathi*. A bird which is shut up, how- 6 ever, should have a spacious portico to which it can go out and bask in the sun; and this should be protected with nets, so that no eagle or hawk can fly in. It is only worth while to go to these expenses and to take these precautions in places where the prices of hens and their produce are high. But in the keeping of fowls, as of all domestic animals, the most important thing is that the man who looks after them should be trustworthy, for, unless he is faithful to his master, the profit from the poultry-house will not surpass the cost. Enough has now been said about the management of hens; we will now pursue the other topics in order.

Of the collection and setting of eggs under the hen.

V. When midwinter is over, this kind of bird is generally wont to lay. In warmer places the most prolific hens begin laying eggs about the first of January, but in colder regions after the 13th of the same month. But their productivity must be en- 2 couraged by suitable food to make them lay earlier. The best food to give them is their fill of half-cooked barley; for it both increases the size of the eggs and makes them lay more often. But this food must be seasoned, as it were, by throwing into it the leaves and seed of shrub-trefoil, which are thought greatly to increase the productivity of birds. The quantity of food, as I have said, should be two *cyathi* of barley per hen if they are allowed to wander freely, but some shrub-trefoil should be mixed with it, or, if this

3 cytisi, vel si id non fuerit, viciae aut milii. Curae autem debebit esse custodi, cum parturient aves, ut habeant quam mundissimis paleis constrata cubilia, eaque [1] subinde converrat, et alia stramenta quam recentissima reponat.[2] Nam pulicibus, atque aliis similibus [3] replentur, quae [4] secum affert avis, cum ad idem cubile revertitur. Assiduus autem debet esse custos et speculari parientes,[5] quod se facere gallinae testantur crebris singultibus interiecta voce
4 acuta. Observare itaque dum edant ova, et [6] confestim circumire oportebit cubilia, ut quae nata sunt recolligantur, notenturque quae quoque die sint edita, et quam recentissima supponantur glucientibus: sic enim appellant rustici aves eas quae volunt incubare; cetera vel reponantur, vel aere mutentur. Aptissima porro sunt ad excludendum recentissima quaeque. Possunt tamen etiam requieta supponi,
5 dum ne vetustiora sint quam dierum decem. Fere autem cum primum partum consummaverunt gallinae, incubare cupiunt ab idibus Ianuariis, quod facere non omnibus permittendum est; quoniam quidem novellae magis edendis, quam excludendis ovis utiliores sunt: inhibeturque cupiditas incubandi [7] pinnula per
6 nares traiecta.[8] Veteranas igitur aves ad hanc rem eligi oportebit, quae iam saepius id fecerint; moresque earum maxime pernosci, quoniam aliae melius

[1] eaque *ac*: que *S*: quae *A*.
[2] reponat *ac*: reponant *SA*.
[3] *post* similibus *add.* animalibus *ac*.
[4] quae *om. SA*.
[5] speculari parientes *ac*: specularientes *SA*.
[6] observare itaque dum edant ova et *ac*: observare dum edant ova itaque dum et *S*: observare idum edant ova itaque dum et *A*.
[7] incubandi *ac*: incubando *SA*.

is not available, vetch or millet. The keeper will 3 have to take care that the hens, when they are breeding, have their nests strewn with the cleanest possible straw, and he must sweep them out from time to time and put in other litter which is as fresh as possible. For the nests become full of fleas and other similar creatures which the hen brings with it when it returns to the same nest. The keeper ought also to be continually on the look-out for hens which are laying, a fact to which they bear witness by frequent cackling interrupted by shrill cries. He will have to watch until 4 they produce eggs and then immediately go round the nests so that the eggs which have been laid may be collected and a record taken to show the number which have been laid each day and that the freshest possible eggs may be put under the clucking hens, for this is what country-folk call those birds which wish to sit. The rest should either be stored or else turned into money. Furthermore, the freshest eggs are most suitable for hatching; those, however, which have been kept for some time can also be set, provided that they are not more than ten days old. Hens which have completed their first clutch of 5 eggs generally want to sit from January the 13th onwards; but they must not all be allowed to do so, since young pullets are more useful for laying eggs than for hatching them, and their desire to sit is checked by passing a small feather through their nostrils. Veteran fowls, therefore, will have to be 6 chosen for the task of sitting, which have already done so frequently, and their disposition must be fully known since some hens are better at hatching the

---

[8] per nares traiecta *ac* : per nasi et a *S* : per nasia et a *A*.

excludunt, aliae editos pullos commodius educant.[1] At e contrario quaedam et sua et aliena ova comminuunt atque consumunt, quod facientem protinus submovere conveniet.[2]

7 Pulli autem duarum aut trium avium exclusi,[3] dum adhuc teneri sunt, ad unam, quae sit melior nutrix, transferri debent, sed primo quoque die, dum mater suos et alienos propter similitudinem dignoscere non potest. Verumtamen servare oportet modum. Neque enim debet maior esse quam triginta capitum. Negant[4] enim hoc ampliorem gregem posse ab una 8 nutriri. Numerus ovorum, quae subiciuntur, impar observatur,[5] nec semper idem. Nam primo tempore, id est mense Ianuario, quindecim, nec unquam plura subici debent: Martio, XIX,[6] nec his pauciora: unum et viginti Aprili:[7] tota deinde aestate usque in calendas Octobris totidem.[8] Postea supervacua est huius rei cura, quod frigoribus exclusi pulli plerum- 9 que intereunt. Plerique tamen etiam ab aestivo solstitio non putant bonam pullationem, quod ab eo tempore etiam si facilem educationem habent, iustum tamen non capiunt incrementum. Verum suburbanis

[1] educant *a* : educent *SAc*.
[2] convenient *S* : conveniet *a* : conveniunt *A* : convenit *c*.
[3] exclusi *edd.* : excusi *SA* : excussi *ac*.
[4] negant *Aa* : necant *S* : negat *c*.
[5] impar observatur *om. S*.
[6] Martis XIX *edd.* : Maio VIII (*aut* novem) *SAac*.
[7] unum et viginti Aprili *edd.* : undecim Aprili *SAa* : unde cum Aprili *c*.
[8] totidem *edd.* : tredecim (*aut* XIII) *SAac*.

chickens and others are more suitable for bringing them up when they have been hatched. Some hens, on the other hand, break and consume both their own and other hens' eggs; any hen which does this will have to be got rid of immediately.

7 The chickens of two or three hens, when they have been hatched and are still very young, should be transferred to one mother, whichever is the best nurse; but this must always be done the very first day while the mother, owing to their similarity, is unable to distinguish her own young and those of other hens. A limit, however, must be observed, which ought not to be more than thirty head; for it is said that a larger flock than this cannot be cared for by a single hen. 8 The rule is observed of putting an uneven number of eggs under a hen, but it is not always the same number. At the first setting, that is, in the month of January, fifteen eggs, and never more, ought to be set, in March nineteen and never less: in April, twenty-one, and the same number throughout the summer until October 1st.[a] After this date any attention given to the matter of hatching is useless, because, owing to the cold, the chickens generally die as soon as they are hatched. 9 Most people, however, do not think that it is good to hatch chickens after the summer solstice, because from that time onwards, even though it is easy to rear them, they never come to their proper growth; but in the

[a] It is clear that the numbers of eggs which should be put under hens at various times of year are wrong in the MSS, according to which fifteen should be set in January, nine in May, eleven in April and thirteen in the summer. This is quite illogical, since obviously more eggs can be given to a hen to sit upon in warm than in cold weather. The readings generally adopted by the editors give the required sense.

locis, ubi a matre pulli non exiguis pretiis veneunt, probanda est aestiva educatio.

Semper autem, cum supponuntur ova, considerari debet, ut luna crescente a decima usque ad quintamdecimam id fiat. Nam et ipsa suppositio per hos fere dies est commodissima; et sic administrandum est, ut rursus cum excluduntur pulli, luna crescat.

10 Diebus quibus animantur ova, et in speciem volucrum conformantur, ter septenis opus est gallinaceo generi. At pavonino[1] et anserino, paulo amplius ter novenis. Quae si quando fuerint supponenda gallinis, prius eas incubare decem diebus fetibus alienigenis patiemur. Tum demum sui generis quattuor ova, nec plura quam quinque fovenda recipient. Sed et haec quam maxima: nam ex pusillis

11 aves minimae[2] nascuntur. Cum deinde quis volet quam plurimos mares excludi, longissima quaeque et acutissima ova subiciet: et rursus cum feminas, quam rotundissima. Supponendi autem consuetudo tradita est ab iis, qui religiosius haec administrant, eiusmodi. Primum quam secretissima cubilia legunt,[3] ne incubantes matrices ab aliis avibus inquietentur: deinde antequam consternant ea, diligenter emundant, paleasque, quas substraturi sunt, sulfure et bitumine atque ardente teda perlustrant, et expiatas

[1] pavonino *ac*: pavone *SA*.
[2] minimae *scripsi*: minima *SA*: minutae *ac*.
[3] legunt *SA*: eligunt *ac*.

neighbourhood of towns, where chickens are sold at a high price straight from their mother's care, summer rearing is to be approved.

When eggs are being put under a hen, care should always be taken that this is done when the moon is increasing, namely, from the tenth to the fifteenth day of the month; for the actual placing of the eggs is most convenient somewhere about this time, and it is necessary to arrange that the moon is increasing again when the chickens are hatched. 10 It takes twenty-one days for the eggs to become quickened and take on the form of birds in the case of farm-yard poultry, but for peacocks and geese rather more than twenty-seven days are required. If ever it should be necessary to put the eggs of the two latter species under ordinary hens, we shall allow them to sit first for ten days on the eggs of these alien birds, and then they will be given four eggs of their own kind to sit upon, and never more than five. These must be as large as possible; for from undersized eggs only very small birds are produced. 11 Next, when anyone wishes as many male chickens as possible to be hatched, he will set the longest and most pointed eggs; if, on the other hand, he wants female chickens, he should set the roundest eggs. The following is the usual method of placing eggs as handed down by those who are most scrupulous in the way they manage such matters. First of all they choose the most retired nesting-boxes, so that the brooding hens may not be disturbed by other fowls; then, before they strew anything in them, they cleanse them carefully and purify the chaff which they are going to put under the hens with sulphur and bitumen and a burning torch, and when they have thus purged it they throw

cubilibus iniciunt, ita factis concavatis nidis, ne advolantibus aut desilientibus evoluta decidant ova.
12 Plurimi etiam infra cubilium stramenta graminis aliquid et ramulos lauri, nec minus alii capita cum clavis ferreis subiciunt: quae cuncta remedio creduntur esse adversus tonitrua, quibus vitiantur ova, pullique semiformes interimuntur antequam toti
13 partibus suis consummentur. Servat autem qui subicit,[1] ne singula ova[2] in cubili manu componat,[3] sed totum ovorum numerum in alveolum ligneum conferat, deinde universum leniter in prae-
14 paratum nidum transfundat. Incubantibus autem gallinis iuxta ponendus est cibus, ut saturae studiosius nidis immorentur, neve longius evagatae refrigerent ova, quae quamvis pedibus ipsae convertant,[4] aviarius tamen cum desilierint matres, circumire debet,[5] ac manu versare, ut aequaliter calore concepto facile animentur, quin etiam si qua unguibus laesa vel fracta sunt, ut removeat. Idque cum fecerit duodeviginti diebus,[6] die undevigesimo animadvertat an pulli rostellis ova pertuderint, et auscultetur, si pipiant. Nam saepe propter crassitudinem puta-
15 minum[7] erumpere non queunt. Itaque haerentes pullos manu eximere oportebit, et matri fovendos subicere, idque non amplius triduo facere. Nam post unum et vigesimum diem silentia ova carent animali-

[1] subicit *acA*²: qui subl *A*²: quis ubi *S*.
[2] ova *om. SAac.*
[3] componat *ac*: componant *SA*.
[4] quamvis pedibus ipse convertant *ac*: quam ipse convert *S*: quam ipsa confere *A*.
[5] debent *SAac*.
[6] duodeviginti diebus *om. ac*.
[7] putaminum *ac*: putaminarum *SA*.

it into the nest-boxes, making the nest hollow so that the eggs may not roll out and fall when the hens fly in or leap down. Very many people also lay a little 12 grass under the litter in the nest-boxes and small branches of bay and also fasten underneath heads of garlic with iron nails, all of which things are regarded as preservatives against thunder by which the eggs are spoilt and the half-formed chickens killed before they can reach complete perfection in all their parts. The man who places the eggs is careful not to place 13 them one by one in the nest-box by hand, but should collect the complete number in a wooden basin and gently pour the whole clutch into the nest ready prepared. Food must be placed near the hens when 14 they are sitting, so that, being well satisfied, they may be more eager to remain on their nests and may not wander too far away and let the eggs grow cold. Though the hens themselves turn the eggs with their feet, the keeper of the poultry, when the hens have leaped down, should go round and turn the eggs by hand, so that they may easily be quickened, receiving heat equally all over, and also that he may remove any eggs which have been damaged or broken by the hen's claws. After doing this for eighteen days, on the nineteenth he should look and see whether the chickens have broken through the eggs with their little beaks and listen whether they are peeping; for often, because of the thickness of the shells, they cannot break their way out. He will, therefore, 15 have to remove with his hand the chickens which are stuck in the shell and put them under their mother to be kept warm, and he should do this for not more than three days, for after the twenty-first day the eggs which are silent have no living creature in them

bus: eaque removenda sunt, ne incubans inani spe diutius detineatur[1] effeta. Pullos autem non oportet singulos, ut quisque natus sit, tollere, sed uno die in cubili sinere cum matre, et aqua ciboque 16 abstinere, dum omnes excludantur. Postero die, cum grex fuerit effectus, hoc modo deponatur.[2] Cribro viciario, vel etiam loliario, qui iam fuerit in usu, pulli superponantur, deinde puleii[3] surculis fumigentur. Ea res videtur prohibere pituitam, 17 quae celerrime teneros interficit. Post haec cavea cum matre claudendi sunt, et farre hordeaceo cum aqua incocto, vel adoreo farre vino resperso modice alendi. Nam maxime cruditas vitanda est: et ob hoc iam tertia die cavea cum matre continendi sunt, priusque quam emittantur ad recentem cibum, singuli tentandi, ne quid hesterni habeant in gutture. Nam nisi vacua est ingluvies, cruditatem significat, 18 abstinerique debent, dum concoquant. Longius autem non est permittendum teneris evagari, sed circa caveam continendi sunt, et farina hordeacea pascendi dum corroborentur: cavendumque ne a serpentibus adflentur, quarum odor tam pestilens est, ut interimat universos. Id vitatur saepius incenso cornu cervino, vel galbano, vel muliebri capillo: quorum omnium fere nidoribus praedicta pestis

[1] retineatur *SAc*: detineatur *a*.
[2] deponatur *SAa*: deponitur *c edd.*
[3] pulei *codd.*: pulegii *edd.*

---

[a] *Mentha pulegium.*
[b] See note on p. 260.

and must be removed, so that the hen may not be kept sitting any longer after the hatching is over, deluded by vain hope. Chickens should not be removed one by one as they are hatched but should be allowed to remain in the nest for one day with their mother and should be kept without water or food until they are all hatched. On the next day, when the brood is complete, it should be brought down from the nest in the following manner. The chickens should be placed in a sieve made of vetch or darnel, which has already been in use, and they should then be fumigated with sprigs of pennyroyal [a]; this seems to prevent the pip, which very quickly kills them when they are young. After this they must be shut up in a coop with their mother and given a moderately large feeding of boiled barley-flour with water or flour of two-grained wheat sprinkled with wine. For above all things indigestion must be avoided, and so on the third day they should be kept in the coop with their mother and before they are let out for fresh food, they should each be examined separately to see if they still have any of the previous day's food in their gorge; for if the crop is not empty, this is a sign of indigestion and they ought to be kept away from food until digestion has taken place. While they are very young, chickens should not be allowed to wander too far but should be kept in the neighbourhood of the coop and fed on barley-meal until they are strong, and care must be taken that they are not breathed upon by snakes, whose odour is so pestilential that it kills them all off. This is prevented by frequently burning hart's-horn or galbanum [b] or women's hair; by the fumes from all these things the aforesaid pest is generally kept

16

17

18

19 submovetur. Sed et curandum erit, ut tepide habeantur. Nam nec calorem nec frigus sustinent. Optimumque est intra officinam clausos haberi cum matre, et post quadragesimum diem potestatem vagandi fieri. Sed primis quasi infantiae diebus pertractandi sunt, plumulaeque sub cauda clunibus [1] detrahendae, ne stercore coinquinatae durescant et
20 naturalia praecludant. Quod quamvis caveatur, saepe tamen evenit, ut alvus exitum non habeat. Itaque pinna pertunditur, et iter digestis cibis praebetur.

Saepe etiam iam [2] validioribus factis, atque ipsis matribus etiam vitanda pituitae [3] pernicies erit. Quae ne fiat, mundissimis vasis et quam purissimam praebebimus aquam: nec minus gallinaria semper fumigabimus, et emundata stercore liberabimus.
21 Quod si tamen pestis permanserit, sunt qui micas [4] alii tepido madefaciunt oleo et [5] faucibus inserant. Quidam hominis urina tepida rigant ora, et tamdiu comprimunt, dum eas amaritudo cogat per nares emoliri pituitae nauseam.[6] Uva [7] quoque, quam Graeci *ἀγρίαν σταφυλὴν* vocant, cum cibo mixta prodest; [8] vel eadem pertrita, et cum aqua potui data.
22 Atque haec remedia mediocriter laborantibus adhibentur. Nam si pituita circumvenit oculos, et iam cibos avis respuit, ferro rescinduntur genae, et coacta sub oculis sanies omnis exprimitur: atque ita paulum

[1] clunibus *a* : crunibus *SA*.
[2] saepe etiam iam *SA* : saepe iam etiam *a* : sed etiam iam *c*.
[3] pituitae *om. SA*.
[4] micas *SA* : spicas *ac*.
[5] madefaciant oleo et *SA* : madefactas oleo *ac*.
[6] nauseam *edd.* : nausa *SA* : nausea *ac*.
[7] uva *edd.* : aqua *SA* : una *ac*.
[8] prodest *ac* : prodent *SA*.

away. Care will also have to be taken that they are kept moderately warm; for they do not bear extreme heat or cold. It is best that they should be kept shut up in the hen-house with their mother and be given full liberty to wander abroad only after forty days. But in the first days of what may be called their infancy they should be held in the hands and the little feathers under their tails should be plucked from their buttocks, lest they become befouled with dung and grow hard and so block the natural passages. It often happens, however, in spite of the precautions taken, that the bowels have no exit; a perforation is, therefore, made and a passage thus opened for the digested food. 19 20

Often too when the chickens have already grown stronger they will have to avoid the fatal disease of the pip, as also will their mothers. To prevent it, we shall give them the purest possible water in the cleanest possible vessels, and we shall also frequently fumigate the hen-houses and keep them cleansed from dung. Some people, if the pestilence persists, moisten morsels of garlic with warm oil and insert them in their throats. Others wet their mouths with warm human urine and keep them closed until the bitter taste of the urine forces them to expel through their nostrils the nauseous matter produced by the pip. The berry also, which the Greeks call the "wild grape," is beneficial mixed with their food, or else pounded up and given them in water to drink. These remedies are given only to those who are suffering just to a slight degree; if the pip surrounds the eyes and the fowl now rejects its food, its cheeks are cut with a lancet and all the diseased matter collected under the eyes is pressed out, and then a little 21 22

23 triti salis vulneribus infricatur.[1] Id porro vitium maxime nascitur cum frigore et penuria cibi laborant aves: item cum per aestatem consistens in cohortibus aqua potatur:[2] item cum ficus aut uva immatura nec ad satietatem permissa est, quibus scilicet cibis abstinendae sunt aves: eosque ut fastidiant efficit uva labrusca de vepribus immatura lecta, quae cum farre triticeo[3] minuto cocta obicitur esurientibus, eiusque sapore offensae aves omnem aspernantur uvam. Similis ratio est etiam caprifici, quae dococta cum cibo praebetur avibus, et ita fici fastidium creat.

24 Mos quoque, sicut in ceteris pecudibus, eligendi quamque optimam et deteriorem vendendi, servetur[4] etiam in hoc genere, ut[5] per autumni tempus omnibus annis, cum fructus carum cessat, numerus quoque minuatur. Submovebimus autem veteres, id est, quae trimatum excesserunt: item quae[6] aut parum fecundae, aut parum bonae[7] nutrices sunt, et praecipue quae ova vel sua vel aliena consumunt: nec minus, quae velut mares[8] cantare atque etiam calcare[9] coeperunt: item serotini pulli, qui ab solstitio nati capere iustum incrementum non potuerunt. In masculis autem non eadem ratio servabitur; sed tamdiu custodiemus generosos,

25 quamdiu feminas implere potuerint. Nam rarior est in his avibus mariti bonitas. Eodem quoque tempore

[1] infricatur *c*: infricantur *Aa*: infriantur *S*.
[2] cohortibus aqua potatur *ac*: cohortibus fuit aqua *SA*.
[3] ordeo triticeo *ac*: hordeo tritico *A*: hordeo trittico *S*.
[4] *post* servetur *add.* ne *SA*.
[5] ut *om. SA*.
[6] itemque aut parum *S*.
[7] parum bonae *om. SA*.
[8] velut mares *ac*: vel mane *SA*.

pounded salt is rubbed into the wounds. Further, 23 this disease chiefly arises when the fowls are suffering from the cold and from poor feeding, and also when, during the summer, water standing in the poultry-yard is drunk, and, again, when they are allowed to eat figs and unripe grapes and not to take their fill of them, foods from which fowls should certainly be kept away. A method of making them loathe them is to pick the wild grapes from the bushes while they are still unripe and put them before them when they are hungry cooked with fine wheat-meal, for being disgusted by the taste the birds refuse every kind of grape. A similar method can be employed also with the wild-fig, which being cooked with their food and given to the birds, creates a distaste for figs also. A 24 practice too, which is employed for all other live-stock, of choosing the better and selling the worse should be observed also in the case of poultry, in order that annually during the autumn, when they cease to be productive, their number may be diminished. We shall get rid of the old hens, that is, those which are more than three years old, also those which are not very prolific or are not very good nurses, and, above all, those which eat their own and other hens' eggs, likewise also those which are beginning to crow like cocks or even to strut about, and also late-born chickens, which have been hatched from the solstice onwards and could not reach their full growth. The same system will not be observed for the cock-birds, but we shall keep those which are well-bred as long as they can impregnate the hens; for good quality in a 25 mating male is rather rare among these birds. Also at

[9] atque etiam calcare *om. SA* : atque calcare *a* : aut etiam calcare *c*.

cum parere desinent aves, id est, ab idibus Novembribus pretiosiores cibi subtrahendi sunt, et vinacea praebenda, quae satis commode pascunt, adiectis interdum tritici excrementis.

VI. Ovorum quoque longioris temporis custodia non aliena est huic curae: quae commode servantur per hiemem, si paleis obruas, aestate, si furfuribus. Quidam prius trito sale sex horis adoperiunt: deinde eluunt, atque ita paleis aut furfuribus obruunt. Nonnulli solida, multi etiam fresa faba coaggerant: alii salibus integris adoperiunt: alii muria tepefacta 2 durant. Sed omnis sal,[1] quemadmodum non patitur putrescere, ita minuit ova, nec sinit plena permanere: quae res ementem deterret. Itaque ne in muriam quidem qui demittunt, integritatem ovorum conservant.

VII. Pinguem quoque facere gallinam, quamvis fartoris, non rustici sit officium, tamen quia non aegre contingit, praecipiendum putavi. Locus ad hanc rem desideratur maxime calidus, et minimi luminis, in quo singulae caveis angustioribus vel sportis inclusae pendeant aves, sed ita coarctatae, ne versari 2 possint. Verum habeant ex utraque parte foramina: unum, quo caput exseratur; alterum, quo cauda clunesque, ut et cibos capere possint et eos digestos sic edere, ne stercore coinquinentur. Substernatur

[1] omnis sal *ac*: omnes salis *S*: omnes es salis *A*[1]: omne sal *A*[2].

the time when the hens cease to lay, that is, from the 13th of November, the more expensive food must be withheld and grape-husks be supplied, which form quite a suitable diet, if refuse from wheat is added from time to time.

VI. The keeping of eggs over a longer period is also germane to the subject which we are now considering. In winter they are conveniently preserved if you bury them in chaff, in summer if you put them in bran. Some people cover them first for six hours with pounded salt; next they wash them and then bury them in chaff or bran. Some people cover them with a heap of whole beans, many with a heap of bruised beans; others bury them in unpounded salt: others harden them in lukewarm brine. But salt in any form, although it does not allow the eggs to rot, shrinks them and prevents them from remaining full: and this is a deterrent to the purchaser. Thus even those who plunge the eggs in brine do not completely preserve their original condition.

Of eggs.

2

VII. Although it is the business of the poulterer rather than of the farmer to fatten hens, yet, since it is not a difficult task, I thought that I ought to give directions on the subject. A spot is required for this purpose which is very warm and has very little light, where the birds may be hung, shut up each separately in rather narrow coops or plaited cages and confined in so close a space that they cannot turn round. They should, however, have holes on either side, one through which they can put out their head and the other through which they can put out their tail and hind-quarters, so that they may be able both to take their food and also get rid of it when it has been digested and so may not be befouled with dung.

On fattening hens.

2

autem mundissima palea, vel molle fenum, id est, cordum. Nam si dure cubant, non facile pinguescunt. Pluma omnis e capite et sub alis atque clunibus detergetur: illic, ne pediculum creet; hic, ne stercore loca naturalia exulceret.

3 Cibus autem praebetur hordeacea farina, quae cum est[1] aqua conspersa et subacta, formantur offae, quibus aves saginantur.[2] Eae[3] tamen primis diebus dari parcius debent, dum plus concoquere consuescant. Nam cruditas vitanda[4] est maxime, tantumque praebendum, quantum digerere possint: neque ante recens admovenda est, quam tentato gutture 4 apparuerit nihil veteris escae remansisse. Cum deinde satiata est avis, paululum deposita cavea[5] dimittitur, sed ita ne vagetur, sed potius, si quid est quod eam stimulet aut mordeat, rostro persequatur. Haec fere communis est cura farcientium. Nam illi qui volunt[6] non solum opimas, sed etiam teneras aves efficere, mulsea recente[7] aqua praedicti generis farinam conspergunt, et ita farciunt: nonnulli tribus aquae partibus unam boni vini miscent, madefactosque triticeo pane obesant avem; quae prima luna (quoniam id quoque custodiendum est) saginari 5 coepta, vicesima pergliscit. Sed si fastidiet cibum, totidem diebus minuere oportebit, quot iam farturae

[1] est *om. Sa.*
[2] saginantur *edd.*: salivatur *codd.*
[3] haec(?) *A*: eae *ac*: ita *S.*
[4] vitanda A^2ac: crudanda SA^1.
[5] deposita cavea *ac*: -ae -ae *SA.*
[6] volunt *ac*: colunt *SA.*
[7] mulsea recente *ac*: multa regenti *S*: multa recentia *A.*

Very clean chaff should be spread under them or soft hay, that is, hay of the second crop; for if their bed is hard they do not easily fatten. All the feathers should be cleared away from their heads and under their wings and hind-quarters, from the head and wings so that they may not breed lice, and from their hind-quarters so that sores may not be caused by dung in the private parts.

Barley-meal is given as food, which, sprinkled with 3 water and kneaded, is formed into pellets with which the birds are crammed. They should, however, be given somewhat sparingly for the first few days, until they become accustomed to digest more of this food; for indigestion must above all things be avoided and only as much given them as they can assimilate; nor ought fresh food be put before them until it is apparent, from feeling the crop, that none of the old food has remained behind. Then, when the bird has 4 had its fill, the coop is lowered a little and the bird is let out, not in order that it may wander at will but rather that it may pursue with its beak anything that stings or bites it. The latter is the common precaution taken by fatteners of birds: but those who wish to make the birds not only plump but also tender, sprinkle meal of the kind already mentioned with fresh honey-water and then cram them with it. Some people mix one part of good wine with three parts of water and fatten the bird with wheaten-bread soaked in it. If the process of cramming is begun at the new moon (for this date too should be observed), the fowl is quite fat by the twentieth day: but, if it takes a dislike to its food, you will have to 5 lessen the amount for the same number of days as the cramming has already proceeded, but only provided

processerint: ita tamen, ne tempus omne opimandi quintam et vicesimam lunam superveniat. Antiquissimum est autem maximam quamque avem lautioribus epulis destinare. Sic enim digna merces sequitur operam et impensam.

VIII. Hac eadem ratione palumbos columbosque cellares pinguissimos facere contingit: neque est tamen in columbis farciendis tantus reditus, quantus in educandis. Nam etiam horum possessio non abhorret a cura boni rustici. Sed id genus minore tutela pascitur longinquis regionibus, ubi liber egressus avibus permittitur: quoniam vel summis turribus, vel editissimis aedificiis assignatas sedes frequentant patentibus fenestris, per quas ad requirendos cibos 2 evolitant. Duobus tamen aut tribus mensibus acceptant conditiva cibaria, ceteris se ipsas pascunt seminibus agrestibus. Sed hoc suburbanis locis facere non possunt, quoniam intercipiuntur variis aucupum insidiis. Itaque clausae intra tectum pasci debent, nec in plano villae loco, nec in frigido: sed in edito fieri tabulatum oportet, quod aspiciat 3 hibernum meridiem. Eiusque parietes, ne iam dicta iteremus, ut in ornithone praecepimus, continuis cubilibus excaventur: vel si non ita competit, paxillis adactis tabulae superponantur, quae vel loculamenta,[1] quibus nidificent aves, vel fictilia columbaria recipiant, praepositis vestibulis, per quae ad cubilia

[1] loculamenta *c* : locum lamenta *SAa*.

[a] *I.e.* due south.

that the whole period of fattening does not go beyond the twenty-fifth day of the lunar period. It is very important that all the biggest fowls should be reserved for the more sumptuous feasts; for thus a worthy recompense attends one's trouble and expense.

VIII. The same method is successfully employed to make wood-pigeons and house-pigeons that live in dovecots very plump; there is, however, not so much profit in cramming pigeons as in just rearing them; for mere possession of them is not unworthy of the attention of a good farmer. The feeding of this kind of bird too requires less supervision in distant parts of the country where they can be allowed free egress, for they frequent the haunts assigned to them on the tops of towers or on very lofty buildings with ever-open windows through which they fly forth to seek their food. Nevertheless for two or three months in the year they welcome food from the store-house, while during the other months they feed themselves on seeds picked up in the fields. But in regions near a city they cannot do this because they are caught by the various snares of the bird-catchers. They ought, then, to be shut up and fed under cover; and on the farm they should not be kept in a part of the farm-house which is level with the ground or cold, but a loft should be constructed for them in an elevated position to face the midday sun in winter;[a] and, that we may not repeat the instructions already given, the walls, as we described in speaking of the hen-house, should be hollowed to form a row of sleeping-places: or, if this is not convenient, pegs should be driven into the walls and boards placed upon them to hold lockers, in which the hens may nest, or earthenware dovecots with porches in front of them through which

Pigeons.

2

3

perveniant. Totus autem locus et ipsae columbarum cellae poliri debent albo tectorio, quoniam eo 4 colore praecipue delectatur hoc genus avium. Nec minus extrinsecus levigari parietes,[1] maxime circa fenestram: et ea sit ita posita, ut maiore parte hiberni diei solem[2] admittat, habeatque appositam satis amplam caveam retibus emunitam, quae excludat accipitres, et recipiat egredientes ad apricationem columbas, nec minus in agros emittat matrices, quae ovis vel pullis incubant, ne quasi gravi perpetuae 5 custodiae servitio contristatae senescant. Nam cum paulum circa aedificia volitaverint, exhilaratae recreantur, et ad fetus suos vegetiores redeunt, propter quos ne longius quidem evagari aut fugere conantur.

Vasa, quibus aqua praebetur, similia esse debent gallinariis, quae colla bibentium admittant, et cupientes lavari propter angustias non recipiant. Nam id facere eas nec ovis nec pullis, quibus plerum-6 que incubant, expedit. Ceterum cibos iuxta parietem conveniet spargi, quoniam fere partes[3] eae columbarii carent stercore. Commodissima cibaria putantur vicia, vel ervum, tum etiam lenticula, miliumque et lolium, nec minus excreta tritici, et si qua sunt alia legumina, quibus etiam gallinae aluntur. Locus autem subinde converri et emundari debet. Nam quanto est cultior, tanto laetior avis conspici-

[1] parietes *Ac*: paries *S*: parientes *a*.
[2] solem *a*: solis *SA*: *om. c*.
[3] quiartesaeae *S*: qui artesae *A*: quoniam fere partes *a*: quam fere parietes *c*.

they may reach their sleeping-quarters. The whole place and the pigeon-cells themselves ought to be finished off with white plaster, since birds of this kind take a special pleasure in that colour; also the walls 4 ought to be made smooth outside, particularly round the window, which should be so placed as to admit the sun for the greater part of a winter's day and should have adjoining it a fairly large pen, protected by nets to keep out hawks, which may accommodate the doves when they come out to bask in the sun; through this also the mother-birds, which are sitting on their eggs or their squabs, can be let out into the fields, so that they may not become prematurely aged through the depression caused by the grievous servitude of perpetual imprisonment; for when they 5 have fluttered about a little round the farm-buildings, they are exhilarated and refreshed and return invigorated to their young, for whose sake they make no attempt to wander far afield or escape by flight.

The vessels in which water is provided should be like those used for fowls, so constructed as to admit the necks of those which drink from them and too narrow to allow the entrance of those which wish to wash in them; for to do so is not good either for the eggs or the young, sitting on which they spend most of their time. It will be found a good plan that 6 their food should be scattered near the wall, since generally those parts of the dove-house are free from dung. Vetch or bitter-vetch and next in order lentils and millet and darnel are considered to be the most suitable foods, likewise the refuse from wheat, also any other kinds of pulse on which hens too are fed. The place ought to be swept and cleaned out from time to time; for the better it is looked after, the more

tur, eaque tam fastidiosa est, ut saepe sedes suas perosa, si detur avolandi potestas, relinquat.[1] Quod frequenter in his regionibus, ubi liberos habent egressus, accidere solet. Id ne fiat, vetus est Democriti praeceptum. Genus accipitris tinnunculum [2] vocant rustici, qui [3] fere in aedificiis nidos facit. Eius pulli singuli fictilibus ollis conduntur, spirantibusque opercula superponuntur, et gypso lita vasa in angulis columbariis suspenduntur: [4] quae res avibus amorem loci sic conciliat, ne unquam deserant.

7

Eligendae vero sunt ad educationem neque vetulae, nec nimium novellae,[5] sed corporis maximi: curandumque, si fieri possit, ut pulli, quemadmodum exclusi sunt, nunquam separentur. Nam fere si sic maritatae [6] plures educant fetus. Sin aliter, certe nec alieni generis [7] coniungantur, ut Alexandrinae et Campanae.[8] Minus enim impares suas [9] diligunt, et ideo nec multum ineunt, nec saepius fetant. Plumae color non semper, nec omnibus idem probatus est: atque ideo qui sit optimus, non facile dictu est. Albus, qui ubique volgo conspicitur, a quibusdam non nimium laudatur; nec tamen vitari debet in his, quae clauso [10] continentur. Nam in vagis maxime

8

9

[1] relinquat *ac* : relinquant *SA*.
[2] tinnunculum *edd.* : titiunculum *codd.*
[3] qui *add. edd.*
[4] suspenduntur *c* : superponuntur *SAa*.
[5] nec nimium novellae *om. SA*.
[6] sic maritate *ac* : si marite *SA*.
[7] alieni generis *Sa* : aliendi generis *A* : alienigene *c*.
[8] alexandrina campane *SA* : alexandrine nec campane *ac*.
[9] impares suas *Ursinus* : pares suos *codd.*
[10] clauso *Aac* : cluso *S*.

cheerful is the appearance of the bird, and so squeamish is it that it often takes a dislike to its own home and abandons it if it is given the opportunity to fly away. This is wont to happen often in districts where the birds are allowed free egress. For the 7 prevention of such an escape, there is an ancient precept of Democritus. There is a kind of hawk which the country-folk call a *tinnunculus* (kestrel) and which generally makes its nest in buildings. The young of this bird are enclosed separately in earthenware pots, and while they are still breathing, lids are put over the pots which are smeared with plaster and hung up in the corners of the pigeon-houses. This induces in the birds such a love for the place that they never desert it.

For the rearing of the young chicks female birds must be chosen which are neither old nor too young, but they should be very large, and care must be taken that, if possible, the chicks should never be separated but be kept together as they were hatched; for if this principle is observed in mating them, they generally rear larger broods. If this is not done, at 8 any rate birds of different breeds, for example the Alexandrine and the Campanian, should not be mated; for they feel less affection for hen-birds unlike themselves and so have little intercourse with them and do not often produce offspring. The same colour of plumage is not approved always or by everybody; it is, therefore, not easy to say which is the best. White, which is generally to be seen 9 everywhere, is not very highly commended by some people; it should not, however, be avoided for birds which are kept in confinement, but for those which wander freely it is much to be con-

est improbandus, quod eum facillime speculatur accipiter.

Fecunditas autem, quamvis longe minor sit quam est gallinarum, maiorem tamen refert quaestum. Nam et octies anno pullos educat, si est bona matrix; et pretiis eorum dominicam[1] complent arcam, sicut eximius auctor M. Varro nobis affirmat, qui prodidit etiam illis severioribus temporibus paria singula 10 milibus singulis sestertiorum solita venire.[2] Nam nostri pudet seaculi, si credere volumus, inveniri qui quaternis milibus nummorum binas aves mercentur. Quamquam vel hos magis tolerabiles putem, qui oblectamenta deliciarum possidendi habendique causa gravi aere et argento pensent, quam illos qui Ponticum Phasim et Scythica stagna Maeotidis eluant.[3] Iam nunc Gangeticas et Aegyptias aves temulenter eructant.

11 Potest[4] tamen etiam in hoc aviario, sicut dictum est, sagina exerceri. Nam si quae steriles aut sordidi coloris interveniunt, similiter ut gallinae farciuntur. Pulli vero facilius sub matribus pinguescunt, si iam firmis, prius quam subvolent, paucas detrahas pinnas, et obteras crura, ut uno loco quiescant, praebeasque copiosum cibum[5] parentibus,[6] quo et se et eos 12 abundantius alant. Quidam leviter obligant crura,

[1] dominicam *SAac* : domini *edd.*
[2] venire *edd.* : veniri *codd.*
[3] eluant *edd.* : heluat *codd.*
[4] potest *ac* : pontes *SA*.
[5] copiosum cibum *ac* : copiosus cibum *S* : copiosus cibus *A*.
[6] parentibus *A* : parientibus *Sac*.

---

[a] *R.R.*, 7. 10.
[b] The Rion, flowing into the Black Sea from the east.

demned, because it is very easily espied by a hawk.

Fecundity in pigeons, though it is much less than in hens, yet brings in greater profit; for a pigeon, if it is a good breeder, rears eight broods in the year, and so pigeons fill the coffers of their owners with the prices which their young command, as that excellent writer Marcus Varro[a] assures us, who has recorded that, even in those more austere times, a single pair used to be sold for 1,000 sesterces. It 10 makes us blush for the present generation, if we are willing to believe that people can be found to pay 4,000 *nummi* for a pair of birds, though I should regard those people who pay great sums in copper and silver for the pleasure which their pets give them merely because they own and possess them, as less insufferable than those who clear of all their birds the river Phasis[b] in Pontus and the pools of Lake Maeotis[c] in Scythia; nay, they are now in their drunkenness belching forth birds brought from the Ganges and from Egypt.

Nevertheless, the fattening process can also be 11 carried out in this pigeon-house, as has already been said; for if any barren or badly-coloured pigeons occur, they are crammed in the same manner as hens. Young pigeons indeed are more easily fattened under their mothers' care, if when they are already strong but before they begin to fly, you pull out a few of their wing-feathers and crush their legs, that they may remain quiet in one spot, and give plenty of food to the parent-birds with which they may feed themselves and their young more abundantly. Some people bind their legs loosely together, 12

[c] The Sea of Azov in South Russia.

quoniam si frangantur, dolorem, et ex eo maciem fieri putant. Sed nihil ista res pinguitudinis efficit. Nam dum vincula exerere conantur, non conquiescunt: et hac quasi exercitatione corpori nihil adiciunt. Fracta crura non plus quam bidui, aut summum tridui dolorem afferunt, et spem tollunt evagandi.

IX. Turturum educatio supervacua est: quoniam id genus [1] in ornithone nec parit nec excludit. Volatura ita ut capitur, farturae destinatur: eoque leviore cura, quam ceterae aves saginatur: verum non omnibus temporibus.[2] Nam per hiemem, quamvis adhibeatur opera, difficulter crescit,[3] et tamen, quia maior est turdi copia, pretium turturum minuitur.

2 Rursus aestate vel sua sponte, dummodo sit facultas cibi, pinguescit. Nihil enim aliud, quam obicitur esca, sed praecipue milium: nec quia tritico vel aliis frumentis minus crassescat;[4] verum quod semine huius maxime delectatur.[5] Hieme tamen offae panis vino madefactae, sicut etiam palumbos, celerius opimant quam ceteri cibi.

3 Receptacula non tanquam columbis loculamenta, vel cellulae cavatae efficiuntur,[6] sed ad lineam mutuli per parietem defixi tegeticulas cannabinas accipiunt, praetentis retibus, quibus prohibeantur volare: quo-

[1] id genus *ac*: ingenuus *SA*.
[2] temporibus *om. SA*.
[3] crescit *SA*: gliscit *ac*.
[4] crassescat *SA*: -ant *a*.
[5] delectatur *SA*: -antur *ac*.
[6] efficiunt *Aac*: fiunt *S*.

because they think that if they are broken, pain, and consequently emaciation, is caused; but doing so does not contribute at all to their fattening, for, while they are trying to get rid of their bonds, they are never at rest, and by this kind of exercise, as it were, they add nothing to their bulk. Broken legs cause pain for not more than two or at most three days and deprive them of all hope of wandering abroad.

IX. The rearing of turtle-doves is of no benefit, because this kind of bird neither lays eggs nor hatches its young in an aviary. A flight of them is ready for cramming in the condition in which it is caught, and can on this account be crammed with less trouble than any other bird, not, however, at every time of year. For in the winter, in spite of all the trouble spent upon them, it is difficult to make them grow, and yet the price of turtle-doves is lessened owing to the greater abundance of thrushes. During the summer, on the other hand, the turtle-dove grows fat even of its own accord, provided it has easy means of getting food. Indeed it is only a question of putting food in its way, especially millet, not that it grows less fat on wheat or other cereals but because it takes the greatest pleasure in millet-seed. In winter, however, pellets of bread soaked in wine fatten turtle-doves as well as wood-pigeons more quickly than any other food.

Turtle-doves.

2

People do not construct either pigeon-boxes or hollow cells as receptacles for turtle-doves as for wood-pigeons, but brackets are fixed in a row along a wall and hold small hempen mats with nets spread in front of them, so that the birds are prevented from flying about, because, if they do so, they lose

3

niam si id faciant, corpori detrahunt. In his assidue pascuntur milio aut tritico, sed ea semina dari nisi sicca non oportet. Satiatque semodius cibi in diebus 4 singulis vicenos et centenos turtures. Aqua semper recens et quam mundissima vasculis, qualibus columbis atque gallinis, praebetur; tegeticulaeque emundantur, ne stercus urat pedes, quod tamen et ipsum diligenter reponi debet ad cultus agrorum arborumque, sicut et omnium avium, praeterquam nantium. Huius avis aetas ad saginam non tam vetus est idonea quam novella. Itaque circa messem, cum iam confirmata est pullities, eligitur.

X. Turdis maior opera et impensa praebetur, qui omni quidem rure, sed salubrius in eo pascuntur, in quo capti sunt. Nam difficulter in aliam regionem transferuntur, quia caveis clausi plurimi despondent: quod faciunt etiam cum eodem momento temporis a rete in aviaria coniecti sunt. Itaque ne id accidat, veterani debent intermisceri, qui ab aucupibus in hunc usum nutriti quasi allectores sint captivorum, maestitiamque eorum mitigent intervolando. Sic enim consuescent et aquam et cibos appetere feri, si 2 mansuetos id facere viderint. Locum aeque munitum et apricum, quam columbi desiderant: sed in eo transversae perticae perforatis parietibus adversis aptantur, quibus insideant, cum satiati cibo requi-

bulk. Here they are constantly fed with millet or wheat; but the grain must not be given them unless it is dry. Half a *modius* of food every day easily satisfies a hundred and twenty turtle-doves. The 4 purest possible water is always provided in vessels such as are used for pigeons and hens. The mats are kept clean so that the dung does not burn their feet, and the dung should itself be carefully set aside for the cultivation of the fields and trees, as also that of all birds except those which swim. This bird is not so suitable for cramming when it is old as when it is young, and so the choice is made about harvest-time when the young brood has already gained strength.

X. Still more labour and expense is spent on thrushes, which are kept in every country district, but, with greater advantage to their health, in that in which they have been caught; for there are difficulties about moving them elsewhere, because, when they are shut up in cages, most of them become despondent; indeed they do so when they are instantaneously hurled from the net into the aviaries. So, to prevent this, some old thrushes ought to be mixed with them which, having been brought up by the fowlers for this purpose, may serve as decoys for the captives and may mitigate their distress by flying in among them. For in this way wild birds will become used to seeking both their water and their food when they have seen the tame birds doing so. They require a place as well protected and as 2 sunny as wood-pigeons need, but transverse poles are fixed in it fitted into holes pierced in the walls which face one another, on which they may perch when they have had their fill of food and wish to rest.

Thrushes.

escere volunt. Eae perticae non altius a terra debent sublevari, quam hominis statura patitur, ut a stante 3 contingi possint. Cibi ponuntur fere partibus his ornithonis, quae super se[1] perticas non habent, quo mundiores permaneant. Semper autem arida ficus diligenter pinsita et permixta polline praeberi debet, 4 tam large quidem ut supersit. Hanc quidam mandunt, et ita obiciunt. Sed istud in maiore numero facere vix expedit, quia nec parvo conducuntur qui mandant, et ab his ipsis aliquantum propter iucunditatem consumitur. Multi varietatem ciborum, ne unum fastidiant, praebendam putant; ea est, cum 5 obiciuntur myrti et lentisci semina; item oleastri et ederaceae baccae,[2] nec minus arbuti.[3] Fere enim etiam in agris ab eiusmodi volucribus haec appetuntur, quae in aviariis[4] quoque desidentium[5] detergent fastidia, faciuntque avidiorem volaturam, quod maxime expedit. Nam largiore cibo celerius pinguescit. Semper tamen etiam canaliculi milio repleti apponuntur, quae est firmissima esca. Nam illa 6 quae supra diximus, pulmentariorum vice dantur. Vasa, quibus recens et munda praebeatur aqua, non dissimilia sint gallinariis.

Hac[6] impensa curaque M. Terentius ternis saepe denariis singulos emptitatos[7] esse significat avorum

[1] se *ac* : *om. SA*.
[2] ederacee bace *ac* : herecee vace *SA*.
[3] arbuti *edd.* : arbusti *codd.*
[4] in aviariis *repetit S*.
[5] desidentium *ac* : sidentur *SA*.
[6] hac *ac* : hanc *SA*.
[7] emptitatos *Sa* : entitatos *A* : eptitatos *c*.

[a] Varro, *R.R.*, III. 2. 15.

These poles ought not to be raised higher from the ground than a man's height allows, so that they may be within his reach when he is standing up. The 3 food is usually placed in those parts of the aviary which have no perches above them, so that it may remain more clean. Dried figs, carefully crushed and mixed with fine flour, ought always to be provided, so abundantly indeed that some is left over. Some people chew a fig and then offer it to the thrushes; 4 but it is scarcely expedient to do this where the number of thrushes is large, because people to chew the figs cost a good deal to hire and themselves eat an appreciable quantity because of the pleasant taste. Many people think that a variety of food ought to be provided, lest the thrushes take a dislike to a single food. This variety consists in putting before them seeds of myrtle and mastic, also wild olive and ivy berries and likewise the fruit of the strawberry-tree, 5 for these are the things for which this kind of bird generally seeks in the fields, and so they do away with the distaste for food which they feel in their idle captivity in the aviaries and make the bird population there more voracious, which is a great advantage; for the more they eat the quicker they get fat. Little troughs, however, full of millet are always placed near them since it is the most solid part of their diet; for the foods which we have mentioned above are given them as relishes. Vessels for the 6 supply of fresh, clean water should be not unlike those for poultry.

Thanks to the expenditure in this way of money and care, so Marcus Terentius informs us,[a] these birds were often bought for three *denarii* a piece in our grandfathers' time, when those who celebrated

temporibus, quibus qui triumphabant populo [1] dabant epulum. At nunc aetatis nostrae luxuria [2] cotidiana fecit haec pretia: propter quae ne rusticis quidem contemnendus sit hic reditus.

Atque ea genera, quae intra saepta villae cibantur, fere persecuti sumus. Nunc de his dicendum est, quibus etiam exitus ad agrestia pabula dantur.

XI. Pavonum educatio magis urbani [3] patrisfamiliae, quam tetrici rustici curam poscit. Sed nec haec tamen aliena est agricolae captantis undique voluptates acquirere, quibus solitudinem ruris eblandiatur. Harum autem decor avium etiam exteros nedum dominos oblectat. Itaque genus alitum nemorosis et parvulis insulis, quales obiacent Italiae, facillime continetur. Nam quoniam nec sublimiter potest nec per longa spatia volitare, tum etiam quia furis ac noxiorum animalium rapinae metus non est, sine custode tuto vagatur, maioremque pabuli partem 2 sibi acquirit. Feminae quidem sua sponte tanquam servitio liberatae studiosius pullos enutriunt: nec curator aliud facere debet, quam ut diei certo tempore, signo dato, iuxta villam gregem convocet, et exiguum hordei concurrentibus obiciat, ut nec avis esuriat, et numerus advenientium recognoscatur.

[1] populo *Aac*: populos *S*.
[2] luxuria *Sc*: luxoriae *A*: luxurie *a*.
[3] urbani *c*: urbanis *SAa*.

triumphs gave a feast to the people. But at the present day luxury has made this their everyday price; wherefore this source of income must not be despised even by farmers.

We have now dealt in general with those kinds of birds which are fed within the precincts of the farm; we must now speak of those which are also given freedom to seek their food in the fields.

Peafowl.

XI. The rearing of peafowl calls for the attention of the city-dwelling householder rather than of the surly countryman; yet it is not alien to the business of the farmer who aims at the acquisition, from every source, of pleasure with which he beguiles the loneliness of country life; and the elegance of these birds delights even strangers, much more their owners. This breed of birds, therefore, can be easily kept on the small wooded islands which lie off the coast of Italy; for since they cannot fly high or over long distances and since too on these islands there is no fear of their being carried off by a thief or by harmful animals, they can safely wander about without anyone to look after them and acquire most of their food for themselves. 2 The hen-birds, finding themselves as it were released from bondage, of their own accord bring up their young with unusual devotion, and the man in charge of them should have nothing to do except, at a fixed time of day, to give the signal and summon the flock to the neighbourhood of the farm and throw down a small quantity of barley before them as they run to meet him, so that the birds may not be hungry and that the number may be verified of those who come to his call.

3 Sed huius possessionis rara conditio est. Quare mediterraneis locis [1] maior adhibenda cura est: eaque sic administretur. Herbidus silvestrisque ager planus sublimi clauditur [2] maceria, cuius tribus lateribus porticus applicantur, et in quarto duae cellae, ut sit altera custodis habitatio, atque altera stabulum pavonum. Sub porticibus deinde per ordinem fiunt arundinea saepta in modum cavearum, quales [3] columbarii tectis superponuntur. Ea saepta distinguuntur velut clatris intercurrentibus calamis, ita 4 ut ab utroque latere singulos aditus habeant. Stabulum autem carere debet uligine, cuius in solo per ordinem figuntur breves paxilli,[4] eorumque partes summae lingulas edolatas habent, quae transversis foratis perticis inducantur.[5] Hae porro quadratae perticae, paxillis superponuntur, ut avem recipiant adsilientem. Sed idcirco sunt exemptiles, ut cum res exigit, a paxillis deductae [6] liberum aditum converrentibus stabulum praebeant.

5 Hoc genus avium, cum trimatum explevit, optime progenerat. Siquidem tenerior aetas, aut sterilis, aut parum fecunda est. Masculus pavo gallinaceam salacitatem habet, atque ideo quinque feminas desiderat. Nam si unam vel alteram fetam saepius compressit, vix dum concepta in alvo vitiat ova, nec ad partum [7] sinit perduci: quoniam immatura geni-

[1] locis *ac*: *om. SA*.
[2] clauditur *a*: cluditur *SAc*.
[3] quales *SA*: qualis *ac*: qualia *Schneider*.
[4] paxilli *ac*: taxilli *SA*.
[5] inducantur *ac*: induantur *SA*.
[6] deductae *Aac*: eductae *S*.
[7] ad partum *ac*: partum *A*: parte *S*.

But the possession of these birds is a rare circumstance and so an unusual amount of care must be exercised in inland districts, and the following procedure must be followed. A flat piece of land covered with grass and trees is enclosed with a high fence to three sides of which galleries are attached, while on the fourth side there are two huts, one for the dwelling-place of the custodian, the other as a peacock-house. Then in the galleries enclosures are made with reeds in a row to form coops such as are placed on the roofs of a pigeon-house. These enclosures are separated from one another by barriers as it were of reeds which run between them, so arranged as to have one entrance on either side. The peacock-house ought to be entirely free from damp, and in the floor short stakes are fixed in a row, the tops of which have carefully hewn tenons for insertion into holes made in the transverse perches. Moreover, these perches which are placed on the top of the stakes are cut square, so that they may give a foothold to a bird when it leaps onto them, but they are made so as to be removable in order that, when it is necessary, they may be detached from the stakes and give free access to those who are sweeping out the peacock-house. 3 4

This kind of fowl, when it has completed its first three years, breeds excellently, but at a tenderer age it is either sterile or not very prolific. The male bird has the salaciousness of the farmyard cock and so requires five hens; for if it frequently covers one or two of them that have been laying, it spoils eggs which are hardly yet formed in the womb and does not allow them to be brought to birth, since they fall out of the genital parts while they are still immature. 5

6 talibus locis excidunt. Ultima parte hiemis concitantibus libidinem cibis utriusque sexus accendenda venus est. Maxime facit ad hanc rem, si favilla levi torreas fabam, tepidamque des ieiunis quinto quoque die. Nec tamen excedas modum sex cyathorum in singulas aves. Haec cibaria non omnibus promiscue[1] spargenda sunt, sed in singulis saeptis, quae arundinibus contexi oportere proposueram, portione[2] servata quinque feminarum et unius maris, ponenda sunt cibaria, nec minus aqua, quae sit idonea potui.
7 Quod ubi factum est, mares sine rixa[3] diducuntur[4] in sua quisque saepta cum feminis, et aequaliter universus grex pascitur. Nam etiam in hoc genere pugnaces inveniuntur masculi, qui et a cibo et a coitu prohibent minus validos, nisi sint hac ratione separati. Fere autem locis apricis ineundi cupiditas exercet mares, cum Favonii spirare coeperunt, id est tempus
8 ab idibus Februariis ante Martium mensem. Signa sunt extimulatae libidinis, cum semetipsum veluti mirantem caudae gemmantibus pinnis protegit: idque cum facit, rotare dicitur.

Post admissurae tempus confestim matrices custodiendae sunt, ne alibi quam in stabulo fetus edant: saepiusque digitis loca feminarum tentanda sunt. Nam in promptu gerunt ova, quibus iam partus appropinquat. Itaque includendae sunt incipientes,[5]
9 ne extra clausum fetum edant: maximeque temporibus his, quibus parturiunt, pluribus stramentis

[1] promiscue *a*: promisce *SA*: permixtae *c*.
[2] proposueram portione *ac*: *om. SA*.
[3] sine rixa *ac*: *om. SA*.
[4] diducuntur *S*: deducuntur *Aac*.
[5] incipientes *SAac*: incientes *Ursinus, Schneider*.

In the last part of the winter the desires of both sexes 6 must be kindled by foods which excite lust. The best means to this end is to toast some beans over embers which are not very hot and give them while still warm to the fowls every fifth day on an empty stomach; but you should not go beyond six *cyathi* to each bird. This food must not be scattered promiscuously to all of them together but must be placed in each of the enclosures, which I had suggested should be made of reeds woven together, a portion having been set aside for five hens and a cock and likewise water which should be suitable for drinking. When this has been done the male birds are driven, 7 without quarrelling, each into its own enclosure together with their hens, and the food is equally distributed over the whole flock. For even among birds of this kind pugnacious males are found which try to deprive those which are weaker than themselves of food and sexual intercourse, if they are not kept apart in this way. Generally in sunny places, when the west winds begin to blow, that is, from the 13th of February until the month of March, a desire for sexual intercourse torments the male birds. It 8 is a sign that a peacock's lust is excited when it covers itself with its bejewelled tail-feathers and seems to be admiring itself; when it does so, it is said to be "forming a wheel."

After the mating season the laying hens must immediately be watched carefully lest they lay their eggs anywhere except in the peacock-house, and the parts of the females must often be felt with the fingers, for, when the time for laying is at hand, they carry their eggs in readiness. When they begin to 9 lay they must be shut up, so that they may not

exaggerandum est aviarium, quo tutius integri fetus excipiantur. Nam fere pavones, cum ad nocturnam requiem venerunt, praedictis perticis insistentes enituntur ova, quae quo propius ac mollius deciderint, illibatam servant integritatem. Quotidie ergo diligenter mane temporibus feturae stabula circumeunda erunt, et iacentia ova colligenda. Quae quanto recentiora gallinis subiecta sunt, tanto commodius excluduntur:[1] idque fieri maxime patris-
10 familias rationi conducit. Nam feminae pavones, quae non incubant, ter anno fere partus edunt: at quae fovent ova, totum tempus fecunditatis aut excludendis aut[2] etiam educandis pullis consumunt. Primus est partus quinque fere ovorum; secundus
11 quattuor; tertius aut trium aut duorum. Neque est quod committatur, ut Rhodiae aves pavoninis incubent, quae ne suos quidem fetus commode nutriunt. Sed veteres maximae quaeque gallinae vernaculi generis eligantur:[3] eaeque novem diebus a primo lunae incremento, novenis ovis incubent, sintque ex his quinque pavonina, cetera gallinacei
12 generis. Decimo deinceps die omnia gallinacea subtrahantur, et totidem recentia eiusdem generis supponantur, ut trigesima luna, quae est fere nova, cum pavoninis excludantur. Sed custodis curam

[1] excuduntur *SAa* : excluduntur *c*.
[2] excludendis aut *edd.* : excudendis aut *ac* : *om. SA*.
[3] eligantur *ac* : religantur *SA*.

produce their eggs outside the enclosure. Above all during the seasons in which they lay, the peacock-house must be piled high with more straw, the better to ensure that the eggs are delivered intact. For usually peahens, having come to seek rest at night, lay their eggs while they are roosting on the perches, which have already been described, and when the eggs have fallen from a lesser height and more softly, they keep their soundness unimpaired. Every day, therefore, during the period of laying you will have to go carefully round the peacock-houses in the early morning and collect the eggs which are lying about, and the fresher they are when they are set under the hen, the better are the prospects of a good hatch, and that this should be done is very much to the householder's advantage. For peahens which do not sit 10 generally produce three lots of eggs during the year, but those which sit spend the whole period of their productivity in either hatching or even rearing their young. The first laying generally consists of five eggs, the second of four, and the third of either three or two. There is no reason for making the mistake 11 of letting Rhodian hens incubate peahens' eggs, since they do not even bring up their own offspring properly; but the biggest veteran farmyard-fowls of our native breed should be chosen and should be put to sit upon nine eggs, five of which should be peahen's and the rest ordinary hen's eggs, nine days after the moon's first increase. Then on the tenth 12 day all the hen's eggs should be removed and the same number of fresh eggs of the same kind substituted, that they may be hatched out with the peahen's eggs on the thirtieth day which is about new moon. But it must not escape the keeper's

non effugiat observare desilientem matricem, saepiusque ad cubile pervenire, et pavonina ova, quae propter magnitudinem difficilius a gallina moventur, versare manu: idque quo diligentius faciat, una pars ovorum notanda est atramento, quod signum habebit

13 aviarius,[1] an a gallina conversa sint. Sed, ut dixi, meminerimus cohortales quam maximas ad hanc rem praeparari. Quae si mediocris habitus sunt, non debent amplius quam terna pavonina et sena generis sui fovere. Cum deinde fecerit pullos, ad aliam nutricem gallinacei debebunt transferri, et subinde qui nati fuerint pavonini ad unam congregari, donec

14 quinque et viginti capitum grex efficiatur. Sed cum[2] erunt editi pulli, similiter ut gallinacei primo die non moveantur: postero die cum educatrice transferantur in caveam: primisque diebus alantur hordeaceo farre vino resperso, nec minus ex quolibet frumento cocta pulticula, et refrigerata. Post paucos deinde dies huic[3] cibo adiciendum erit concisum porrum Tarentinum, et caseus mollis vehementer expressus; nam serum nocere pullis manifestum est.

15 Locustae quoque pedibus ademptis utiles cibandis pullis habentur, atque his pasci debent usque ad sextum mensem: postmodum satis est hordeum de manu praebere. Possunt autem post quintum et trigesimum diem quam nati sunt, etiam in agrum

[1] habiarius *S*. [2] cum *ac*: *om. SA*.
[3] huic *ac*: hic *S*: hoc *A*.

attention to mark the mother-hen when she leaps down and to visit the nest-box frequently and with his hand to turn the peahen's eggs, which on account of their size are more difficult for the farmyard-hen to move; and so that he may carry out this task with greater care, one side of the eggs should be marked with ink and the poultry-man will then have a means of knowing whether the eggs have been turned by the hen. But, as I have said, we must remember that 13 farmyard hens of the greatest possible size are provided for this purpose; and if they are of only moderate build, they ought not to sit upon more than three peahen's eggs and six of their own kind. When the hen has hatched the chickens, the farmyard chickens will have to be transferred to another nurse, and any young peafowls which are hatched from time to time should be collected round one nurse until a flock of twenty-five head is made up. But when the young 14 peafowls are hatched out, on the first day, like farmyard chickens, they should not be moved, but on the following day they should be transferred to a coop with the hen that is to bring them up, and during the first days they should be fed on barley-meal sprinkled with wine and with gruel made from any kind of cereal and allowed to grow cold. Then after a few days a Tarentine leek cut up small should be added to their diet and soft cheese which has been pressed out with great force, for whey is obviously harmful to chickens. Locusts too, whose feet have been re- 15 moved, are regarded as useful for feeding the peachicks and they ought to eat them until the sixth month; afterwards it is enough to give them barley from the hand. After the thirty-fifth day following their birth they may even be quite safely taken out

satis tuto educi, sequiturque grex velut matrem gallinam singultientem. Ea cavea clausa [1] fertur in agrum a pastore, et emissa ligato pede longa linea custoditur, ad quam [2] circumvolant pulli. Qui cum ad satietatem pasti sunt, reducuntur in villam perse- 16 quentes, ut dixi, nutricis singultus.[3] Satis autem convenit inter auctores, non debere alias gallinas, quae pullos sui generis educant, in eodem loco pasci. Nam cum conspexerunt pavoninam prolem, suos pullos diligere desinunt, et immaturos relinquunt, perosae videlicet, quod nec magnitudine, nec specie pavoninis pares sint. Vitia quae gallinaceo generi [4] nocere solent, eadem has aves infestant: sed nec remedia traduntur alia, quam quae gallinaceis adhibentur. Nam et pituita et cruditas, et si quae aliae sunt pestes, iisdem remediis, quae proposuimus, 17 prohibentur. Septimum deinde mensem cum excesserunt, in stabulo cum ceteris ad nocturnam requiem debent includi. Sed erit curandum, ne humi maneant. Nam qui sic cubitant, tollendi sunt, et supra perticas imponendi, ne frigore laborent.

XII. Numidicarum eadem est fere quae pavonum educatio. Ceterum silvestres gallinae, quae rusticae appellantur, in servitute non fetant: et ideo nihil de his praecipimus, nisi ut cibus ad satietatem praebeatur, quo sint conviviorum epulis aptiores.

[1] ea cavea clausa *ac* : ex causam causaque *SA*.
[2] ad quam *ac* : aquam *SA*.
[3] singultus *ac* : singuli *SA*.
[4] gallinaceo generi *ac* : gallinacei generis *SA*.

into a field, and the flock follows the clucking hen as though it were their mother. The latter is shut up in a coop and taken out to the field by the man who feeds them, and when it is let out it is secured by a long line attached to its foot. The chicks flutter round it and, when they have eaten their fill, they are brought back to the farm, following the clucking of their foster-mother, as I have already described. The authorities are pretty well agreed that the other hens which are bringing up chickens of their own kind ought not to be fed in the same place; for when they have seen the little peachicks, they cease to care for their own chickens and abandon them before they reach maturity, evidently hating them because they do not equal the little peachicks either in size or in beauty. 16

The same diseases as usually harm farmyard fowls attack these birds also, and no remedies are applied to them other than those which are administered to ordinary cocks and hens; for the pip and indigestion and any other plagues which occur are checked by the same remedies as we have prescribed. When they have passed the seventh month, they should be shut up with the others in the peacock-house for their night's rest; but care will have to be taken that they do not remain on the ground. Those who go to sleep in this position must be picked up and placed on the perches, so that they may not suffer from the cold. 17

Guinea-fowls and "rustic"-cocks.

XII. The rearing of guinea-fowls is almost the same as that of peacocks. But woodland hens, which are called "rustic"-fowls, do not breed in captivity, and, therefore, we have no instructions to give about them except that they must be given their fill of food, so that they may be better suited for feasts to which guests are invited.

XIII. Venio nunc ad eas aves, quas Graeci vocant ἀμφιβίους, quia non tantum terrestria, sed aquatilia quoque desiderant pabula, nec magis humo quam stagno consueverunt. Eiusque generis anser praecipue rusticis gratus est, quod nec maximam curam poscit, et solertiorem custodiam quam canis praebet. 2 Nam clangore prodit insidiantem, sicut etiam memoria tradidit in obsidione Capitolii, cum adventum[1] Gallorum vociferatus est, canibus silentibus.[2] Is autem non ubique haberi potest, ut existimat verissime Celsus, qui sic ait: anser neque sine aqua, nec sine multa herba facile sustinetur, neque utilis est locis consitis, quia quicquid rerum[3] contingere 3 potest, carpit. Sicubi vero flumen aut lacus est, herbaeque copia, nec nimis iuxta satae fruges, id quoque genus[4] nutriendum est. Quod etiam nos facere censemus, non quia magni sit fructus, sed quia minimi oneris. Attamen praestat ex se pullos atque plumam, quam non, ut in ovibus lanam, semel demetere, sed bis anno, vere et autumno vellere licet. Atque ob has quidem causas, si permittit locorum conditio, vel paucos utique oportet educare, singulisque maribus ternas feminas destinare. Nam propter gravitatem plures inire non possunt. Quinetiam intra cohortem, ut protecti sint, secretas singulis

[1] adventum *SAac* : adventu *edd.*
[2] silentibus *ac* : *om. SA.*
[3] rerum *SA* : tenerum *ac.*

XIII. I now come to those birds which the Greeks call "amphibious," because they require not only food produced from the earth but also that which comes from the water, and have accustomed themselves quite as much to standing water as to the land. Of this type of bird the goose is particularly acceptable to farmers, because it does not demand very much attention and keeps watch more cleverly than a dog, since by its cackling it betrays the presence of anyone who is lying in wait, just as (so history has informed us) when during the siege of the Capitol it was the goose which loudly announced the approach of the Gauls while the dogs kept silence. The goose, however, cannot be kept everywhere, an opinion which Celsus expresses with much truth when he says: "A goose cannot easily be maintained without plenty of water and plenty of grass and is not profitable in closely planted land because it plucks at anything which it can reach; but wherever there is a river or a lake and an abundance of grass and there are not sown crops too near at hand, this kind of bird also should be reared." We, furthermore, are in favour of keeping geese not because it brings a large profit but because it gives very little trouble. Yet it produces goslings and feathers; the latter you may gather not merely once a year, like wool from sheep, but you can pluck twice, in spring and in autumn. Indeed for these reasons, if local conditions permit, you should rear at any rate a few geese and assign three female birds to one male; for because of their weight they cannot couple with more. Moreover, so that they may have protection, separate goose pens should be made for each inside

Amphibious birds.

2

3

[4] id quoque genus *ac*: *om. SA.*

haras faceret oportet,[1] in quibus cubitent et fetus ubi edant.

XIV. Qui vero greges nantium possidere student, chenoboscia [2] constituunt,[3] quae tum demum vigebunt, si fuerint ordinata ratione tali. Cohors ab omni cetero pecore secreta clauditur alta novem pedum maceria, porticibusque circumdata, ita ut in aliqua parte sit cella custodis. Sub porticibus deinde quadratae harae [4] caementis vel etiam laterculis extruuntur: quas singulas satis est habere quoquoversus pedes ternos, et aditus singulos firmis ostiolis munitos: quia per feturam diligenter claudi debent.

2 Extra villam deinde non longe ab aedificio si est stagnum vel flumen, alia non quaeratur aqua: sin aliter, lacus piscinaque manu fiant, ut sint quibus inurinare possint aves. Nam sine isto primordio non magis quam sine terreno recte vivere queunt.[5] Palustris quoque,[6] sed herbidus ager destinetur, atque alia pabula conserantur, ut vicia, trifolium, faenum Graecum, sed praecipue genus intubi, quod *σέριν* [7] Graeci appellant. Lactucae quoque in hunc usum semina vel maxime serenda sunt, quoniam et mollissimum est olus, et libentissime ab his avibus appetitur. Tum etiam pullis utilissima est esca.

3 Haec cum praeparata sunt, curandum est, ut mares feminaeque quam amplissimi corporis et albi coloris

[1] quinetiam—oportet *Schneider*: quin et etiam in rutectis circa chortem secretis angulis haras (aras *A*) facere *SA*: quin etiam intra cohortem protecti sint secretis anglis haras facere *a*: intra cohortem pretecti secretis angulis raras facere *c*.

[2] chenoboscia nam *A*: XHNOBOC nam *S*: *om. ac.*

[3] statuunt *A*.

[4] harae *Aa*: are *c*: habere *S*.

[5] queunt *ac*: nequeunt *SA*.

[6] quodque *S*.

the poultry-yard [a] in which they can rest and where they can lay their eggs.

The housing and feeding of geese.

XIV. Those who desire to possess flocks of swimming birds establish goose-pens, which then will flourish only if they are arranged in the following manner. A yard remote from any other livestock is enclosed by a wall nine feet high and surrounded by porticos so arranged that the keeper's hut may be in some part of them. Then under the porticos square pens are built of unhewn stones or even small bricks. It is enough if each pen measures three feet each way and has a single entrance fitted with strong little doors, because the pens ought to be kept shut when the geese are laying or sitting. 2 If there is a pool or river outside the farm and not far from the building, no other water need be looked for; otherwise a lake and fish-pond should be artificially constructed, so that the geese may have water into which to dive; for they can no more live properly without the element of water than they can without the element of earth. A marshy field too which is also grassy should be set aside for them, and other foods be sown such as vetch, trefoil, fenugreek and above all the kind of endive which the Greek call *seris*.[b] Lettuce seeds in particular should also be sown for this purpose, since it is a very tender vegetable and is also much sought after by these birds; also it is a very useful food for goslings.

3 Having made all these preparations, you must take care that the male and female birds which you choose are of the largest possible size and of a white colour;

[a] The text here is uncertain but the meaning is clear.
[b] Dioscorides, II. 132.

[7] σέριν *edd.*: caepim *S*: cepi *A*: *om. ac.*

eligantur. Nam est aliud genus varium, quod a fero mitigatum domesticum factum est. Id neque aeque fecundum est, nec tam pretiosum: propter quod

4 minime nutriendum est. Anseribus admittendis tempus aptissimum est a bruma; mox ad pariendum, et ad incubandum a Calen. Februariis vel Martiis usque ad solstitium, quod fit ultima parte mensis Iunii. Ineunt autem non, ut priores aves, de quibus diximus, insistentes humi: nam fere in flumine aut piscinis id faciunt: singulaeque ter anno pariunt, si prohibeantur fetus suos excudere,[1] quod magis

5 expedit, quam quum ipsae suos fovent. Nam et a gallinis melius enutriuntur, et longe maior grex efficitur. Pariunt autem singulis fetibus ova, primo quina, sequenti quaterna, novissimo terna: quem partum nonnulli permittunt ipsis matribus educare, quia reliquo tempore anni vacaturae sunt a fetu. Minime autem concedendum est feminis extra saeptum parere, sed cum videbuntur sedem quaerere, comprimendae sunt atque tentandae. Nam si appropinquant partus, digito tanguntur ova, quae

6 sunt in prima parte locorum genitalium. Quamobrem perduci ad haram debent, includique ut fetum edant: idque singulis semel fecisse satis est, quoniam unaquaeque recurrit eodem, ubi primo peperit. Sed novissimo fetu cum volumus ipsas incubare, notandi erunt uniuscuiusque partus, ut suis matribus

[1] excudere *codd.*: excludere *edd.*

for there is another kind which is of various colours and, originally wild, has been tamed and become a domestic bird, but it is not so prolific and commands a lower price, and so should certainly not be reared. 4 The most suitable time for coupling geese is from the height of winter onwards, and then for laying eggs and sitting on them from the first of February or March until the summer solstice, which falls in the last part of the month of June. They couple not standing on the ground, like the birds of whom we dealt before, but generally in a river or pond; and each hen-bird lays a clutch of eggs three times a year if prevented from hatching them out, which is a better plan than if they sit on their own eggs; for the young 5 are better reared by ordinary hens and also the result is a much larger flock. At each laying they produce the following numbers of eggs, at the first five, at the next four and at the last three. Some people allow the geese themselves to rear the last clutch, because for the rest of the year they will be taking a holiday from laying. The female birds must not on any account be allowed to lay outside the enclosure, but, when they seem to be looking for a nesting-place, they must be stopped and must be examined; for if they are near laying, the eggs, which are in the nearest part of the genital organs, can be felt with the finger. Wherefore they ought to be 6 taken to the goose-pen and shut up there so that they may lay their eggs; and it is enough to have done this once with each of them since every one of them returns to the place where it first laid an egg. But, after the last laying, when we wish the geese themselves to sit, the eggs of each will have to be marked so that they may be put under those which

subiciantur: quoniam negatur anser aliena excudere ova, nisi subiecta sua quoque habuerit. Supponuntur autem gallinis huius generis ova, sicut pavonina, plurima quinque, paucissima tria: ipsis 7 autem anseribus paucissima VII, plurima XV. Sed custodiri debet, ut ovis subiciantur herbae urticarum, quo quasi remedio medicantur, ne noceri possit excusis [1] anserculis, quos enecant urticae, si teneros pupugerint. Pullis autem formandis excudendisque triginta diebus opus est, cum sunt frigora: nam tepidis XXV satis est. Saepius tamen anser trigesimo 8 die nascitur. Atque is dum exiguus est, decem primis diebus pascitur in hara clausus [2] cum matre: postea cum serenitas permittit, producitur in prata, et ad piscinas. Cavendumque est, ne aut aculeis urticae compungatur, aut esuriens mittatur in pascuum: sed ante concisis intubis vel lactucae foliis saturetur. Nam si est adhuc parum firmus indigens ciborum pervenit in pascuum, fruticibus aut solidioribus herbis obluctatur ita pertinaciter, ut collum abrumpat. Milium quoque aut etiam triticum mixtum cum aqua recte praebetur. Atque ubi paulum se confirmavit, in gregem coaequalium compellitur, et hordeo alitur: quod et matricibus 9 praebere non inutile est. Pullos autem non expedit plures in singulas haras quam vicenos adici; nec rursus omnino cum maioribus includi, quoniam vali-

[1] excusis *edd.*: excussis *codd.*
[2] clausus *ac*: clausum *SA*.

laid them; for it is said that a goose does not hatch another's eggs unless she has some of her own also beneath her. Goose eggs, like those of peahens, are put under ordinary hens, the maximum numbers being five and the minimum three, whereas a minimum of seven and a maximum of fifteen are put under the geese themselves. But care must be 7 taken, when stalks of nettle (which are used as a remedy to cure disease) are placed under the eggs, that they may not possibly hurt the goslings when they are hatched; for nettles kill them if they sting them when they are quite young. Thirty days are required for the forming and hatching of the goslings when the weather is cold; for when it is warm, twenty-five days are enough, but more often the gosling is hatched on the thirtieth day. While it is 8 quite small, for the first ten days it is shut up with its mother in the pen and fed there; afterwards, when calm weather allows, it is taken out into the meadows and to the ponds. Care must be taken that it is not stung by the prickles of the nettle or sent out hungry to pasture, but that it has had its fill beforehand of chopped endive or lettuce leaves; for if it is still not very strong and arrives hungry at the pasture-ground, it struggles so persistently with shrubs or the tougher plants that it breaks its neck. It is also well to provide it with millet or even wheat mixed with water. When it has become a little stronger, it is driven out to join a flock of birds of its own age and fed on barley, the provision of which for laying geese also is not without advantage. It is not 9 expedient to assign more than twenty goslings to each goose-pen, nor, again, must they be shut up at all with birds older than themselves, since the stronger

dior enecat infirmum. Cellas, in quibus incubitant, siccissimas esse oportet, substratasque habere paleas: vel si eae non sunt, crassissimum[1] quodque[2] faenum. Cetera eadem, quae in aliis generibus pullorum servanda sunt, ne coluber, ne vipera, felesque, aut etiam mustela possit aspirare: quae fere pernicies ad internecionem prosternunt teneros.

10 Sunt qui hordeum maceratum incubantibus apponant, nec patiantur matrices saepius nidum relinquere. Deinde pullis excusis primis quinque diebus polentam vel maceratum far,[3] sicut pavonibus obiciunt. Nonnulli etiam viride nasturtium consectum minutatim cum aqua praebent, eaque eis est esca iucundissima.[4] Mox ubi quattuor mensium facti sunt, farturae maximus quisque destinatur, quoniam tenera aetas praecipue habetur ad hanc rem aptissima: et est

11 facilis harum avium sagina:[5] nam praeter polentam et pollinem ter die nihil sane aliud dari necesse est, dummodo large bibendi potestas fiat, nec vagandi facultas detur, sintque calido et tenebricoso loco: quae[6] res ad creandas adipes multum conferunt. Hoc modo duobus mensibus pinguescunt etiam

[1] crassissimum *SA* : gratissimum *ac*.
[2] quodque *edd.* : quoque *codd.*
[3] maceratum far *ac* : carata fari *S* : caratam farris *A*.

kills the weaker. The coops in which they sleep must be very dry and have chaff spread on the floor, or, if this is not available, the coarsest possible hay. For the rest, the same precautions must be taken as for other kinds of young birds to prevent a grass-snake or a viper or a cat or even a weasel from being able to catch them; for these pestilential creatures generally lay them low and destroy them while they are young and tender.

Some people put barley soaked in water by the side 10 of geese which are sitting and do not allow them to leave the nest too often; then, when the goslings have been hatched, for the first five days they put before them pearl-barley or meal soaked in water, as they also give to peahens. Others give them green cress cut up very small with water—a food which is very agreeable to them. Then when they have become four months old, all the biggest goslings are set aside for fattening, since a tender age is regarded as especially suitable for this process. Indeed the cramming of these birds is an easy matter; for 11 besides pearl-barley and wheat-flour three times a day, absolutely nothing else need be given them, provided that they have facilities for drinking freely and are not allowed to wander about and are kept in a warm, shady place; for all these precautions contribute greatly to the formation of fat. In this manner even the older birds grow fat in two months, for the tenderest young brood

---

[4] iocundissima *ac* : iucundissimum *SA*.

[5] sagina *c* : saginam *SAa*.

[6] tenebricoso loco quae *Ac* : tenebroso *a* : tenebricosolo quoque *S*.

maiores. Nam tenerrima pullities[1] saepe XL diebus opima[2] redditur.

XV. Nessotrophii cura similis, sed maior impensa est. Nam clausae pascuntur anates, querquedulae, boscides, phalerides,[3] similesque volucres, quae stagna et paludes rimantur. Locus planus eligitur, isque munitur sublimiter pedum quindecim maceria: deinde clatris superpositis, vel grandi macula retibus contegitur, ne aut evolandi sit potestas domesticis avibus, aut aquilis vel accipitribus involandi. Sed ea tota maceries opere tectorio levigatur extra intraque, 2 ne feles, aut viverra perrepat. Media[4] deinde parte nessotrophii lacus defoditur in duos pedes altitudinis, spatiumque longitudini[5] datur et latitudini quantum loci[6] conditio permittit.

3 Ora lacus ne corrumpantur violentia restagnantis undae, quae semper influere debet, opere signino consternuntur, eaque non in gradus oportet erigi, sed paulatim clivo subsidere, ut tamquam e litore descendatur in aquam. Solum autem stagni per circuitum, quod sit instar modi totius duarum partium, lapidibus inculcatis ac[7] tectorio muniendum est, ne possit herbas evomere, praebeatque nantibus aquae[8]

[1] pinguescunt etiam maiores. Nam tenerrima pullities *Schneider*: pinguescunt etiam patriminam pullities *SA*: pinguescunt etiam propter nimiam pullutiem (polliciem *c*) *a*.

[2] opima *edd.*: optima *codd.*

[3] phalerides *edd.*: plargides *S*: philagrides *ac*: *om.* *A*.

[4] media *ac*: medio *SA*.

[5] longitudini *edd.*: longitudinis *codd.*

[6] loci *ac*: locis *SA*.

[7] ac *ac*: ad *SA*.

[8] ad quam *a*: aquae *c*: aquam *S*: ad aquam *A*.

---

[a] The text of this passage is undoubtedly corrupt. Schneider's restoration certainly gives the right sense, since

is often brought to a plump condition in forty days.[a]

XV. A place for rearing ducks requires similar attention but is more costly. For mallard, teal, pochard and coots and similar birds, which root about in pools and marshes, can be kept in captivity. A level space is chosen and is provided with a wall fifteen feet high; then it is covered in by having lattice-work or nets of a large mesh placed over it, so that there may be no opportunity for the tame birds to fly away or for eagles or hawks to fly in. The whole of the wall is made smooth by plastering it inside and outside, so that no cat or ferret may creep through it. Then in the middle of the duck-yard a pond is dug, two feet deep, and as much space is assigned to its length and width as the local conditions permit.

Ducks.

2

The edges of the pond are paved with plaster, so that they may not be damaged by the violence of the water when it overflows (for it ought to be always running in), and they should not be raised in the form of steps but should slope down gradually, so that there may be an easy descent as if from the shore into the water. The floor of the pond along the circumference to the extent of about two-thirds of its whole dimension must be constructed with stones well rammed down and plaster, so that it may not be able to put forth any vegetation and may keep the surface of the water clear for the fowls which swim

3

the passage is imitated by Palladius, *R.R.*, Chapter XXX: *melius pinguescunt in tenera aetate. Polenta dabitur in die ter. Large vagari licentia prohibetur. Loco obscuro claudentur et calido. Sic maiores etiam secundo mense pinguescunt; nam parvuli saepe die trigesimo.*

4 puram superficiem. Media rursus terrena pars esse debet, ut colocasiis conseratur, aliisque familiaribus aquae [1] viridibus, quae inopacant avium receptacula. Sunt enim quibus cordi est vel in silvulis tamaricum, aut scirporum frutectis immorari. Nec ob hanc tamen causam totus locus silvulis occupetur, sed ut dixi, per circuitum vacet, ut sine impedimento, cum apricitate [2] diei gestiunt aves, nandi velocitate con-
5 certent. Nam quemadmodum desiderant esse quo [3] irrepant, et ubi delitescentibus fluvialibus [4] animalibus insidientur, ita offenduntur, si non sunt libera spatia, qua permeent. Extra lacum deinde per vicenos undique [5] pedes gramine ripae vestiantur: sintque post hunc agri modum circa maceriam lapide fabricata et expolita tectoriis pedalia in quadratum cubilia, quibus innidificent aves: eaque contegantur intersitis buxeis aut myrteis fruticibus, qui non excedant altitudinem parietum.

6 Statim deinde perpetuus canaliculus humi depressus construatur, per quem quotidie mixti cum aqua cibi decurrant: sic enim pabulatur [6] id genus avium. Gratissima est esca terrestris leguminis panicum et milium, necnon et hordeum: sed ubi copia est, etiam glans ac vinacea praebentur. Aquatilis autem cibi si sit facultas, datur cammarus, et

[1] atque *SAac*.
[2] apricitate *ac* : apricitatem *SA*.
[3] quo *edd.* : qui *codd.*
[4] fluvialibus *SA* : fluviaticis *ac*.
[5] undique *ac* : undequi *SA*.
[6] pabulatur *Aac* : ambulatur *S*.

---

[a] *Nelumbium speciosum*, a plant of the lily kind which grows in the lakes and marshes of Egypt.

upon it. On the other hand, the middle part of the pond should be made of earth, so that it may be sown with the Egyptian bean [a] and other green stuff which generally grows in the water and provides shade for the haunts of the waterfowl. Some of them take pleasure in lingering in little plantations of tamarisk and thickets of club-rushes. Nevertheless the whole space should not for this reason be occupied by little plantations, but, as I have said, should be left free all round the circumference, so that, as they are cheered by a day of sunshine, the water fowl may vie with one another to see which swims the fastest. 4 For just as they require to be where there are holes into which they can creep and where they can lie in wait for fresh-water creatures which are in hiding, so they are displeased if there are no open spaces in which they can roam freely. The banks of the pond should be clothed with grass to a distance of twenty feet all round and beyond this space round the wall there should be nest-boxes one foot square made of stone and covered with a smooth layer of plaster in which the birds may lay their eggs. These nest-boxes should be protected by bushes planted between them of box and myrtle which should not exceed the walls in height. 5

6 Next a continuous channel should be constructed, sunk into the ground, along which the food may be carried down every day mingling with the water, for this is how birds of this kind get their food. The foods grown on dry land which they like best are panic-grass and millet and also barley; but, where there is abundance of them, acorns and grape-husks are also provided. If there is food which grows in the water available, they are given fresh-water crayfish and small

rivalis alecula, vel si qua sunt incrementi parvi fluviorum animalia.

7 Tempora concubitus eadem quae ceterae silvestres alites observant Martii, sequentisque mensis: per quos festucae[1] surculique in vivariis[2] passim spargendi sunt, ut colligere possint aves, quibus nidos construant. Sed antiquissimum est, cum quis nessotrophion constituere volet, ut praedictarum avium circa paludes, in quibus plerumque fetant, ova colligat, et cohortalibus gallinis subiciat. Sic enim excussi educatique pulli deponunt ingenia silvestria, clausique vivariis haud dubitanter progenerant. Nam si modo captas aves, quae consuevere libero[3] victu, custodiae tradere velis, parere cunctantur in servitute. Sed de tutela nantium volucrum satis dictum est.

XVI. Verum opportune, dum meminimus aquatilium animalium, ad curam pervenimus piscium, quorum reditum quamvis alienissimum agricultoribus putem (quid enim tam contrarium est, quam terrenum fluido?), tamen non omittam: nam et harum studia rerum maiores nostri celebraverunt, adeo quidem, ut etiam dulcibus aquis marinos clauderent[4] pisces, atque eadem cura mugilem scarumque[5] nutrirent, qua nunc muraena et lupus educatur. 2 Magni enim aestimabat vetus illa Romuli et Numae rustica progenies, si urbanae vitae compararetur villatica, nulla parte copiarum defici. Quamobrem

[1] festucae *ac*: fetu *SA*.
[2] vivariis *S*: aviariis *Aac*.
[3] consuevere libero *om. SA*.
[4] dulcibus aquis marinos clauderent *edd.*: dulcibus aquibus a fluviatilis cludent *SA*: duabus aquis fluviales clauderent *a*: dulcibus aquis fluviales clauderent *c*.

pickled river-fish and any other river animals which grow only to a small size.

7 They observe the same seasons for coupling as other wild birds, namely, March and the following month. During these months stalks and twigs should be scattered about everywhere in the bird-pens, so that the birds may be able to collect them and use them to build their nests. But it is most important, when anyone wishes to establish a place for rearing ducks, to collect the eggs of the said fowls in the region of the marshes, where they usually lay, and set them under farm-yard hens. For when they are hatched and reared in this way they lay aside their wild nature and undoubtedly breed shut up in the bird-pens. If you want to hand over to custody birds which have only just been caught and have been used to a life of liberty, they are slow to begin to lay in captivity. But enough has now been said about the care of fowls which swim.

Fishes.

XVI. In dealing with aquatic animals we come in due course to the management of fishes, the profitable nature of which, though I regard it as far removed from the business of farmers—for what things are so contrary to one another as dry land and water?—I will nevertheless not pass over. Our ancestors carried their zeal for this pursuit to such a pitch that they even imprisoned salt-water fish in fresh water and fed the grey mullet and parrot wrasse with the same care with which the lamprey and the sea-pike are now reared. 2 The country-bred descendants of Romulus and Numa of old prided themselves greatly on the fact that, if life on the farm were compared with that in the town, it did not fall short of it in abundance of any kind;

[5] squalumque *SAa*: scalumque *c*.

non solum piscinas, quas ipsi construxerant, frequentabant: sed etiam quos rerum natura lacus fecerat, convectis marinis seminibus replebant. Inde Velinus, inde etiam Sabatinus, item Volsiniensis, et Ciminius lupos auratasque procreaverunt, ac si qua sunt alia piscium genera dulcis undae tolerantia.
3 Mox istam curam sequens aetas abolevit, et lautitiae locupletum maria ipsa Neptunumque clauserunt, ut[1] iam tum avorum[2] memoria circumferretur Marcii Philippi velut urbanissimum, quod erat luxuriosi[3] factum atque dictum. Nam is forte Casini cum apud hospitem cenaret, appositumque e vicino flumine lupum degustasset atque expuisset, improbum factum dicto prosecutus: Peream, inquit, nisi piscem
4 putavi. Hoc igitur periurium multorum subtiliorem fecit gulam, doctaque et erudita palata fastidire docuit fluvialem lupum, nisi quem Tiberis adverso torrente defatigasset. Itaque Terentius Varro: Nullus est, inquit, hoc saeculo nebulo,[4] ac minthon,[5] qui non iam dicat, nihil sua interesse, utrum eiusmodi
5 piscibus, an ranis frequens habeat vivarium. Ac tamen iisdem temporibus, quibus hanc memorabat Varro luxuriem, maxime laudabatur severitas Catonis, qui nihilo minus et ipse tutor Luculli grandi aere

[1] ut *edd.*: et *ac*: *om. SA*.
[2] avorum *ac*: quorum *SA*.
[3] luxuriosi *scripsi*: luxuriose *SAa*: luxuriosissime *c*.
[4] nebulus *ac*: nebullus *SA*.
[5] minthon *ac*: mintho *SA*.

[a] The Lago di Piedi di Luco in Umbria.
[b] The Lago Bracciano about 35 miles N.W. of Rome.
[c] The Lago di Bolseno about 70 miles N.W. of Rome.
[d] The Lago di Vico near Viturbo about 45 miles N.W. of Rome.

they, therefore, not only stocked the fish-ponds which they had themselves constructed, but also filled the lakes which nature had formed, with fish-spawn brought from the sea. Hence the Veline [a] and Sabatine [b] lakes, also the Volsinian [c] and Ciminian [d] lakes produced basse and gilt-head, and all the fishes to be found anywhere which can live in fresh water. 3 Then an age followed which abandoned this method of keeping fish and the extravagance of the wealthy enclosed the very seas and Neptune himself, so that within the memory of our grandfathers the action and speech of Marcius Philippus [e] was on everyone's lips as being very witty, whereas it was the action and speech of a luxurious man. For once when he happened to be dining at a friend's house at Casinum,[f] and after having tasted a pike from a neighbouring river which was set before him had spit it out, he followed this opprobrious action with the words: "Plague take me if I did not think that it was a fish." 4 This oath caused many people to put more refinement into their gluttony and has taught learned and educated palates to loathe the basse unless it were one which had been wearied by struggling against the current of the Tiber. Therefore Terentius Varro says: [g] There is no paltry or foppish fellow in these days who does not now declare that he cares not whether he has a fish-pond crowded with this sort of fish or with frogs. 5 Yet in the very times to which Varro ascribed this luxury, the austerity of Cato was highly commended, who, nevertheless, himself as the guardian of Lucullus sold his ward's fish-ponds for the

[e] This story is borrowed from Varro, *R.R.* III. 3. 9.
[f] The modern Monte Cassino in the north of Campagna.
[g] *Loc. cit.*

sestertium[1] milium quadringentorum piscinas pupilli sui venditabat. Iam enim celebres erant deliciae popinales, cum ad mare deferrentur[2] vivaria, quorum studiosissimi, velut ante devictarum gentium Numantinus et Isauricus, ita Sergius Orata, et Licinius Muraena captorum piscium laetabantur vocabulis.

6 Sed quoniam sic mores occalluere, non ut haec usitata, verum ut maxime laudabilia et honesta iudicarentur, nos quoque ne videamur tot seculorum seri castigatores, hunc etiam quaestum villaticum patrisfamilias demonstrabimus. Qui sive insulas,[3] sive maritimos agros mercatus,[4] propter exilitatem soli, quae plerumque litori vicina est, fructus terrae percipere non 7 poterit,[5] ex mari reditum constituat. Huius autem rei quasi primordium est, naturam loci contemplari, quo piscinas facere constitueris. Non enim omnibus litoribus omne genus piscium haberi potest. Limosa regio planum educat piscem, velut soleam, rhombum, passerem. Eadem quoque maxime idonea est conchyliis:[6] purpurarum muricibus, tum concharum 8 ostreis,[7] pectunculis, balanis, vel sphondylis.[8] At arenosi gurgites planos quidem non pessime,[9] sed

[1] sestertium *Aac* : sestertius *S*.
[2] defertur *SA* : deferantur *ac*.
[3] insulas *ac* : in insula *SA*. [4] mercatus *a* : mercatur *SA*.
[5] precipere non potuerit *ac* : percipuerit ut *SA*.
[6] conchiliis *ac* : conciliis *SA*.
[7] *Warmington*: muricibus et ostreis, purpurarumque tum concharum *prior. edd.*
[8] pectunculi, balani vel sphondili *codd.*
[9] pessime *a* : spessime *S* : spissime *A* : proxime *c*.

[a] Scipio Africanus Minor. [b] P. Servilius Vatia.
[c] According to Pliny, IX. § 168, Sergius Orata established the first oyster-beds at Baiae near Naples.

immense sum of 400,000 sesterces. For culinary delicacies were already in great demand when fish-ponds were made to communicate with the sea and, just as at an earlier date Numantinus [a] and Isauricus [b] rejoiced in names taken from conquered nations, so Sergius Orata (goldfish) [c] and Licinius Muraena (lamprey),[d] who made fish-ponds their chief interest, rejoiced in the names of the fish they had captured.

6 But since men's moral sense has become so blunted that such behaviour is reckoned not only as customary but also as highly laudable and honourable, we too, lest we should seem to be only out-of-date critics of so many past generations, will show that the fish-pond is also a source of profit which the head of a household can gain from his country estate. He who has bought either islands or land near the sea and is unable, owing to the poverty of the soil which is generally found near the coast, to gather the fruits of the earth, should establish a source of revenue from the sea. 7 The first step in this direction is to examine the nature of the ground where you have decided to construct your fish-ponds, for every kind of fish cannot be kept on every coast. A muddy stretch of shore is the place for rearing flat fish, such as the sole, the turbot and the flounder;[e] it is also very suitable for testaceous animals: of purple-producing shell-fish, the true purple fish; and also, of other molluscs, the oyster, small scallops, barnacles or *sphondyli*.[f] 8 But the sandy whirlpools are not bad feeding-grounds for flat-fish—better, however,

[d] Licinius Muraena according to Pliny (*loc. cit.* § 170) invented fish-ponds for all sorts of fish.

[e] Or dab; the identification is doubtful.

[f] Apparently another kind of mussel, perhaps *spondylus gaedaropus*.

pelagios melius pascunt, ut auratas ac dentices, Punicasque et indigenas umbras: verum conchyliis minus apti. Rursus optime saxosum mare nominis sui pisces nutrit, qui scilicet, quod in petris stabulentur, saxatiles dicti sunt, ut merulae turdique, nec minus melanuri. Atque ut litorum sic et fretorum differentias nosse oportet, ne nos alienigenae pisces decipiant. Non enim omni mari potest omnis esse, ut helops, qui Pamphilio profundo nec alio pascitur: ut Atlantico faber, qui et generosissimis piscibus adnumeratur, in nostro Gadium municipio eumque prisca consuetudine[1] zeum appellamus: ut scarus, qui totius Asiae Graeciaeque litoribus Sicilia tenus frequentissimus exit, nunquam in Ligusticum, nec per Gallias enavit ad Hibericum mare. Itaque ne si capti quidem perferantur in nostra vivaria, diuturni queant possideri. Sola ex pretiosis piscibus muraena, quamvis Tartesii Carpathiique pelagi, quod est ultimum, vernacula,[2] quovis hospes freto peregrinum mare sustinet. Sed iam de situ piscinarum dicendum est.

9

10

XVII. Stagnum censemus eximie optimum, quod sic positum est, ut insequens maris unda priorem submoveat, nec intra conseptum sinat remanere veterem. Namque id simillimum est pelago, quod agitatum ventis assidue renovatur, nec concalescere potest: quoniam gelidum ab imo fluctum revolvit in

[1] consuetudine *Aac*: consuetudinem *S*.
[2] vernacula *c*: vernaculo *SAa*.

---

[a] *Oblata melanurus.*
[b] Off the S. coast of Asia Minor.
[c] Between Corsica and the Italian Riviera.
[d] *I.e.* Gallia Cisalpina and Gallia Transalpina.
[e] Between Spain and the Balearic Islands.

for deep-sea fish such as gilt-head and sea-braize and the Carthaginian and our own Italian maigres, but they are less suitable for shell-fish. On the other hand a rocky sea provides excellent nourishment for fishes which bear its name, that is, are called rock-fish because they find shelter among the rocks, such as merles and wrasse and likewise "black tails."[a] We must also know the different qualities both of shores and of seas, lest we be deceived about foreign fish; for every fish cannot exist in every sea, the sturgeon for example, which feeds in the depths of the Pamphylian Sea[b] and nowhere else, and the dory in the Atlantic which in our municipality of Gades is numbered amongst the noblest of fishes and which by an ancient custom we call *zeus*, and the parrot wrasse which is produced in great numbers on the coasts of the whole of Asia Minor and Greece as far as Sicily but has never swum into the Ligurian[c] sea nor past the Gauls[d] into the Iberian Sea;[e] therefore, even if they were captured and conveyed to our fish-ponds, they could not long remain in our possession. Alone of the valuable fish the lamprey, although a native of the Tartessian and the Carpathian Sea, which is very far away, in whatever sea it finds itself a guest can thrive in strange waters. But the time has come to speak of the situation of fish-ponds.

9

10

Fish-ponds.

XVII. We consider that incomparably the best pond is one which is so situated that the incoming tide of the sea expels the water of the previous tide and does not allow any stale water to remain within the enclosure; for a pond most resembles the open sea if it is stirred by the winds and its waters constantly renewed and it cannot become warm, because it keeps rolling up a wave of cold water from the

partem superiorem. Id autem stagnum vel exciditur[1] in petra, cuius rarissima est occasio, vel in litore 2 construitur opere signino. Sed utcunque fabricatum est, si semper influente gurgite riget, habere[2] debet specus iuxta solum, eorumque alios simplices, et rectos, quo secedant squamosi greges, alios in cochleam retortos, nec nimis spatiosos, in quibus muraenae delitescant; quamquam nonnullis commisceri eas cum alterius notae piscibus non placet: quia si rabie vexantur, quod huic generi velut canino solet accidere, saepissime persequuntur squamosos, 3 plurimosque mandendo consumunt; itineraque, si loci natura permittit, omni lateri piscinae dari convenit. Facilius enim vetus submovetur unda, cum quacunque parte fluctus urget, per adversam patet exitus. Hos autem meatus fieri censemus per imam consepti partem, si loci situs ita competit, ut in solo piscinae posita libella septem pedibus sublimius esse maris aequor ostendat: nam piscibus stagni haec in altitudinem gurgitis mensura abunde est. Nec dubium, quin quanto magis imo mari venit unda, tanto sit frigidior, quod est aptissimum nantibus. 4 Sin autem locus, ubi vivarium constituere censemus, pari libra cum aequore maris est, in pedes novem defodiatur piscina, et infra duos a[3] summa parte cuniculis rivi perducantur; curandumque est, ut

[1] exciditur *ac* : excitur *SA*. [2] haberi *SAac*.
[3] a *ac* : ad *SA*.

[a] *Cf.* Plautus, *Rud.* 4. 3. 5.

bottom to the uppermost part. The pond is either hewn in the rock, which only rarely occurs, or built of plaster on the shore; but in whatever way it is 2 constructed, if it is kept cold by the swirl of water which is constantly flowing in, it ought to contain recesses near the bottom, some of them simple and straight to which the "scaly flocks"[a] may retire, others twisted into a spiral and not too wide, in which the lampreys may lurk. Some people, however, hold that lampreys should not be mixed with fishes of another kind, because, if they are seized with madness, which sometimes happens to this sort of fish just as it happens to dogs, they very often pursue their scaly companions and chew them up and devour great numbers of them. If the nature of the ground 3 permits, channels should be provided for the water on every side of the fish-pond; for the old water is more easily carried away if there is an outlet on the side opposite to that from which the wave forces its way in. We are of opinion that these passages, if the lie of the ground is suitable, should be made along the lowest part of the enclosure, so that a plummet placed on the bottom of the pond may show that the level of the sea is seven feet higher; for this measurement in the depth of the water is fully enough for the fish in the pond, and there is no doubt that, the greater the depth of the sea from which the water comes, the colder it is, and this suits the swimming fishes very well. But if the place 4 where we think of constructing the fish-pond is on a level with the surface of the sea, the pond should be excavated to the depth of nine feet, and two feet below the top streams of water should be conducted along small channels, and care must be taken that

largissime veniant, quoniam modus ille aquae iacentis infra libram maris non aliter exprimitur, quam si 5 maior recentis freti vis incesserit. Multi putant in eiusmodi stagnis longos [1] piscibus recessus et flexuosos in lateribus specus esse fabricandos, quo sint opaciores aestuantibus latebrae. Sed si recens mare non semper stagnum permeat, id facere contrarium est. Nam eiusmodi receptacula nec facile novas admittunt aquas, et difficulter veteres emittunt: plusque nocet putris unda, quam prodest opacitas. 6 Debent tamen similes velut cellae parietibus excavari, ut sint, quae protegant refugientes ardorem solis, et nihilominus facile, quam conceperint aquam, remittant. Verum meminisse oportebit, ut rivis, per quos exundat piscina, praefigantur [2] aenei foraminibus exiguis cancelli, quibus impediatur fuga piscium. Si vero laxitas permittit, e litore scopulos, qui praecipue verbenis [3] algae vestiuntur, non erit alienum per stagni spatia disponere, et quantum comminisci valet hominis ingenium, repraesentare faciem maris, 7 ut clausi quam minime custodiam sentiant. Hac ratione stabulis ordinatis aquatile pecus inducemus; sitque nobis antiquissimum meminisse etiam in fluviatili negotio, quod in terreno praecipitur: Et

[1] longos *ac* : longis *SA*.

[2] praeficantur *S* : praeficentur *A* : prefingetur *a* : praefigentur *c*.

[3] verbenis algae *ac* : velvenis algae *S* : velvenis ac leve *A*.

---

[a] It is doubtful whether *verbenis* can bear the meaning of "vegetation" in general and the reading is perhaps wrong.

the flow is very abundant, since the quantity of water which lies below the level of the sea is only forced out by the greater violence of the fresh sea water rushing in. Many people think that in the 5 sides of ponds of this kind deep recesses and winding caves should be constructed for the fishes, so that there may be shadier places of retreat for them when they feel the heat. But if a change of sea water is not continually passing through the pond, the result is to cause a contrary condition, for lurking-places of this kind do not easily admit a change of water and only with difficulty get rid of the stale water, and more harm results from the putrid water than benefit from the shade. There ought, however, to be 6 excavated in the sides of the pond what may be described as a series of similar cells which may serve to protect the fish when they want to avoid the heat of the sun and yet at the same time let the water, which they have received, easily flow out again. It will be well to remember that gratings made of brass with small holes should be fixed in front of the channels through which the fish-pond pours out its waters, to prevent the fish from escaping. If space allows, it will not be amiss to place in various parts of the pond rocks from the sea-shore, especially those which are covered with bunches of sea-weed [a] and, as far as the wit of man can contrive, to represent the appearance of the sea, so that, though they are prisoners, the fish may feel their captivity as little as possible.

Having arranged "stalls" for them on this 7 principle, we shall introduce our "water flock" into it, and it should be our prime concern to recall also in our dealings with rivers the advice given for our business with dry land: "And consider well what every

quid quaeque ferat regio.[1] Neque enim si velimus, ut in mari non nunquam conspeximus, in vivario multitudinem mullorum [2] pascere queamus, cum sit mollissimum genus, et servitutis indignantissimum.
8 Raro itaque unus aut alter de multis milibus claustra patitur: at contra frequenter animadvertimus intra septa pelagios greges, inertis mugilis et rapacis lupi. Quare, ut proposueram, qualitatem litoris nostri contemplemur; et si videmus scopulosam, probemus. Turdi complura genera, merulasque et avidas mustelas, tum etiam sine macula (nam sunt et varii) lupos includemus, item plautas,[3] quae maxime probantur, muraenas, et si quae sunt aliae saxatilis notae, quarum pretia vigent. Nam vile ne captare quidem, nedum alere conducit. Possunt ista eadem
9 genera etiam litoris arenosi stagnis contineri. Nam quae limo caenoque lita sunt,[4] ut ante iam dixi, conchyliis magis et iacentibus apta sunt animalibus. Neque est eadem lacus positio, quae recipit cubantes: neque [5] eadem praebentur cibaria prostratis piscibus et rectis.[6] Namque soleis ac rhombis et similibus animalibus humilis in duos pedes piscina deprimitur in ea parte litoris, quae profundi recessu [7] nunquam
10 destituitur. Spissi deinde clatri marginibus infiguntur, qui super aquam semper emineant, etiam cum maris aestus intumuerit. Mox praeiaciuntur in gyrum moles, ita ut complectantur sinu suo, et tamen excedant stagni modum. Sic enim et maris atrocitas

[1] *post* regio *add.* *SA* : oportet si quis in eo.
[2] mulorum *codd.*
[3] plautas *SA* (= Greek πλωτάς) : flutas *ac.*
[4] lita sunt *S* : litescunt *Aac.* [5] atque *SAac.*
[6] rectis *SA* : erectis *ac.*
[7] profundo recessu *c* : frondi recensu *A* : frondi recessu *S* : frondi recente *a.*

place will bear." For we cannot, if we should wish to do so, feed in a fish-pond a multitude of red mullet, such as we have very often seen in the sea, since it is a very delicate kind of fish and most intolerant of captivity, and so only one or two out of many 8 thousands can on rare occasions endure confinement, while, on the contrary, we frequently notice in closed waters shoals of those deep-sea fish: the sluggish grey mullet and the greedy basse. Therefore, as I have already suggested, let us consider the quality of our sea-shore and, if we find it rocky, let us be content with it. We shall imprison in our ponds several kinds of wrasse and sea-merles and greedy sea-weasels and also basse which have no spots (for there is also a mottled kind), also floating lampreys, which are much esteemed, and any other lampreys of the rock-dwelling kind which command a high price; for it does not pay to catch, much less to keep, anything which is cheap. These same kinds 9 of fish can also be kept in ponds on a sandy shore; for shores which are covered with slime and mud are, as I have already said, better suited to shell-fish and animals which lie at the bottom. A different position too is required for ponds which harbour those fish which lie down, nor is the same food provided for prostrate as for upright fish. For soles and turbots and similar creatures a shallow pond is sunk two feet in that part of the shore which is never left high and dry by ebbing of deep water. Next close barriers 10 are fixed along the edges of the pond, so that they always stand out of the water even when the tide of the sea is at its highest; then dams are thrown up all round so as to encompass the pond in their embrace and at the same time to rise above its level. For in

obiectu crepidinis frangitur, et in tranquillo consistens piscis sedibus suis non exturbatur, neque ipsum vivarium repletur algarum congerie, quam 11 tempestatibus eructat pelagi violentia. Oportebit autem nonnullis locis moles intercidi more Maeandri parvis sed angustis itineribus, quae quantalibet hiemis saevitia mare sine fluctu transmittant.

Esca iacentium mollior esse debet, quam saxatilium, nam quia dentibus carent, aut lambunt cibos, aut integros hauriunt, mandere quidem non possunt. 12 Itaque praeberi convenit tabentes aleculas,[1] et salibus exesam chalcidem, putremque sardinam, nec minus scarorum[2] branchias, vel quicquid intestini pelamis[3] aut lacertus gerit: tum scombri, carcharique et elacata[4] ventriculos[5] et ne per singula enumerem, salsamentorum omnium purgamenta, quae cetariorum[6] officinis everruntur. Nos autem plura nominavimus genera, non quia cuncta cunctis litoribus exeunt, sed ut ex his aliqua, quorum erit 13 facultas, praebeamus. Facit etiam ex pomis viridis adaperta ficus; et mitis digitis infractus unedo; nec minus elisum molle sorbum, quique sunt cibi sorbilibus proximi, ut e mulctra recens caseus, si loci conditio vel lactis annona permittit. Nulla tamen aeque,[7] quam praedictae salsurae pabula commode dantur,

[1] halleculas *ac*: halleculam *SA*.
[2] scaurorum *ac*: aurorum *SA*.
[3] pelamis *a*: pelanus *c*: palemis *SA*.
[4] elacatae *edd.*: lacte *a*: lapte *SAc*.
[5] venterculos *SAac*.
[6] ceterarum *SA*: ceterum *ac*.
[7] aeque *ac*: quae *SA*.

this way the violence of the sea is broken by the barriers of a bank, and the fish, keeping in calm water, are not driven out of their usual haunts nor is the pond itself filled with a collection of sea-weed which the force of the sea throws up in stormy weather. It 11 will, however, be necessary that cuts should be made in the moles at some points, forming small but narrow passages with meandering course, so that, however fierce a winter storm is raving, they may let the sea-water pass in without creating a wave.

Diet for fish

The diet of flat fish ought to be softer than that of rock-fish, for, lacking teeth, they either lick up their food or swallow it whole, being unable to chew it. It is, therefore, fitting that decaying pilchards or 12 over-salted herrings or rotten sardines, also the gills of parrot wrasse and any part of the intestines of a young tunny or lizard-fish, also the entrails of a mackerel, a dog-fish or a spindle-fish,[a] and, not to go into further details, the refuse of any salted fish which is swept out of fishmongers' shops. We have named several kinds, not because they are all produced on every coast, but in order to mention some of those which it will be possible to provide. Of fruits too the 13 green fig cut open is suitable and a ripe arbutus-berry crushed by the fingers, likewise a soft sorb-apple squeezed out and any foods which most closely resemble things which can be easily swallowed, such as curds fresh from the milk-pail, if local conditions and the cheap price of milk make this possible. No food, however, is so suitable for giving them as the diet of salt fish already mentioned, since it has a

[a] The readings of the MSS. give no sense here, but the name of a fish is clearly intended and *elacatae* is suggested by the reading of *a*. Warmington suggests *elacatenis* (ἠλακατῆνος).

14 quoniam odorata sunt. Omnis enim iacens piscis magis naribus escam, quam oculis vestigat. Nam dum supinus semper cubat, sublimiora [1] aspectat, et ea quae in plano sunt dextra laevaque non facile pervidet. Itaque cum salsamenta obiecta sunt, eorum sequens odorem, pervenit ad cibos.

Ceteri autem saxatiles aut pelagici [2] satis ex his, sed recentibus melius pascuntur. Nam et alecula modo capta, et cammarus exiguusque gobio, quisquis denique est incrementi minuti piscis, maiorem alit. 15 Siquando tamen hiemis saevitia non patitur eius generis escam dari, vel sordidi panis offae, vel siqua sunt temporis poma concisa praebentur. Ficus quidem arida semper obicitur, eximie si sit, ut Baeticae Numidiaeque regionibus, larga. Ceterum illud committi non debet, quod multi faciunt, ut nihil praebeant, quia semetipsos etiam clausi diu tolerare possint. Nam nisi piscis domini cibariis saginatur, cum ad piscatorium [3] forum perlatus est, macies indicat eum non esse libero mari captum, sed de custodia elatum, propter quod plurimum pretio detrahitur.

16 Atque haec villatica pastio finem praesenti disputationi [4] faciat, ne immodico volumine lector [5] fatigetur. Redibimus autem sequenti exordio ad curam silvestrium pecorum, cultumque apium.

[1] sublimiora *S* : sublimior *A* : sublimius *a* : sublimus *c*.
[2] pelasci *SA*¹ : pelasgi *A*² : pelagici *ac*.
[3] piscatorium *edd.* : piscatoris *SAac*.
[4] disputationi *ac* : disputationis *SA*.
[5] lector *S*²*ac* : delector *S*¹*A*.

strong odour; for every flat fish tracks down its food 14 rather by scent than by sight. For lying constantly on its back it looks towards what is above it and does not easily see things which are on a level with itself on the right or left. When, therefore, salted fish is put in its way, it follows the scent of it and so reaches its food.

The other kinds of fish, namely those which live among the rocks and in the open sea, can quite well be fed on this diet, but still better on fresh food. For a newly caught pilchard, crayfish or small goby, in a word any fish of minute growth serves as food for a larger fish. If, however, the violence of the 15 winter does not allow this kind of food to be given, bits of coarse bread or any fruits that are in season are cut up and given. Dried figs indeed are always offered to them, an excellent thing to do if they are abundant as they are in the regions of Baetica and Numidia. But the mistake ought not to be made, which many people make, of providing no food at all on the ground that the fish can maintain themselves for a long time even when they are shut up; for unless a fish is fattened with food provided by its owner, when it is brought to the fish-market, its leanness shows that it has not been caught in the open sea but brought out of a place of confinement, and on this account a large sum is knocked off the price.

Let this account of the method of feeding fish on 16 the farm-estate bring our present discourse to a close, lest the reader be wearied with the immoderate length of this volume. In the next book we will return to the management of wild stock and the culture of bees.

BOOK IX

LIBER IX

PRAEFATIO

Venio nunc ad tutelam pecudum silvestrium et[1] apium educationem: quas et ipsas, Publi Silvine, villaticas pastiones iure dixerim; siquidem mos antiquus lepusculis capreisque, ac subus feris iuxta villam plerumque subiecta dominicis habitationibus ponebat vivaria, ut et conspectu suo clausa[2] venatio possidentis oblectaret oculos, et cum exegisset usus epularum, velut e cella promeretur. Apibus quoque dabatur[3] sedes adhuc nostra memoria vel in ipsis villae parietibus excisis, vel in protectis porticibus ac pomariis. Quare quoniam tituli, quem praescripsimus huic disputationi, ratio reddita est, ea nunc quae proposuimus singula persequamur.

I. Ferae pecudes, ut capreoli, damaeque, nec minus orygum cervorumque genera et aprorum, modo lautitiis ac voluptatibus dominorum serviunt, modo quaestui ac reditibus. Sed qui venationem voluptati suae claudunt, contenti sunt, utcunque competit proximus aedificio loci situs, munire viva-

[1] *post* silvestrium *om.* et *SA*.
[2] suo clausa *ac* : sui clausa *A* : sui classa *S*.
[3] dabantur *SAac*.

BOOK IX

PREFACE

I now come to the care of wild cattle and the rearing of bees, which also, Publius Silvinus, I can justly place among creatures which are fed on the farm, since ancient custom placed parks for young hares, wild goats and wild boars near the farm, generally within the view of the owner's dwelling-place, so that the sight of their being hunted within an inclosure might delight the eyes of the proprietor and that when the custom of giving feasts called for game, it might be produced as it were out of store. Also within our own memory accommodation for bees was provided either in holes cut in the actual walls of the farm-building or in sheltered galleries and orchards. So, since we have assigned a reason for the title which we have prefixed to this discourse, let us now proceed to deal, one by one, with the topics which we have proposed. Preface.

I. Wild creatures, such as roebucks, chamois and also various kinds of antelopes, deer and wild boars sometimes serve to enhance the splendour and pleasure of their owners, and sometimes to bring profit and revenue. Those who keep game shut up for their own pleasure are content to construct a park, on any suitable site in the neighbourhood of Wild creatures.

rium, semperque de manu cibos et aquam praebere: qui vero quaestum reditumque desiderant, cum est vicinum villae nemus (id enim refert non procul esse ab oculis domini) sine cunctatione praedictis animali-
2 bus destinant.[1] Et si naturalis defuit aqua, vel inducitur [2] fluens, vel infossi lacus signino consternuntur, qui receptam pluviatilem contineant.[3]

Modus silvae pro cuiusque [4] facultatibus occupatur; ac si lapidis et operae vilitas suadeat, haud dubie [5] caementis et calce formatus circumdatur murus: sin
3 aliter, crudo latere [6] ac luto constructus. Ubi vero neutrum patrifamiliae conducit, ratio postulat vacerris includi: sic enim appellatur genus clatrorum: idque fabricatur ex robore querceo, vel subereo. Nam oleae rara est occasio. Quidquid denique sub iniuria pluviarum magis diuturnum est, pro condicione regionis ad hunc usum eligitur. Et sive teres [7] arboris truncus, sive ut crassitudo postulavit, fissilis stipes compluribus locis per latus efforatur, et in circuitu vivarii certis intervenientibus spatiis defixus erigitur: deinde per transversa laterum cava [8] transmittuntur ramices, qui exitus ferarum obserent.
4 Satis est autem vacerras inter pedes octonos defigere, serisque transversis ita clatrare, ne spatiorum laxitas, quae foraminibus intervenit, pecudi praebeat fugam. Hoc autem modo licet etiam latissimas regiones tractusque montium claudere, sicuti Galliarum nec-

[1] destinatur *SAac*.
[2] si naturalis—inducitur *om. S*.
[3] contineant *c*: contineat *SAa*.
[4] pro cuiusque *ac*: ut pro cuius *S*: ut pro cuiusque *A*.
[5] haud dubie *c*: id haud dubie *a*: ita ut dubiis *SA*.
[6] crudo latere *c*: crudo lateri *SA*: crudeliter *a*.
[7] teres *Gesner*: teris *SA*: veteris *ac*.
[8] cuca *SA*: cavea *ac*.

the farm buildings, and always give them food and water by hand. Those on the other hand who look for profit and revenue, when there is a wood near the farm (for it is important that it should not be far out of sight of the owner), reserve it without hesitation for the above-mentioned animals, and if there 2 is no natural supply of water, either running-water is introduced or else ponds are dug and lined with mortar to receive and hold the rain-water.

The extent of wood involved is in proportion to the size of each man's property and, if the cheapness of stone and labour make it advisable, certainly a wall built with unhewn stone and lime is put round it; otherwise it is made with unburnt brick and clay. When neither of these methods serves the purpose 3 of the master of the house, reason requires that they should be shut up with a post fence; for this is the name given to a certain kind of lattice made of oak or cork-wood, since olive-wood is only rarely obtainable; in a word, according to local conditions, any kind of wood is chosen for this purpose which resists injury from rain better than any other. Whether it be the round trunk of a tree or cleft into stakes, as its thickness demands, it has several holes bored through its side and is erected firmly in the ground at fixed intervals all round the park; then bars are put across through the holes in the sides of the posts to prevent the passage of the wild beasts. It is enough to fix 4 the posts at intervals of eight feet and to fasten them to the cross-bars in such a way that the width of space which occurs where holes are left may not offer the animals a means of escape. In this manner you can even enclose very wide regions and tracts of mountains, as the vast extent of ground permits in

non et in aliis quibusdam provinciis locorum vastitas patitur. Nam et fabricandis ingens est vacerris materiae[1] copia, et cetera in hanc rem feliciter suppetunt; quippe crebris fontibus abundat solum, 5 quod est maxime praedictis generibus salutare: tum etiam sua sponte pabula feris benignissime subministrat[2] praecipueque saltus eliguntur, qui et terrenis fetibus et arboribus abundant. Nam ut graminibus ita frugibus roburneis opus est[3]: maximeque laudantur, qui sunt feracissimi querneae glandis et iligneae, nec minus cerrea,[4] tum[5] et arbuti, ceterorumque pomorum silvestrium, quae diligentius persecuti sumus, cum de cohortalibus subus disputaremus. Nam eadem fere sunt pecudum 6 silvestrium pabula, quae domesticarum. Contentus tamen non debet esse diligens paterfamilias cibis, quos suapte natura terra gignit, sed temporibus anni, quibus silvae pabulis carent, condita messe clausis succurrere, hordeoque alere, vel adoreo farre aut faba, primumque[6] etiam vinaceis, quicquid denique vilissime constiterit, dare. Idque ut[7] intelligant ferae praeberi, unam vel alteram domi mansuefactam conveniet immittere, quae pervagata totum vivarium cunctantes ad obiecta cibaria pecudes 7 perducat. Nec solum istud per hiemis penuriam

[1] vacerris materie *c*: materiae vacerriis *SA*: variis materie *a*.
[2] subministrat *a*: -ant *SAc*.
[3] opus est *SA*: opus habet *c*: robor est *a*.
[4] acerreae *SAa*: ceree *c*.
[5] cum *SAac*.
[6] primumque *SAa*: plurimumque *c*.
[7] idque un *c*: ut *SA*: itaque ut *a*.

[a] Book VII. Chapter 9. § 6 ff.

the provinces of Gaul and in certain others; for there is both a great abundance of timber for making posts and everything else which is needed for the purpose is in abundant supply. The soil abounds in frequent 5 springs, which is particularly wholesome for the above-named kinds of animals; then too it furnishes wild creatures with fodder most liberally even of its own accord. Woodlands are chiefly chosen which abound in the fruits of the ground and also in trees; for as these creatures have need of grass, so too they require the fruits of oak-trees, and those woods are most highly commended which are most productive of the acorn of the ordinary oak and of the evergreen oak and likewise of the Turkey-oak, also of the fruit of the strawberry-tree and the other wild fruits which we described in great detail when we were discussing farm-yard pigs.[a] For the fodder of wild cattle is almost the same as that of domestic animals.

Nevertheless the careful head of a household ought 6 not to be content with the foods which the earth produces by its own nature, but, at the seasons of the year when the woods do not provide food, he ought to come to the help of the animals which he has confined with the fruits of the harvest which he has stored up, and feed them on barley or wheat-meal or beans, and especially, too, on grape-husks; in a word, he should give them whatever costs the least. Also in order that the wild creatures may understand that provision is being made for them, it will be a good plan to send among them one or two animals which have been tamed at home, and which, roaming through the whole park, may direct the hesitating creatures to the fare offered to them. It is advisable 7 that this should be done not only during the scarce

fieri expedit, sed cum etiam fetae partus ediderint, quo melius educant[1] natos. Itaque custos vivarii frequenter speculari debebit, si iam effetae sint, ut manu datis sustineantur frumentis. Nec vero patiendus est oryx, aut aper, aliusve quis ferus ultra quadrimatum senescere. Nam usque in hoc tempus capiunt incrementa, postea macescunt senectute. Quare dum viridis aetas pulchritudinem corporis 8 conservat, aere mutandi[2] sunt. Cervus tamen compluribus annis sustineri potest. Nam diu iuvenis possidetur, quod aevi longioris vitam sortitus est. De minoris autem incrementi[3] animalibus, qualis est lepus, haec praecipimus,[4] ut in iis vivariis, quae maceria munita sunt, farraginis et olerum, ferae intubi lactucaeque semina parvulis areolis per diversa spatia factis iniciantur. Itemque Punicum cicer, vel hoc vernaculum,[5] nec minus hordeum, et cicercula condita ex horreo promantur, et aqua caelesti macerata obiciantur. Nam sicca non nimis ab lepus-9 culis appetuntur. Haec porro animalia vel similia his, etiam silente me, facile intelligitur, quam non expediat conferre in vivarium, quod vacerris circumdatum est: siquidem propter exiguitatem corporis facile clatris subrepunt, et liberos nactae[6] egressus fugam moliuntur.

II. Venio nunc ad alvorum curam, de quibus neque diligentius quidquam praecipi potest, quam ab

[1] educant *SA* : -ent *ac*.
[2] aere mutandi *ac* : remutandi *A* : remuttendi *S*.
[3] incrementis *SAac*.
[4] praecipiemus *SAac*.
[5] vernaculo *SAac*.
[6] nacte *Aac* : nancte *S*.

---

[a] See Book I. 1. 13 and note.

season of winter but also when those which were with young have brought them forth, so that they may rear them better. And so the park-keeper will have frequently to be on the watch and see if they have borne their young, in order that their strength may be sustained by cereals given them by hand. But neither the antelope nor the wild boar nor any other wild creature should be allowed to live to a greater age than four years. For up to that time they advance in growth, after it they grow old and lean; and so they should be turned into cash while a vigorous time of life preserves their bodily comeliness. The deer, however, may be kept for many 8 years, for it long remains young in your possession, because it has been allotted a life of longer duration. But as regards animals of lesser growth, such as the hare, our advice is that, in those parks surrounded by a wall, the seeds of mixed cereals and of the pot-herbs, wild endive and lettuce, should be thrown upon small beds of earth made at different intervals apart. Also the Carthaginian and our own native chick-pea, and barley too and chickling should be produced out of store and put before them after having been soaked in rain-water; for dry food is not much sought after by hares. Moreover, it is easily understood even 9 without my mentioning it, concerning these animals and others like them, how inexpedient it is to introduce them into a park which is surrounded by railings, since owing to the small size of their bodies they can easily creep under the bars and, having obtained free exit, manage to escape.

II. I come now to the management of bee-hives, Bees. about which no instructions can be given with greater care than in the words of Hyginus,[a] more

Hygino iam dictum est, nec ornatius quam Vergilio, nec elegantius quam Celso. Hyginus veterum auctorum placita [1] secretis dispersa monimentis industrie collegit: Vergilius poeticis floribus illuminavit: Celsus utriusque memorati adhibuit mo-
2 dum. Quare ne attemptanda quidem nobis fuit haec disputationis materia, nisi quod consummatio susceptae professionis hanc quoque sui partem desiderabat, ne universitas inchoati operis nostri, velut membro aliquo reciso, mutila [2] atque imperfecta [3] conspiceretur. Atque ea, quae Hyginus fabulose tradita de originibus apum non intermisit, poeticae magis licentiae quam nostrae fidei concesserim.
3 Nec sane rustico dignum est sciscitari, fuerit ne mulier pulcherrima specie Melissa,[4] quam Iuppiter in apem convertit, an (ut Euhemerus poeta dicit) crabronibus et sole genitas apes, quas nymphae Phryxonides educaverunt, mox Dictaeo specu Iovis [5] extitisse nutrices, easque pabula munere dei sortitas quibus ipsae parvum [6] educaverant alumnum. Ista enim, quamvis non dedeceant poetam, summatim tamen et uno tantummodo versiculo leviter attigit Vergilius, cum sic ait:

Dictaeo caeli regem pavere sub antro.

[1] placita *ac* : plagita *SA*.
[2] mutila *SA* : mutilata *ac*.
[3] imperfecta *ac* : in infectu *SA*.
[4] mellisam *SA* : mellissa *a* : melisa (?) *c*.
[5] iovis *ac* : io quis *SA*.
[6] ipse parvum *ac* : ipsa et arvom *SA*.

---

[a] See Book I. 1. 14 and note.

ornately than by Vergil, or more elegantly than by Celsus.[a] Hyginus has industriously collected the opinions of ancient authors dispersed in their different writings; Vergil has embellished the subject with the flowers of poetry; and Celsus has applied the method of both the above-mentioned authors. 2 Therefore, we ought never to have even attempted to discourse on this subject, did not the fulfilment of the promise which we made call for the treatment of this part of our subject also, lest the body of the work begun, looked at as a whole, should appear mutilated and imperfect, as if a limb had been cut off. The tradition of the fabulous origin of the bees which Hyginus has not passed over I would rather attribute to poetic licence than submit to the test of our belief; nor indeed is it a fit question for a husbandman to 3 ask whether there ever existed a woman of surpassing beauty called Melissa, whom Jupiter changed into a bee, or whether (as Euhemerus[b] the poet says) the bees were bred from hornets and the sun, and that the nymphs, the daughters of Phryxon,[c] reared them, and that soon after they became the nurses of Jupiter in the Dictaean Cave[d] and that, by the gift of the god, they had allotted to them the food with which they themselves had reared their little foster-child. Upon this story, though not unworthy of a poet, Vergil touched briefly and lightly in a single line when he said:

'Neath Dicte's cave they fed the king of heaven.[e]

[b] A Greek writer who flourished about 300 B.C. and wrote a work *Hiera Anagraphe*, which rationalized mythology and which was translated into Latin by Ennius.

[c] This name is not otherwise mentioned in Latin literature.

[d] In Crete. [e] *Georg.* IV. 152. Dicte is Mount Sethia in Crete.

4 Sed ne illud quidem pertinet ad agricolas, quando et in qua regione primum natae sint: utrum in Thessalia sub Aristaeo, an in insula Cea, ut scribit Euhemerus, an Erechthei temporibus in monte Hymetto, ut Euthronius; an Cretae Saturni temporibus, ut Nicander: non magis quam utrum examina, tanquam cetera videmus animalia, concubitu subolem procreent, an heredem generis sui floribus eligant, quod affirmat noster Maro: et utrum evo-
5 mant liquorem mellis, an alia parte reddant. Haec enim et his similia magis scrutantium rerum naturae latebras, quam rusticorum est inquirere. Studiosis quoque literarum gratiora sunt ista in otio legentibus, quam negotiosis agricolis: quoniam neque in opere neque in re familiari quidquam iuvant.

III. Quare revertamur ad ea, quae alveorum cultoribus magis apta sunt quot [1] genera sunt apium et quid [2] ex his optimum.[3] Peripateticae sectae conditor Aristoteles in iis libris, quos de animalibus conscripsit, apium [4] examinum genera complura demonstrat, earumque alias [5] vastas sed glomerosas,

[1] quot *SA*²: quod *A*¹.
[2] quid *scripsi*: quod *A*: quot *S*.
[3] quot—optimum *om. ac.*
[4] apium sive *c*: *om. SAa.*
[5] aliaque *SAac.*

---

[a] Son of Apollo and Cyrene, also said to have planted the first olive-tree.

[b] Cea, or in Greek Ceos, an island, one of the Cyclades, near Cape Sunium.

[c] A mountain in Attica near Athens.

[d] Mythical king of Athens.

But it does not even concern husbandmen when and in what country bees first came into existence, whether in Thessaly under Aristaeus,[a] or in the island of Cea,[b] as Euhemerus writes, or on Mount Hymettus[c] in the time of Erechtheus,[d] as Euthronius[e] says, or in Crete in the time of Saturn, as Nicander[f] says. All this no more concerns farmers than the question whether the swarms of bees produce their offspring, as we see the other animals do, by copulation, or whether they pick up the heir of their race from the flowers, as our own poet Maro[g] affirms, and whether they vomit the liquid honey from their mouths or yield it from some other part. The inquiry into these and similar questions concerns those who search into the hidden secrets of nature rather than husbandmen. They are subjects more agreeable to the students of literature, who can read at their leisure, than to farmers who are busy folk, seeing that they are of no assistance to them in their work or in the increase of their substance. 4 5

III. Therefore let us return to topics which are more suitable to those who have charge of bee-hives, namely, how many kinds of bees there are and which of them is the best. Aristotle, the founder of the Peripatetic School, in the books which he wrote about animals,[h] shows that there are several kinds

The different kinds of bees and which is best.

[e] This name is not otherwise mentioned in Latin literature. We should perhaps read Euphonius; two agricultural writers of this name are mentioned by Varro (I. 1. 8), one of Athens and the other of Amphipolis.

[f] Physician, poet and grammarian of Colophon in Asia Minor; he flourished about 150 B.C. His *Theriaca* and *Alexipharmaca* have survived.

[g] Vergil, *Georg.* IV. 197 ff.

[h] *Hist. anim.* V. 22 (553[b]. 22 ff.).

easdemque nigras et hirsutas apes habent: alias[1] minores quidem, sed aeque[2] rotundas et fusci[3] coloris 2 horridique pili: alias[4] magis exiguas, nec tam rotundas, sed obesas tamen et latas, coloris meliusculi: nonnullas[5] minimas gracilesque, et acuti alvi, ex aureolo varias atque leves: eiusque[6] auctoritatem sequens Vergilius, maxime probat parvulas, oblongas, leves, nitidas,

Ardentes auro, et paribus lita corpora guttis,

moribus etiam placidis. Nam quanto grandior 3 apis, atque etiam rotundior, tanto peior. Si vero saevior, maxime pessima est. Sed tamen iracundia notae melioris apium facile delenitur[7] assiduo interventu eorum qui curant.[8] Nam cum saepius tractantur,[9] celerius mansuescunt, durantque si diligenter excultae sunt, in annos decem; nec ullum examen hanc[10] aetatem potest excedere, quamvis in demortuarum locum quotannis pullos substituant. Nam fere decimo ad internecionem anno gens universa 4 totius alvei consumitur. Itaque ne hoc in toto fiat apiario, semper propaganda erit soboles, observandumque vere cum se nova profundent examina, ut excipiantur, et domiciliorum numerus augeatur. Nam saepe morbis intercipiuntur, quibus quemadmodum mederi oportet, suo loco dicetur.

[1] alias *Aac*: alia *S*.
[2] aeque *ac*: neque *SA*.
[3] fusci *SA*: infusci *ac*.
[4] alia *SA*: alas *a*: alias *c*.
[5] nonnullas *ac*: nonnulla *SA*.
[6] eiusque *ac*: eius *SA*.
[7] delenitur *scripsi*: delinitur *ac*: denitur *SA*.
[8] curant *ac*: currant *AS*.
[9] tractantur *ac*: tractatur *SA*.
[10] hanc *Aac*: hac *S*.

of swarms of bees, some of them having bees huge and globular in shape and at the same time black and hairy; others smaller but equally round and of a 2 dusky colour and with bristling hairs; others still smaller but not so round, but nevertheless fat and broad and of rather a better colour; some very small and slender with bellies which end in a point, striped of a golden colour and quite smooth. Vergil, following Aristotle as his authority, approves most of bees which are very small, oblong, smooth and shining,

> Burning with gold, their bodies stained with spots of equal size,[a]

calm, too, in disposition; for the larger and rounder a bee is, the worse it is, and if it is unusually fierce, it is 3 by far the worst kind of all. However, the irascibility of the better kind of bees is easily soothed by the frequent intervention of those who look after them; for when they are often handled, they quickly become tame. If they are carefully looked after, they live for ten years; but no swarm can exceed this age, even if young stock is substituted yearly in place of those which have died; for usually in the tenth year all the population of the whole hive is destroyed and exterminated. In order, therefore, that this may 4 not be the fate of the whole apiary, fresh stock must be continually propagated and care must be taken in the spring, when the fresh swarms issue forth, that they are intercepted and the number of dwelling-places increased; for bees are often overtaken by diseases. The methods by which these ought to be cured will be dealt with in their proper places.

[a] *Georg.* IV. 99.

IV. Interim per has notas, quas iam diximus, probatis apibus destinari debent pabulationes, eaeque sint secretissimae, et ut[1] noster praecipit Maro, viduae[2] pecudibus, aprico et minime procelloso caeli statu:

Quo neque sit ventis aditus; nam pabula venti
Ferre domum prohibent: neque oves haedique petulci
Floribus insultent, aut errans bucula campo
Decutiat rorem, et surgentes atterat herbas.

2 Eademque regio fecunda sit fruticis exigui, et maxime thymi aut origani, tum etiam thymbrae, vel nostratis cunilae, quam satureiam[3] rustici vocant. Post haec frequens sit incrementi maioris surculus, ut rosmarinus,[4] et utraque cytisus. Est enim sativa et altera suae spontis. Itemque semper virens pinus, et minor ilex: nam prolixior ab omnibus improbatur. Ederae quoque non propter bonitatem recipiuntur, 3 sed quia praebent plurimum mellis. Arborum[5] vero sunt probatissimae, rutila atque alba ziziphus, nec minus tamarices,[6] tum etiam amygdalae, persicique, ac pyri, denique pomiferarum pleraeque, ne singulis immorer. Ac silvestrium commodissime faciunt glandifera robora, quin etiam terebinthus, nec dissimilis huic lentiscus[7] ac tiliae. Solae[8] ex omnibus nocentes

[1] ut et *SAac*.
[2] viduae *ac* : vide *SA*.
[3] satureiam *ac* : satyram etiam *S* : satyratis *A*.
[4] marinum *SAac*.
[5] arborum *SAa* : arbores *c*.
[6] tamarices *edd.* : amaracus *SAac*.
[7] *post* lentiscus *add.* et odorata cedrus *Aac* : *om. S*.
[8] solae *A* : sole *ac* : sola *S*.

IV. Meanwhile, when you have chosen your bees in accordance with the points which we have just mentioned, feeding-grounds ought to be assigned to the bees of which you approve. These should be as retired as possible and, as our Maro[a] directs, void of cattle and with a sunny aspect as little as possible exposed to storms,

Feeding-grounds for bees, and which is best.

> Where winds may not approach; for winds prevent
> The bees from bearing home their food; nor sheep,
> Nor frisky kids must trample down the flowers,
> Nor heifers wandering o'er the plain shake off
> The dews or crush the rising blades of grass.

The region should also be rich in small clumps, 2 especially thyme and marjoram and also in Greek savory and our own Italian savory, which the country-folk call *satureia*. Next let there be plenty of shrubs of larger growth, such as rosemary and both kinds of trefoil (for there is one variety which is sown and another which grows of its own accord), also the ever-green pine and the lesser holm-oak (for the taller variety is universally condemned). Ivy, too, is admitted not for its other good qualities but because it provides a large quantity of honey. Of 3 trees the following are very highly commended, the red and white jujube-trees, likewise tamarisks, also almond-trees and peach-trees and pear-trees, in a word, so as not to waste time in naming each kind, the majority of the fruit-bearing trees. Of woodland trees the most suitable are the acorn-bearing oaks, also terebinths and mastic-trees, which closely resemble them, and lime-trees. Of all the trees of this class

[a] Vergil, *Georg.* IV. 8–12.

4 taxi repudiantur. Mille praeterea semina vel crudo cespite virentia, vel subacta sulco, flores amicissimos apibus creant, ut sunt in virgineo [1] solo frutices amelli, caules acanthini, scapus asphodeli, gladiolus narcissi. At in hortensi lira consita nitent candida lilia, nec his sordidiora leucoia,[2] tum puniceae [3] rosae luteolaeque, et Sarranae violae, nec minus caelestis luminis hyacinthus, Corycius item Siculusque bulbus croci 5 deponitur, qui coloret odoretque mella.[4] Iam vero notae vilioris innumerabiles nascuntur herbae cultis atque pascuis regionibus, quae favorum ceras exuberant: ut vulgares lapsanae, nec his pretiosior armoracia, rapistrique olus, et intubi silvestris ac nigri papaveris flores, tum agrestis pastinaca, et eiusdem nominis edomita, quam Graeci σταφυλῖνον [5] 6 vocant. Verum ex cunctis, quae proposui, quaeque omisi temporis [6] compendia sequens (nam inexputabilis erat numerus) [7] saporis praecipui mella reddit thymus.[8] Eximio deinde proximum thymbra, serpyllumque et origanum. Tertiae notae, sed adhuc generosae, marinus [9] ros et nostras cunila, quam dixi satureiam. Mediocris deinde gustus tamaricis [10] ac ziziphi flores, reliquaque, quae [11] proposuimus, cibaria. 7 Sed ex sordidis deterrimae notae mel habetur [12]

[1] virgineo *SAac* : irriguo *edd.*

[2] sordidiora leucoia *ac* (*in marg. A*) : sordido la reucolatum *SA*.

[3] puniceae *edd.* : punice *SAac*.

[4] mella *ac* (*in marg. A*) : *om. S.*

[5] σταφυλῖνον *A marg.* : σταφυαει non *A* : σταφυλει non *S*.

[6] temporis *c* : temporum *SAa*.

[7] erat numerus *ac* : et enumeri *SA*.

[8] thymus *ac* : thymum *SA*.

[9] marinum *SAac*.

[10] tamaricis *edd.* : amarachinia *SA* : amaranchini *ac*.

[11] reliqua quae *SAa* : reliquaque *c*.

yews only are excluded as being hurtful. Moreover 4 a thousand seeds, which flourish in uncultivated turf or are turned up in the furrow, produce flowers which are much loved by bees, for example shrubs of starwort [a] in virgin soil, stalks of bear's foot,[b] stems of asphodel and the sword-like leaf of the narcissus. White lilies sown between the furrows in the garden make a brilliant show and the gilliflowers have no less pure a colour; then there are red and yellow roses and purple violets and sky-blue larkspur; also the Corycian [c] and Sicilian saffron-bulbs are planted to give colour and scent to the honey. Moreover, 5 countless herbs of a baser kind spring up on cultivated land and pasture which supply an abundance of wax for the honey-combs, such as the common charlock and the horse-radish, which is no more precious, the mustard-herb, and flowers of wild endive and black poppy, also the field parsnip, and the cultivated variety which bears the same name and which the Greeks call *staphylinos* (carrot). But of 6 all the plants which I have suggested and of those which I have not mentioned so as to save time (for their number could not be computed), thyme yields honey with the best flavour; the next best are Greek savory, wild thyme and marjoram. In the third class, but still of high quality, are rosemary and our Italian savory, which I have called *satureia*. Next the flowers of the tamarisk and the jujube-tree and the other kinds of fodder which I suggested have only a mediocre flavour. The honey which is considered 7

[a] *Aster amellus*. [b] *Acanthus mollis*.
[c] Corycus was in Cilicia in southern Asia Minor.

---

[12] habetur *c*: habentur *SAa*.

nemorense, quod sparto atque arbuto[1] provenit: villaticum, quod nascitur in oleribus.[2] Et quoniam situm pastionum atque etiam genera pabulorum exposui, nunc de ipsis receptaculis et domiciliis examinum loquar.

V. Sedes apium collocanda est contra brumalem meridiem procul a tumultu, et coetu hominum ac pecudum, nec calido loco, nec frigido: nam utraque re infestantur. Haec autem sit ima parte vallis, et ut vacuae cum prodeunt pabulatum apes, facilius editioribus advolent, et collectis utensilibus cum onere per proclivia non aegre devolent.

Si villae situs ita competit, non est dubitandum quin aedificio iunctum apiarium maceria circumdemus, sed in ea parte, quae tetris latrinae ster-
2 quiliniique et a balinei libera est odoribus. Verum[3] si positio repugnabit, nec maxima tamen incommoda congruent,[4] sic quoque magis expediet sub oculis domini esse apiarium. Sin autem cuncta fuerint inimica, certe vicina vallis occupetur, quo saepius descendere non sit grave possidenti. Nam res ista maximam fidem desiderat; quae quoniam rarissima est, interventu domini tutius custoditur. Neque ea

[1] arbusto *SAac*.

[2] *post* oleribus *add.* et stercorosis herbis *a*: *om. SA*: et stercoris herbis *c* (*in marg.*) *A*.

[3] verum A^2ac: vel et *S*: vellet A^1.

[4] congruent A^2ac: congluent SA^1.

of the poorest quality is the woodland honey which comes from dirty feeding-grounds and is produced from broom-trees and strawberry-trees, and the farm-house honey which comes from vegetables. Now that I have described the situation of the feeding-grounds and also the various kinds of food, I will next speak of the arrangement for receiving and housing the swarm.

On the best situation for an apiary.

V. A position must be chosen for the bees facing the sun at midday in winter, far from the noise and the assemblage of men and beasts and neither hot nor cold, for bees are troubled by both these conditions. It should be situated in the bottom of a valley, that the empty bees, when they go forth to feed, may be able more easily to fly up to the higher ground, and also, when they have collected what they require, they may fly with their burden on a down-hill course without any difficulty.

If the situation of the farm permits, we ought not to hesitate to join the apiary to a building and surround it with a wall, but it must be on the side of the house which is free from the foul odours which come from the latrines, the dunghill and the bath-room. 2 If, however, this position has drawbacks, but yet the worst disadvantages are not all present, even under these conditions it will be more expedient for the apiary to be under the master's eye. If, however, everything is unfavourable, at all events a valley should be pitched upon close at hand, so that the owner may be able to go down rather often and visit it without grave inconvenience; for in bee-keeping perfect honesty is necessary, and since this is very rare, it is better secured by the intervention of the master. Not only is an overseer who

curatorem fraudulentum tantum, sed etiam immundae segnitiae perosa est. Aeque enim dedignatur, si minus pure habita est, ac si tractetur fraudulenter.

3 Sed ubicumque fuerint alvearia [1] non editissimo claudantur muro. Qui si metu praedonum sublimior placuerit, tribus elatis ab humo pedibus, exiguis in ordinem fenestellis apibus sit pervius: iungaturque tugurium, quod et custodes habitent, et quo [2] condatur instrumentum: sitque maxime repletum praeparatis alveis [3] ad usum novorum examinum, nec minus herbis salutaribus, et siqua sunt alia, quae languentibus adhibentur.

4 Palmaque vestibulum aut ingens oleaster obumbret,
Ut cum prima novi [4] ducent examina reges,
Vere suo, ludetque favis emissa iuventus:
Vicina invitet decedere ripa calori,
Obviaque hospitiis teneat frondentibus arbos.

Tum perennis aqua, si est facultas, inducatur, vel
5 extracta [5] manu detur, sine qua neque favi neque mella nec pulli denique figurari queunt. Sive igitur, ut dixi, praeterfluens unda vel putealis canalibus

[1] alvearia A^2ac : albaria SA^1.
[2] quo *om.* *SAac.*
[3] alveis A^2ac : alubis *SA*.
[4] prima novi *edd.* : vere novo *SAac*.
[5] extracta SA^1 : extructo A^2ac.

is fraudulent abhorrent to the business but also one whose laziness causes filthy conditions; for bee-keeping revolts alike against a lack of cleanliness and against fraudulent management.

Wherever the hives are placed, they should not be enclosed within very high walls. If, through fear of robbers, a rather lofty wall is thought desirable, passages through it should be made for the bees in the form of a row of little windows three feet above the ground, and there should be an adjoining cottage in which the keepers may live and the apparatus may be stored. The store-house should be chiefly occupied by hives ready for the use of new swarms and also by health-giving herbs and any other remedies which may be applied to bees when they are sick. 3

And let a palm or vast wild-olive tree 4
O'ershade the porch, that when new kings lead forth
The infant swarms and the young bees make sport
In their own spring, from honey-combs set free;
Then let the neighbouring bank invite retreat
From mid-day heat, and let the sheltering tree
Hold them in leafy hospitality.[a]

Next let ever-flowing water, if it is available, be introduced or drawn by hand and provided, without which neither combs nor honey nor even young bees can be formed. Whether, therefore, as I have said, it be running water which has been conveyed in channels or well-water, it should con- 5

[a] Vergil, *Georg.* IV. 20 ff.

immissa fuerit, virgis ac lapidibus aggeretur apium causa,

> Pontibus ut crebris possint consistere, et alas
> Pandere ad aestivum solem, si forte morantis
> Sparserit, aut praeceps Neptuno immerserit Eurus.

6 Conseri deinde circa totum apiarium debent arbusculae incrementi parvi, maximeque propter salubritatem (nam sunt etiam remedio languentibus) cytisi, tum deinde casiae atque pini et rosmarinus:[1] quin etiam cunilae et thymi frutices, item violarum, vel quaecunque[2] utiliter deponi patitur qualitas terrae. Gravis et tetri odoris non solum virentia sed et quaelibet res prohibeantur, sicuti cancri nidor, cum est ignibus adustus, aut odor palustris caeni. Nec minus vitentur cavae rupes aut vallis argutiae, quas Graeci vocant ἠχοῦς.[3]

VI. Igitur ordinatis sedibus, alvearia[4] fabricanda sunt pro conditione regionis. Sive illa ferax est suberis, haud dubitanter utilissimas alvos[5] faciemus ex corticibus, quia nec hieme frigent,[6] nec candent aestate; sive ferulis exuberat, iis quoque, quod sunt naturae corticis similes, aeque commode vasa texuntur. Si neutrum aderit, opere textorio[7] salicibus connectentur: vel si nec haec suppetent,

[1] marinum *SAac*.
[2] quaecunque *A²ac* : quae *SA¹*.
[3] ἠχοῦς *om. ac*.
[4] alvearia *ac* : albaria *SA*.
[5] alvos *A²c* : albos *SA¹* : alveos *a*.
[6] frigent *SAa* : rigent *c*.
[7] opere textorio *edd.* : opererio *S* : operario *Aa* : opere vitorio *c*.

tain heaps of sticks and stones for the use of the bees,

> That upon frequent bridges they may rest
> And spread their wings to catch the summer sun,
> If swift east winds have caught them loitering
> And rained on them or plunged them in the deep.[a]

Next, round the whole apiary, little trees of small 6 growth ought to be planted and in particular shrub-trefoils on account of their health-giving properties (for they are a remedy for bees when they are listless); also wild cinnamon and pines and rosemary, and clumps of marjoram and thyme and violets and whatever else the nature of the ground allows to be profitably planted. Not only growing things but also anything whatsoever which has a disagreeable and noisome odour should be kept away from the apiary, such as the smell of a crab when it is burnt on the fire or the odour of mud taken from a marsh. Likewise let hollow rocks and shrill noises produced by valleys, which the Greeks call echoes, be avoided.

VI. When, therefore, the sites have been arranged, beehives must be constructed in accordance with local conditions. If the place is rich in cork-trees, we shall certainly make the most serviceable hives from their bark, because they are neither cold in winter nor hot in summer; or if it grows plenty of fennel-stalks, with these too, since they resemble the nature of bark, receptacles can be quite as conveniently made by weaving them together. If neither of these materials is at hand, the hives can be made by plaiting withies together; or, if these are not available either, they will have to be made with

On the choice of beehives.

[a] Vergil, *Georg.* IV. 27 ff.

ligno cavae[1] arboris aut[2] in tabulas desectae fabricabuntur. 2 Deterrima est conditio fictilium, quae et accenduntur aestatis vaporibus, et gelantur hiemis frigoribus. Reliqua sunt alvorum genera duo, ut vel ex fimo fingantur,[3] vel lateribus extruantur: quorum alterum iure damnavit Celsus, quoniam maxime est ignibus obnoxium; alterum probavit, quamvis incommodum eius praecipuum non dissimulaverit, quod, si res postulet, transferri non possit. 3 Itaque non assentior ei, qui putat nihilo minus eius generis habendas esse alvos: neque enim solum id repugnat rationibus domini, quod immobiles sint, cum vendere aut alios agros instruere velit; (hoc enim commodum pertinet ad utilitatem solius patrisfamilias) sed, quod ipsarum apium causa[4] fieri debet, cum aut morbo aut sterilitate et penuria locorum vexatas conveniet[5] in aliam regionem mitti, nec propter praedictam causam moveri poterunt,[6] hoc 4 maxime vitandum est. Itaque quamvis doctissimi viri auctoritatem reverebar, tamen ambitione submota, quid ipse censerem, non omisi. Nam quod maxime movet Celsum, ne sint stabula vel igni vel furibus obnoxia, potest vitari opere lateritio circumstructis alvis, ut impediatur rapina praedonis, et contra flammarum violentiam protegantur:[7] easdemque, cum fuerint movendae, resolutis structurae compagibus, licebit transferre.

[1] cave *SA* : cavatae *ac*.
[2] aut *ac* : *om. SA*.
[3] fingantur *ac* : finguntur *SA*.
[4] causa *ac* : coriosa *SA*.
[5] conveniet *SAa* : -at *c*.
[6] poterint *S* : -ant *Aac*.
[7] proteguntur *SAac*.

wood of a tree either hollow or cut up into boards. Those made of earthenware have the worst qualities 2 of all, since they are burnt by the heat of summer and frozen by the cold of winter. Two kinds of hives remain to be described, those which are either made of dung or built of bricks. Celsus was right in condemning the former because it is very liable to catch fire; the latter he approved, although he made no secret of its chief disadvantage, namely, that if occasion should arise, it cannot be moved to another site. I do not agree with him who thinks that hives 3 of this kind ought to be used in spite of this drawback, for it is not only against the interests of the owner that they should be immovable when he wants to sell them or furnish another site with hives (for these considerations concern the convenience of the owner alone), but the question arises as to what ought to be done for the sake of the bees themselves, when it is advisable that they should be sent to another district because they are suffering from disease or from the barrenness and poverty of the locality and yet cannot be moved for the reason mentioned above—a state of affairs which ought above all things to be avoided. So, though holding in respect the 4 authority of a learned man, yet, without seeking to set myself up against him, I have not omitted to express my own opinion. For Celsus' chief anxiety, lest the bees' quarters should be exposed to fire or thieves, can be avoided by building a brick wall round the hives to prevent the plundering of robbers and to give protection against the violence of fire, and, when the hives have to be moved it will be possible to take apart the framework of the structure and move the hives elsewhere.

VII. Sed quoniam plerisque videtur istud operosum, qualiacunque vasa placuerint, collocari debebunt. Suggestus lapideus extenditur per totum apiarium in [1] tres pedes altitudinis [2] extructus, isque diligenter opere tectorio levigatur, ita ne ascensus lacertis, aut anguibus, aliisve noxiis animalibus 2 praebeatur. Superponuntur deinde, sive, ut Celso placet, lateribus facta domicilia, sive, ut nobis, alvearia, praeterquam a tergo [3] circumstructa: seu, quod paene omnium in usu est, qui modo diligenter ista curant, per ordinem vasa disposita ligantur, vel laterculis, vel caementis, ita ut singula binis parietibus angustis contineantur, liberaeque frontes utrimque sint. Nam et qua procedunt, nonnunquam patefaciendae sunt,[4] et multo magis a tergo, quia subinde curantur ex-3 amina. Sin autem nulli parietes alvis intervenient, sic tamen collocandae erunt, ut paulum altera ab altera distet, ne,[5] cum inspiciuntur, ea, quae in curatione tractatur, haerentem sibi alteram concutiat, vicinasque apes conterreat, quae omnem motum imbecillis ut cereis [6] scilicet operibus suis tamquam ruinam timent. Ordines quidem vasorum superinstructos in altitudinem tres esse abunde est, quoniam summum sic quoque parum commode curator in-4 spicit. Ora cavearum, quae praebent apibus vestibula, proniora sint quam terga, ut ne influant imbres,

[1] in *ac*: per *SA*.
[2] *post* altitudinis *add.* totidemque crassitudinis *ac*: *om. SA.*
[3] *ante* tergo *add.* et frontibus *SAa*: *om. edd.*
[4] sunt *Aac*: sint *S*.
[5] nec *SAac*.
[6] cereis *ac*: ceteris *SA*.

VII. But since most people regard all this as involving too much trouble, whatever kind of receptacles take their fancy will have to be arranged thus. A bank made of stones built three feet high is stretched across the apiary and carefully smoothed over with plaster, so that no chance of climbing it may be offered to lizards and snakes or other harmful creatures; then on the top of it are placed either bee-houses made with bricks, which Celsus prefers, or, as we prefer, hives walled round except at the back; or else—and this is the practice of almost all those who are careful in these matters—receptacles arranged in a row are fastened together either with small bricks or with unhewn stones in such a way that each is contained within two narrow walls and the two sides, at the back and at the front, are left free; for the sides on which they issue forth have sometimes to be opened and this is especially necessary at the back because the swarms have to be attended to from time to time. If there are no partitions between the hives, they will, nevertheless, have to be so placed as to be at a little distance from one another, so that, when they are being inspected, one which is handled in the course of being attended to may not shake another which is closely joined to it, and alarm the neighbouring bees, which are afraid of every movement as threatening ruin to their structures which are frail, being of wax. It is quite enough to have three rows of hives one above the other, since even so the man who looks after them cannot very conveniently inspect the top row. The fronts of the hives, which afford entries for the bees, should slope down more than their backs, so that the rain may not flow in,

On the position of beehives.

2

3

4

et si forte tamen incesserint,[1] non immorentur, sed per aditum effluant. Propter quos convenit alvearia porticibus supermuniri; sin aliter, luto Punico frondibus inlimatis adumbrari, quod tegmen cum frigora et pluvias, tum et aestus arcet. Nec tamen ita nocet huic generi calor aestatis ut hiemale frigus.[2] Itaque semper aedificium sit post apiarium, quod Aquilonis excipiat iniuriam, stabulisque praebeat 5 teporem. Nec minus ipsa domicilia, quamvis aedificio protegantur,[3] obversa tamen ad hibernum orientem componi debebunt, ut apricum habeant apes matutinum egressum, et sint experrectiores. Nam frigus ignaviam creat; propter quod etiam foramina, quibus exitus aut introitus datur, angustissima esse debent, ut quam minimum frigoris admittant: eaque satis est ita forari, ne possint [4] capere plus unius apis incrementum. Sic nec venenatus stellio, nec obscaenum scarabaei [5] vel papilionis genus, lucifugaeque blattae, ut ait Maro, per laxiora spatia ianuae favos 6 populabuntur. Atque utilissimum est pro frequentia domicilii duos vel tres aditus in eodem operculo distantes inter se fieri contra fallaciam lacerti, qui velut custos vestibuli [6] prodeuntibus inhians [7] apibus affert exitium, eaeque pauciores intereunt, cum

[1] incesserit *SA* : ingesserint *ac*.

[2] calor aestatis ut hiemale frigus *Gesner ex Palladii citatione* : caloris ut hiemalitus *SA* : caloris ut hiemis alitus *a* : caloris aut hiemis estus *c*.

[3] protegantur *ac* : -untur *SA*.

[4] possint *ac* : possit *SA*.

[5] scarabei *c* : -ri *SAa*.

[6] vestibuli *SAac*.

[7] inhians *ac* : in hanc *SA*.

---

[a] The text here is doubtful but the sense clear.

[b] Vergil, *Georg.* IV. 243.

and that, if by chance it does find its way in, it may not remain there but flow out through the entrance. Also, on account of the rain, the hives should be protected above with colonnades, or, failing these, they should be overshadowed by green foliage daubed over with Carthaginian clay, forming a covering which keeps off both the cold and rain and also the heat. However the heat of summer is not so harmful to this kind of creature as the cold of winter,[a] and so there should always be a building behind the apiary to intercept the violence of the north wind and provide warmth for the hives. Likewise the bees' dwelling-places, although they 5 are protected by buildings, ought to be so arranged as to face the south-east, in order that the bees may enjoy the sun when they go out in the morning and may be more wide-awake; for cold begets sloth. For the same reason, too, the holes through which they go in and out ought to be very narrow, so as to admit as little cold as possible; indeed it is enough that they should be so bored that they cannot admit the bulk of more than one bee at a time. Thus neither the poisonous gecko nor the foul race of beetles and butterflies and the cockroaches that shun the day-light, as Maro says,[b] will not lay waste the honey-combs by having too wide an entrance to pass through. It is also a most useful 6 device to have made in proportion to the number of bees in the hive, two or three entrances in its outer covering at a distance from one another to defeat the craftiness of the lizard, which standing like a door-keeper at the entry, with open mouth, brings destruction upon the bees as they come forth, and fewer of them perish when they are at liberty to

licet[1] vitare pestis obsidia per aliud volantibus[2] effugium.

VIII. Atque haec de pabulationibus, domiciliisque[3] et sedibus eligendis abunde diximus: quibus provisis, sequitur ut examina desideremus. Ea[4] porro vel aere parta, vel gratuita contingunt. Sed quas pretio comparabimus, scrupulosius praedictis comprobemus notis, et earum frequentiam prius quam 2 mercemur, apertis alvearibus consideremus: vel si non fuerit inspiciendi facultas, certe id quod contemplari licet, notabimus:[5] an in vestibulo ianuae complures consistant, et vehemens sonus intus murmurantium exaudiatur. Atque etiam si omnes intra domicilium silentes forte conquiescent, labris foramini aditus admotis, et inflato spiritu ex respondente earum subito[6] fremitu poterimus aestimare vel multitudinem, vel paucitatem.

3 Praecipue autem custodiendum est, ut ex vicinia potius, quam peregrinis regionibus petantur, quoniam solent caeli novitate lacessiri. Quod si non contingit, ac necesse habuerimus longinquis itineribus advehere, curabimus[7] ne salebris solicitentur, optimeque noctibus collo portabuntur. Nam diebus requies danda est, et infundendi sunt grati apibus 4 liquores, quibus intra clausum alantur. Mox cum perlatae domum fuerint, si dies supervenerit, nec

[1] licet *c* : liceant *SA* : liceat *a*.
[2] volantibus *S* : ulantibus *A* : vadentibus *ac*.
[3] domiciliisque *Sa* : domiciliis et *c* : domicilibusque *A*.
[4] ea *ac* : *om. SA*.
[5] notabimus *ac* : notavimus *SA*.
[6] subito *ac* : sumito *SA*.
[7] curabimus *c* : curavimus *SA* : *om. a*.

avoid the pest which lies in wait for them by flying out by another passage.

On the purchase of bees and the taking of swarms.

VIII. We have now said enough about the choice of feeding-grounds, dwelling-places and their sites. These having been provided, the next things that we require are swarms of bees. These come to us either by purchase or without being paid for. Those which we are going to buy we shall test with particular care by means of the points already mentioned, and we must consider how numerous they are before we purchase them, by opening the hives; or if there are 2 no facilities for inspecting them, we shall at any rate take note of what we are allowed to see, namely, whether a goodly number of bees are standing in the entrance-porch and whether a loud noise is to be heard of bees buzzing inside. Also if it so happens that they are all silent and at peace within their dwelling-place, we shall be able to estimate their great or small number from the sudden noise on the part of the bees as a result of our applying our lips to the hole by which they enter and blowing into it.

But we must be particularly careful that the 3 swarms are brought from the neighbourhood rather than from distant regions, since they are usually irritated by a change of climate. But if this is impossible and we are obliged to convey them over long distances, we shall be careful that they are not disturbed by the roughness of the road, and they will be best carried on the shoulders and at night; for they must be given rest in the day-time, and liquids which they like must be poured into the hives, so that they may be fed while remaining shut up. Then 4 when they have arrived at their destination, if daylight has come on, the hive must be neither opened

aperiri nec collocari oportebit alvum, nisi vesperi, ut apes placidae mane post totius requiem[1] noctis egrediantur: specularique debebimus[2] fere triduo, numquid universae se profundant. Quod cum faciunt, fugam meditantur. Ea remediis quibus debeat inhiberi, mox praecipiemus.

5 At quae dono vel aucupio contingunt, minus scrupulose probantur: quamquam ne sic quidem velim nisi optimas possidere, cum et impensam et eandem operam custodis postulent bonae atque improbae: et quod[3] maxime refert,[4] non sunt degeneres intermiscendae, quae infament generosas. Nam minor fructus mellis respondet, cum segniora 6 interveniunt examina. Verumtamen quoniam interdum propter conditionem locorum vel mediocre pecus (nam malum nullo quidem modo) parandum est, curam vestigandis examinibus hac ratione adhibe-7 bimus. Ubicunque saltus sunt idonei, mellifici, nihil antiquius apes, quam, quibus utantur, vicinos eligunt fontes. Eos itaque convenit plerumque ab hora secunda obsidere, specularique quae turba sit aquantium. Nam si paucae admodum circumvolant (nisi tamen complura capita rivorum diductas faciunt rariores) intelligenda est earum penuria, propter quam locum quoque non esse mellificum suspica-8 bimur. At si commeant frequentes, spem quoque

[1] requiem *ac* : quiem *SA*.
[2] debebimus *SA* : debemus *ac*.
[3] quod *ac* : quoque *SA*.
[4] refert *Aac* : referunt *S*.

nor placed in position until evening comes, so that the bees may go forth quietly in the morning after a whole night's rest, and we shall need to watch carefully for about three days to see whether they all sally forth in a body; for when they do this, they are meditating escape. We will presently prescribe what remedies we ought to apply to prevent this.

5 Bees which come to us by gift or by capture are accepted less scrupulously, although even in these circumstances I would not care to possess any but the best, since good and bad bees require the same expenditure and the same labour on the part of their keeper; also (and this is especially important) inferior bees should not be mixed with those of high quality, since they bring discredit upon them; for a smaller yield of honey rewards your efforts when the idler swarms take part in the gathering of it. 6 Nevertheless, since sometimes, owing to local conditions, an indifferent set of bees has to be procured (though never on any account should a bad one be acquired), we shall exercise care in seeking out swarms by the following method. 7 Wherever there are suitable woodlands where honey can be gathered, there is nothing that the bees would sooner do than make choice of springs near at hand for their use. It is a good plan, therefore, usually to frequent these springs from the second hour onwards and watch how many bees come to them for water. For if only a few are flying about (unless there are several sources of water which attract them and cause them to be more widely dispersed) we must conclude that there is a scarcity of them, which will make us suspect that the place will not produce much honey. 8 But if they come and go in large numbers, they inspire greater

aucupandi examina maiorem faciunt; eaque sic inveniuntur. Primum quam longe sint explorandum est, praeparandaque [1] in hanc [2] rem liquida rubrica: qua cum festucis illitis contigeris apium terga fontem libantium, commoratus [3] eodem loco facilius redeuntes agnoscere poteris; ac si non tarde id [4] facient, scies eas in vicino consistere: sin autem serius, pro 9 morae tempore [5] aestimabis distantiam loci. Sed cum animadverteris celeriter redeuntes, non aegre persequens iter volantium ad sedem perduceris examinis. In iis autem quae longius meare videbuntur, solertior adhibebitur cura, quae talis est. Arundinis internodium cum suis articulis exciditur, et terebratur ab latere talea [6] per quod foramen exiguo melle vel defruto [7] instillato, ponitur iuxta fontem. Deinde cum ad odorem dulcis liquaminis complures apes irrepserunt,[8] tollitur talea, et apposito [9] foramini pollice non emittitur, nisi una, quae cum evasit, fugam suam demonstrat observanti: atque is, dum sufficit, persequitur evolantem. 10 Cum [10] deinde conspicere possit [11] apem, tum [12] alteram emittit: et si eandem petit [13] caeli partem, vestigiis prioribus inhaeret. Si minus, aliam atque aliam foramine adaperto patitur egredi; regionemque

[1] praeparandumque *SAac.*
[2] hac *SA* : hanc *ac.*
[3] commoratur *SAc* : -os *a.*
[4] id *ac* : in *SA.*
[5] temporis *SAac.*
[6] alba ter et alia *S* : alvatere talea *A* : ab latere talea *a* : ab latera talea *c.*
[7] defriti *SA* : defruto *ac.*
[8] irrepserunt *ac* : inperserunt *SA.*
[9] apposito *ac* : imposito *SA.*
[10] cum *ac* : *om. SA.*
[11] desit *ac* : possit *SA.*

hopes of our catching swarms of them; and the following is the method of finding them. First we must try to discover how far away they are, and for this purpose liquid red-ochre must be prepared; then, after touching the backs of the bees with stalks smeared with this liquid as they are drinking at the spring, waiting in the same place you will be able more easily to recognize the bees when they return. If they are not slow in returning, you will know that they dwell in the neighbourhood; but if they are late in doing so, you will calculate the distance by the period of their delay. 9 If you notice them returning quickly, you will have no difficulty in following the course of their flight and will be led to where the swarm has its home. As regards those who apparently go farther away, a more ingenious plan will be adopted, as follows. The joint of a reed with the knots at either end is cut and a hole bored in the side of the rod thus formed, through which you should drop a little honey or boiled-down must. The rod is then placed near a spring. Then when a number of bees, attracted by the smell of the sweet liquid, have crept into it, the rod is taken away and the thumb placed on the hole and one bee only released at a time, which, when it has escaped, shows the line of its flight to the observer, and he, as long as he can keep up, follows it as it flies away. 10 Then, when he can no longer see the bee, he lets out another, and if it seeks the same quarter of the heavens he persists in following his former tracks. Otherwise he opens the hole and allows them to

---

[12] apem tum *ac* : apertum *SA*.
[13] petit *ac* : cepit *SA*.

notat, in quam plures revolent, et eas persequitur, donec ad latebram perducatur examinis.

Quod sive est abditum specu, fumo elicietur, et cum erupit, aeris strepitu coercetur. Nam statim sono territum vel in frutice vel in editiore silvae fronde considet, et a vestigatore praeparato vase[1] recon-
11 ditur. Sin autem sedem habet arboris cavae, et aut extat ramus, quem obtinent, aut sunt in ipsius arboris trunco,[2] tunc, si[3] mediocritas patitur, acutissima serra, quo celerius id fiat, praeciditur primum superior pars, quae ab apibus vacat; deinde inferior, quatenus videtur inhabitari. Tum recisus utraque parte mundo vestimento contegitur, quoniam hoc quoque plurimum refert, ac si quibus rimis hiat illitis ad locum perfertur: relictisque[4] parvis, ut iam dixi, foraminibus,[5] more ceterarum alvorum collocatur.[6]
12 Sed indagatorem convenit matutina tempora vestigandi eligere, ut spatium diei habeat, quo exploret commeatus apium. Saepe enim, si serius coepit eas denotare, etiam cum in propinquo sunt, iustis operum peractis se recipiunt, nec remeant ad aquam: quo evenit ut vestigator ignoret, quam longe a fonte

[1] vase *ac*: vaso *SA*.
[2] in ipsius arboris trunco *edd.*: aut ipsius truncis (trucis *A*) *SA*: aut ipsius trunci *c*: ipsius trunci *a*.
[3] in eo sunt *SA*: si in eo *ca*.
[4] relictisque *ac*: relictis *SA*.
[5] foraminibus *ac*: certaminibus *S*: certaminis *A*.
[6] collocatur *ac*: -antur *SA*.

emerge one after another, and marks the direction in which most of them fly home, and pursues them until he is led to the lurking-place of the swarm.

If it is hidden in a cave, the swarm will be driven out with smoke, and when it has sallied forth, it is checked by the noise of brass being beaten; for, terrified by the sound, it will immediately settle on a shrub or on a higher kind of foliage, that of a tree, and is enclosed in a vessel prepared for the purpose by the man who has tracked down the bees. But if the 11 swarm has its home in a hollow tree and either the branch which the bees occupy stands out from the tree or they are inside the trunk of the tree itself, then, if the small size of the branch or trunk allows, first the upper part, which is empty of bees, is cut through with a saw which should be very sharp so that the operation may be more quickly carried out, and then the lower part so far as it seems to be inhabited. Then, when it has been cut off at both ends, it is covered with a clean garment (for this too is very important), and if there are any gaping holes,[a] they are daubed over, and it is carried to the place where the bees are kept, and, small holes being left in it, as I have said, it is put in position like the rest of the hives. The searcher for swarms should 12 choose the morning for his search, so that he may have the whole day to spy out the comings and goings of the bees. For often, if he is too late in beginning to observe them, when they have finished their usual tasks, they go home and do not return to the water, even though they are near at hand, with the result that the man who is searching for them does not know how far away the swarm is from the

[a] *I.e.* in the vessel.

13 distet examen. Sunt qui per initia veris apiastrum, atque, ut ille vates ait,

Trita melisphylla et cerinthae ignobile gramen,

aliasque colligant similes herbas, quibus id genus animalium delectatur, et ita alvos perfricent, ut odor et succus vasis inhaereat: quae deinde mundata exiguo melle respergant, et per nemora non longe a fontibus disponant, eaque cum repleta sunt exami-
14 nibus, domum referant. Sed hoc nisi locis, quibus abundant apes, facere non expedit. Nam saepe vel inania vasa nacti, qui forte praetereunt, secum auferunt: neque est tanti vacua perdere complura, ut uno vel altero potiare [1] pleno. At in maiore copia, etiam si multa [2] intercipiuntur, plus est quod in repertis apibus acquiritur. Atque haec est ratio capiendi silvestria examina.

IX. Deinceps talis altera est vernacula retinendi.[3] Semper quidem custos sedule circumire debet alvearia. Neque enim ullum tempus est, quo non curam desiderent; sed eam postulant diligentiorem, cum vernant et exundant novis fetibus, qui nisi curatoris obsidio protinus excepti sunt, diffugiunt. Quippe talis est apium natura, ut pariter quaeque plebs generetur cum regibus; qui ubi evolandi vires adepti sunt, consortia dedignantur vetustiorum, multoque magis

[1] potiare *ac*: patiore *SA*.
[2] si multa *Sa*: simulata *Ac*.
[3] *post* retinendi *add.* quem ad modum vernacula nova examina observentur et in alvos condantur *SA*: *om. ac*.

[a] Vergil, *Georg.* IV. 63.

fountain. There are some people who during the 13 early spring collect wild parsley and, in the words of the great poet,

Bruised balm and wax-flower's lowly greenery,[a]

and other similar herbs in which this kind of creatures takes delight, and rub the hives thoroughly with them, so that the scent and juice stick to them; then, after cleaning them, they sprinkle them with a little honey and place them here and there in the woods not far from the springs and, when they are full of swarms, they carry them back home. It is 14 not, however, expedient to do this except in places where there is an abundance of bees, because it often happens that chance passers-by, finding the hives empty, carry them off with them, nor is the possession of one or two full of bees enough to compensate for the loss of several empty hives. But where bees are more plentiful, even if many hives are carried off, their loss is made up by the bees which are obtained. Such is the method of catching wild swarms of bees.

The treatment of bees bred in the home apiary.

IX. Next there is another method of retaining the swarms produced from our own apiaries. The keeper ought always diligently to go round the hives, for there is no time when they do not need his care; but they demand still more careful attention when the bees feel the approach of spring and the hives overflow with new offspring, which, unless they are promptly intercepted by the constant watchfulness of the keeper, fly off in different directions. For such is the nature of bees that each brood of ordinary bees is generated together with its king and, when they have acquired enough strength to fly away, they despise the society of their elders and even more the

imperia: quippe cum rationabili generi[1] mortalium, tum magis egentibus consilii mutis[2] animalibus, nulla 2 sit regni societas. Itaque novi duces procedunt cum sua iuventute, quae uno aut altero die in ipso domicilii vestibulo glomerata consistens, egressu suo propriae desiderium sedis ostendit; eaque tanquam[3] patria contenta est, si a procuratore[4] protinus assignetur. Sin autem defuit custos, velut iniuria 3 repulsa[5] peregrinam regionem petit. Quod ne fiat, boni curatoris est vernis temporibus observare[6] alvos[7] in octavam fere diei horam, post quam[8] non temere se nova proripiunt agmina;[9] eorumque[10] egressus diligenter custodiat. Nam quaedam solent, cum subito evaserunt, sine cunctatione se proripere. 4 Poterit exploratam fugam praesciscere vespertinis temporibus aurem singulis alveis admovendo. Siquidem fere ante triduum, quam eruptionem facturae sint, velut militaria[11] signa moventium tumultus ac murmur exoritur: ex quo, ut verissime dicit Vergilius,

Corda licet vulgi praesciscere namque morantes
Martius ille aeris rauci canor increpat,[12] et vox
Auditur fractos sonitus imitata tubarum.

[1] genere *SAac*.
[2] mutis *ac* : muti *S* : multa *A*.
[3] tanquam *SA* : velut *ac*.
[4] procurator *S* : procuratori *A* : procurati *c*.
[5] repulsa *c* : -am *SA*a.
[6] observari *SA* : -e *ac*.
[7] alveos *ac* : alvis *SA*.
[8] postquam horam *SAac*.
[9] agmina *ac* : -e *SA*.

orders which they give; for as the human race, which possesses reason, allows no partnership of the kingly power, much less do the dumb animals who are lacking in understanding. Therefore the new chieftains 2 come forth with their following of young bees, which, remaining in a mass for one or two days at the very entrance of their abode, by their coming out show their desire for a home of their own, and if the man in charge immediately assigns it to them, are as content with it as if it were their native place. If, however, the keeper has been away, they make for some strange region as if they had been driven away unjustly. To 3 prevent this, it is the duty of a good overseer in spring-time to keep an eye upon the hives until about the eighth hour of the day (after which the new battalions of bees do not take to impetuous flight), and carefully watch their departures, for some of them, when they have broken out, usually immediately rush away. He will be able to find out beforehand their 4 decision to escape by putting his ear to each of the hives in the evening; for about three days before they intend to break out an uproar and buzzing arises like that of an army setting out on the march. From this, as Vergil very truly says,

> You can foreknow the purpose of the herd;
> The martial roar of the hoarse brass reproves
> The loiterers, and a voice is heard whose notes
> The broken sound of trumpets imitates.[a]

[a] *Georg.* IV. 70 ff.

---

[10] eoque ut eorumque egressus *c*: eoque regressus *SA*: eorumque egressus eoque regressus *a*.
[11] militaria *S*²*ac*: milia *S*¹*A*.
[12] increpat *ac*: invocat *SA*.

5 Itaque maxime observari debent, quae istud faciunt, ut sive ad pugnam eruperint, nam inter se tanquam civilibus bellis, et cum alteris quasi cum exteris gentibus proeliantur, sive fugae causa se proripuerint, praesto sit ad utrumque casum paratus[1] 6 custos. Pugna quidem vel unius inter se dissidentis vel duorum examinum discordantium facile compescitur: nam ut idem ait,

Pulveris exigui iactu compressa quiescit:

aut aqua mulsea[2] passove, aut aliquo liquore simili[3] respersa,[4] videlicet familiari dulcedine saevientium iras mitigante. Nam eadem mire etiam dissidentes reges conciliant. Sunt enim saepe plures unius populi duces, et quasi procerum seditione plebs in partes diducitur: quod frequenter fieri prohibendum est, quoniam intestino bello totae gentes consu- 7 muntur. Itaque si constat principibus gratia, manet[5] pax incruenta. Sin autem saepius acie dimicantes notaveris, duces seditionum interficere curabis: dimicantium vero proelia praedictis remediis sedantur. Ac deinde cum agmen glomeratum in proximo frondentis arbusculae ramo consederit, animadvertito, an totum examen in speciem unius uvae dependeat: idque signum erit aut unum regem

[1] paratus *Sac* : -os *A*.
[2] aqua mulsea *SA* : aqua mulsa *a* : aut mulsa *c* : mulso *edd.*
[3] simili *ac* : simplici *SA*.
[4] respersa *SA* : -am *ac*.
[5] maneat *SAac*.

[a] *Georg.* IV. 87.

The bees, therefore, which behave like this ought 5 especially to be kept under observation, so that, whether they sally forth to battle (for they wage a kind of civil war amongst themselves and as it were foreign wars with other swarms) or break out in order to escape, the keeper may be at hand, ready for either event. Fighting either of the bees of one swarm 6 quarrelling amongst themselves or of two swarms at variance with one another is easily quelled; for, as the same poet says,

> By casting of a little dust the strife
> Is stayed and laid to rest,[a]

or else by sprinkling over them honey-water or raisin-wine or some similar liquid, that is to say the sweet taste of things familiar to them, abates their wrath. The same expedients too are wonderfully efficacious for reconciling king-bees when they are at enmity; for there are often several leaders of one people, and the common herd is as it were divided into factions by the quarrels of its chiefs. This must be prevented from happening often, since whole nations are destroyed by civil war. And so, if good 7 feeling exists between the princes, peace continues and no blood is shed. If, however, you have often noticed them fighting a pitched battle, you will take care to put to death the leaders of the factions; but when they are actually fighting, their battles can be calmed by the above-mentioned remedies. Next, when a host of bees has settled in a mass on the neighbouring branch of a leafy shrub, you should take notice whether the whole swarm hangs down in the form of a single bunch of grapes. This will be a sign either that there is only one king-bee in it

inesse, aut certe plures bona fide reconciliatos; quos sic[1] patieris,[2] dum in suum revolent[3] domicilium. 8 Sin autem duobus aut etiam compluribus velut uberibus diductum[4] fuerit examen, ne dubitaveris et plures proceres et adhuc iratos esse. Atque in iis partibus, quibus maxime videris apes glomerari, requirere duces debebis. Itaque succo praedictarum herbarum, id est, melissophylli vel apiastri manu illita, ne ad tactum diffugiant, leviter inseres digitos, et diductas apes scrutaberis, donec auctorem pugnae reperias.

X. Sunt autem hi reges maiores paulo et oblongi magis quam ceterae apes, rectioribus cruribus, sed minus amplis pinnis, pulchri coloris et nitidi, levesque ac sine pilo, sine spiculo, nisi quis forte pleniorem quasi capillum, quem in ventre gerunt, aculeum putat, quo et ipso tamen ad nocendum non utuntur. Quidam etiam infusci atque hirsuti reperiuntur, quorum pro habitu damnabis ingenium.

2 Nam duo sunt regum facies, ita corpora plebis.
Alter erit maculis auro squalentibus ardens
insignis et ore
Et rutilis clarus squamis.

[1] quos sic *om. SAac.*
[2] patieris *a* : paterisque *SA* : petierunt(?) *c.*
[3] revolet *SAac.*
[4] deductum *ac* : ductum *SA.*

or, at any rate, that, if there are several, they are reconciled and on good terms with one another, in which case you will leave them as they are until they fly back to their abode. If, however, the swarm is 8 divided into two or even more clusters, you need have no doubt that there are several chiefs and that they are still in an angry mood, and you will have to search for the leaders in the parts of the clusters where you see the bees most closely massed together. Having, then, smeared your hand with the juice of the herbs already named, that is, balm and wild parsley, lest they fly away at your touch, you will gently insert your fingers and, after separating the bees from one another, you will search until you find the author of the quarrel.

X. Now the king-bees are slightly larger and more oblong in shape than the other bees, with straighter legs but less ample wings, of a beautiful shining colour and smooth, without any hair, and stingless, unless one regards as such the coarser hair-like object growing on their belly, of which, however, they do not make use to inflict a hurt. Some, too, are found of a dusky colour and hairy, of whose disposition you will form an unfavourable opinion judging from their bodily appearance.

The king bee.

As two-fold are the features of the kings, 2
So are the bodies of their subjects; one
Will gleam with markings rough with gold, and bright
With ruddy scales, and of a comely mien.[a]

[a] Parts of Vergil, *Georg.* IV. 91–7.

Atque hic maxime probatur, qui est melior: nam deterior, sordido sputo similis, tam foedus est,

quam pulvere ab alto
Cum venit et sicco terram spuit ore viator.

Et, ut idem ait,

Desidia latamque trahens inglorius alvum.

Omnes igitur duces notae deterioris
Dede neci, melior vacua sine regnet in aula.

3 Qui tamen et ipse spoliandus est alis, ubi saepius cum examine suo conatur eruptione facta profugere. Nam velut quadam compede retinebimus erronem ducem detractis alis, qui fugae destitutus praesidio, finem regni non audet excedere, propter quod ne ditionis quidem suae populo permittit longius evagari.

XI. Sed nonnunquam idem necandus est, cum vetus alveare numero apium destituitur, atque infrequentia eius alio[1] examine[2] replenda est. Itaque cum primo vere in eo vase nata est pullities, novus rex eliditur[3] ut multitudo sine discordia cum parentibus suis conversetur. Quod si nullam progeniem tulerint favi, duas[4] vel tres alvorum plebes in unum contribuere licebit, sed prius respersas dulci liquore: tum demum includere, et posito cibo, dum

[1] alio *scripsi*: aliquo *codd.*
[2] examine *ac*: -a *SA*.
[3] eligitur *SAac*.
[4] duas *ac*: dius *SA*.

That is why this one is especially approved, being superior; for the inferior kind, like dirty spittle, is as foul as

> The wayfarer who comes from depth of dust
> And from his parchèd mouth the dirt spits forth:[a]

And as the same writer says,

> With sloth inglorious his wide paunch he drags.[b]

Therefore all the leaders of the baser kind

> Give them to death, and let the better prince
> Rule in the empty hall.[c]

3 Nevertheless he too must be despoiled of his wings, when he oft-times attempts to break out with his swarm and fly away; for, if we strip him of his wings, we shall keep the vagrant chieftain as though in fetters chained, who, deprived of the resource of flight, ventures not to leave the confines of his realm and, for this reason, does not allow even the people under his sway to wander further than he is able.

How to keep up the proper complement of the apiary.

XI. But sometimes the king-bee has to be put to death when an old hive falls short of its proper complement of bees, and its want of numbers must be made up from another swarm. Therefore, when in the early spring a young brood is born in the hive, the new king-bee is squeezed to death, so that the multitude of bees may live with their parents without discord. But if the combs have produced no offspring, it will be open to you to bring together the population of two or three hives into one, but only after they have been sprinkled with sweet liquid; then you can shut them up and, after placing food for them, keep

[a] Parts of Vergil, *Georg.* IV. 96 f.
[b] *Ib.* 94. [c] *Ib.* 90.

conversari consuescant, exiguis spiramentis relictis
2 triduo fere clausas habere. Sunt qui seniorem potius regem submoveant, quod est contrarium: quippe turba vetustior, velut quidam senatus, minoribus parere non censent, atque imperia validiorum contumaciter
3 spernendo[1] poenis ac mortibus afficiuntur.[2] Illi quidem incommodo, quod iuniori[3] examini solet accidere, cum antiquarum apium relictus a nobis rex senectute defecit, et tanquam domino mortuo familia nimia licentia discordat, facile occurritur. Nam ex iis alvis, quae plures habent principes, dux unus eligitur: isque translatus ad eas, quae sine imperio sunt, rector constituitur.

Potest autem minore molestia in iis domiciliis, quae aliqua peste vexata sunt, paucitas apium
4 emendari. Nam ubi cognita est clades frequentis alvi, si quos habet favos, oportet considerare: tum deinde cerae eius quae semina pullorum continet, partem recidere, in qua regii generis proles animatur. Est autem facilis conspectu, quoniam fere in ipso fine cerarum velut papilla uberis apparet eminentior et laxioris fistulae[4] quam sunt reliqua foramina,
5 quibus popularis notae pulli detinentur. Celsus quidem[5] affirmat in extremis favis transversas fistulas esse, quae contineant regios pullos. Hyginus quoque auctoritatem Graecorum sequens negat ex vermiculo,[6] ut ceteras apes, fieri ducem, sed in circuitu favorum paulo maiora, quam sunt plebeii

[1] spernendo *SAa* : -os *c*.
[2] afficiuntur *c* : afficitur *SA* : afficiunt *a*.
[3] iuveniori *SA* : iuniori *ac*.
[4] apparet eminentior et laxioris *ac* : *om. SA*.
[5] quidem *ac* : quae quidam *SA*.
[6] vermiculo *Gesner* : vernaculo *SAac*.

them enclosed for about three days, leaving only small breathing-holes, until they are accustomed to live together. There are some people who prefer 2 to get rid of a king-bee that is old, but this is harmful; for the crowd of older bees, who form a kind of senate, do not think fit to obey the juniors and, through obstinately despising the orders of those who are stronger than themselves, are visited with punishment and death. The trouble, indeed, which usually 3 befalls a younger swarm, when the king of the old bees whom we have left in power has failed through old age and wild discord arises through lack of control (just as happens in a family when its head dies), can easily be met. For one leader is chosen from those hives which have several chiefs and is transferred to those which have no one to govern them, and set up as ruler.

In those quarters which are afflicted by some pestilence the lack of bees can be remedied with less trouble; for when the disaster to the crowded hive 4 is recognized, you must examine any combs which it contains. You must then next cut away, from the wax which holds the seeds, that part in which the offspring of the kingly race comes to life. It is easy to see this, since almost at the very end of the wax there appears as it were the nipple of a breast projecting somewhat and with a wider cavity than the rest of the holes, in which the young bees of the common kind are enclosed. Celsus indeed declares that there are 5 transverse cavities in the outermost combs which contain the royal progeny. Hyginus, too, following the authority of the Greeks, says that the ruler is not formed, like the rest of the bees, from a small worm, but that, on the circumference of the combs, straight holes are to be found somewhat larger than those

seminis, inveniri recta foramina repleta quasi sorde rubri coloris, ex qua protinus alatus rex figuretur.

XII. Est et illa vernaculi examinis cura, si forte praedicto tempore facta eruptione[1] patriam[2] fastidiens[3] sedem longiorem fugam denuntiavit. Id autem significat, cum sic apis evadit vestibulum, ut nulla intra revolet, sed se confestim levet sublimius. 2 Crepitaculis aeris[4] aut testarum plerumque vulgo iacentium terreatur fugiens iuventus: eaque vel pavida cum repetierit alvum maternam, et in eius aditu glomerata pependerit, vel statim se ad proximam frondem contulerit, protinus custos novum loculamentum in hoc praeparatum perlinat intrinsecus praedictis herbis: deinde guttis mellis respersum admoveat: tum manibus, aut etiam trulla con- 3 gregatas apes recondat: atque, uti debet, adhibita cetera cura, diligenter compositum et illitum vas interim patiatur in eodem loco esse, dum advesperascat. Primo deinde crepusculo transferat, et re- 4 ponat in ordinem reliquarum alvorum. Oportet autem etiam vacua domicilia collocata in apiariis habere. Nam sunt nonnulla examina, quae cum processerint,[5] statim sedem sibi quaerant in proximo, eamque[6] occupent quam vacantem reperiunt. Haec fere acquirendarum, atque etiam retinendarum apium traditur cura.

[1] eruptione *ac*: -em *SA*.
[2] patriam *ac*: -ae *SA*.
[3] *post* fastidiens *add.* sedens *SA*.
[4] aeris *c*: eris *SA*: aereis *a*.
[5] processerint *A*: -unt *Sac*.
[6] eaque *SA*: eandemque *a*: eademque *c*.

which hold the bees of common birth, filled with a kind of dirt of a red colour from which the winged king-bee is immediately formed.

How to capture a swarm and prevent its escape.

XII. Care must also be taken of the home-bred swarm, if by chance, taking a dislike to their paternal abode, they break forth at the time already mentioned and announce their intention of taking a more distant flight. This the swarm intimates when the bees so completely avoid the entrance to the hive that not a single one flies back again into it, but immediately rises high into the sky. 2 The young bees who are escaping should be frightened by the rattling of brass or potsherds, which are usually to be found lying about; and when in their alarm they have returned to the maternal hive and hang in a mass at the entrance to it or betake themselves immediately to the nearest foliage, the keeper should immediately besmear the inside of a new receptacle prepared for the purpose with the herbs mentioned above, and then, after sprinkling it with drops of honey, bring it near and gather the mass of bees together with his hands or with a scoop; 3 and, after taking every proper precaution, he should let the hive, after it has been carefully adjusted and besmeared inside, remain in the same place until evening begins to fall. Then at first twilight he should remove it and replace it in a row with the other hives. 4 But you should also have empty hives placed in the apiary; for there are some swarms which, as soon as they have come forth, immediately seek a home for themselves nearby and occupy one which they find empty. You now have a practically complete account of the measures to be taken for acquiring bees and keeping them in your possession.

XIII. Sequitur ut morbo vel pestilentia laborantibus remedia desiderentur. Pestilentiae rara in apibus pernicies, nec tamen aliud, quam quod in cetero pecore praecepimus, quid fieri possit[1] reperio, nisi ut longius alvi transferantur. Morborum autem facilius et causae dispiciuntur, et inveniuntur medi-
2 cinae. Maximus autem annuus[2] earum labor est initio veris, quo tithymali floret frutex, et quo[3] amara ulmi semina sua promunt. Nam quasi novis pomis, ita his primitivis floribus illectae avide vescuntur post hibernam famem, alioqui[4] citra satietatem[5] tali non[6] nocente cibo: quo[7] cum se affatim repleverunt, profluvio alvi, nisi celeriter succurritur, intereunt. Nam et tithymalus maiorum quoque animalium ventrem solvit, et proprie ulmus apium. Eaque causa est, cur in regionibus Italiae,[8] quae sunt eius generis[9] arboribus consitae, raro frequentes durent
3 apes. Itaque veris principio si medicatos cibos praebeas, iisdem remediis et provideri[10] potest, ne tali peste vexentur, et cum iam laborant, sanari. Nam illud quod Hyginus antiquos secutus auctores prodidit, ipse non expertus asseverare non audeo:[11]
4 volentibus tamen licebit experiri. Siquidem prae-

[1] possit *Sac*: potest *A*.
[2] maximus autem annuus *Schneider*: maximumque vel minimum annuus *S*: maximusque vel minimus annuus *Aa*.
[3] et quo (quos *a*) amara ulmi *ac*: quo samaras ulmis *SA*.
[4] alioqui *SA*: alioquin *ac*. [5] sacietatem *S*: satietatem *Aac*.
[6] non *om. SAac*. [7] quo *om. SAac*.
[8] in regionibus Italiae *ac*: in geniobus; taliae *SA*.
[9] generis *SAac*. [10] provideri *ac*: -ere *SA*.
[11] audeo *SA*: audet *ac*.

[a] Minor troubles, distinct from *pestilentia*, which is what is now called 'bee-pest' or 'foul brood.'
[b] Now called 'dysentery.'

Remedies for the diseases of bees.

XIII. The next thing is that remedies are needed for those which are suffering from disease or pestilence. The ruinous disease of 'pestilence' is rare in bees, nor can I find anything which ought to be done other than what we have prescribed in the case of the other animals (except that the hives should be moved far away); but the causes of common ailments [a] in bees are more easily diagnosed and remedies found for them. 2 The most serious is their annual distemper at the beginning of spring, when the spurge-bush flowers and the elms put forth their bitter blossoms; for as by fresh apples, so are they allured by these early flowers and eat greedily of them after their winter hunger, such food not being hurtful when not eaten beyond satiety, but when they have gorged themselves abundantly with it, they die from a flux of the belly, unless help is quickly given. For spurge produces looseness of the bowels in the larger animals also, but elm has this effect particularly on bees. This is the reason why bees rarely continue numerous in the districts of Italy which are planted with trees of this kind. 3 And so at the beginning of spring, if you supply them with medicated food, by means of the same remedies it is possible both to provide against their being troubled by plague [b] of this kind and also to cure them when they are already suffering from it. Now I myself do not venture to insist on the treatment which Hyginus, following ancient authorities, has recorded, since I have not tried it; but it is open to those who wish to do so to test it. 4 For his instructions are: when a plague of this kind has attacked the bees, and the bodies are found for dead in heaps under the honeycombs, lay them aside in a dry place through the winter, and, at

cipit apium corpora, quae cum eiusmodi pestis incessit, sub favis acervatim enectae[1] reperiuntur, sicco loco per hiemem reposita circa aequinoctium vernum, cum clementia diei suaserit, post horam tertiam in solem proferre, ficulneoque cinere obruere. Quo[2] facto, affirmat intra duas horas cum vivido halitu caloris animatae sunt, resumpto spiritu, si 5 praeparatum vas obiciatur, irrepere. Nos magis ne intereant, quae deinceps dicturi sumus, aegris examinibus exhibenda[3] censemus. Nam vel grana mali Punici[4] tunsa et vino Amineo conspersa,[5] vel uvae passae cum rore Syriaco[6] pari mensura[7] pinsitae et austero vino insuccatae[8] dari debent: vel si per se ista frustrata sunt, omnia eadem aequis ponderibus in unum levigata, et fictili vase cum Amineo vino infervefacta, mox etiam refrigerata, 6 ligneis canalibus apponi. Nonnulli rorem[9] marinum aqua mulsa decoctum, cum gelaverit, imbricibus infusum praebent libandum. Quidam bubulam vel hominis urinam, sicut Hyginus affirmat, alvis appo- 7 nunt. Nec non etiam ille morbus maxime est conspicuus, qui horridas contractasque carpit, cum frequenter aliae mortuarum corpora[10] domiciliis efferunt, aliae intra tecta, ut in publico luctu, maesto silentio torpent. Id cum accidit, arundineis infusi canalibus

[1] enectae *Aac* : -a *S*.
[2] qui *SA* : quo *ac*.
[3] ad exhibenda *SAac*.
[4] Punici *ac* : -a *SA*.
[5] consparsam *SA* : consparsa *a* : conspersa *c*.
[6] sutorio *SAac*.
[7] mensura *SAac*.
[8] insucatae *ac* : insucae *SA*.
[9] ros *ac* : robore *SA*.
[10] corpora *ac* : -is *SA*.

about the time of the spring equinox, when the mildness of the day invites us, bring them out into the sunshine, after the third hour, and cover them with fig-wood ashes. If this is done, he declares that within two hours, brought to life by the quickening breath of the heat, they begin to breathe again and crawl into a vessel provided for this purpose, if it is placed in their way. We 5 rather, that they may not perish, are of opinion that the diet, which we will forthwith describe, should be put before the swarms when they are sick. For they ought to be given either seeds of pomegranate, bruised and sprinkled with Aminean [a] wine, or raisins with an equal quantity of Syrian sumach [b] and soaked in rough wine; or, if these are without effect taken separately, all the same ingredients should be pounded in equal quantities into a single mass and boiled in an earthenware vessel with Aminean wine and then allowed to cool right away and placed before the bees in wooden troughs. Some 6 people boil rosemary in honey-water and, when it has cooled, pour it into troughs and give it to the bees to sip. Others put the urine either of oxen or of human beings near the hives, as Hyginus declares. Moreover also, that disease is particularly remarkable 7 which makes them hideous and shrunken and consumes them, when some often carry out from their abodes the bodies of those which have died, while others remain listless within their dwellings in sad silence, as though in time of public mourning. When this

[a] From a district of Picenum (Vergil, *Georg.* II. 97).

[b] *Ros* or, more correctly, *rhus Syriacus* is said by Pliny, *N.H.* XIII. § 55, to be used as a drug, which shows that *Syriacus* is the right reading here.

offeruntur cibi, maxime dococti mellis, et cum galla [1] vel arida rosa detriti. Galbanum etiam, ut eius odore medicentur, incendi convenit, passoque et 8 defruto vetere fessas sustinere. Optime tamen facit amelli radix, cuius est frutex luteus purpureus flos: ea cum vetere Amineo vino decocta exprimitur, et ita liquatus eius succus datur. Hyginus quidem in eo libro, quem de apibus scripsit, Aristomachus, inquit, hoc modo succurrendum laborantibus existimat: primum, ut omnes vitiosi favi tollantur, et cibus ex [2] integro recens ponatur; deinde ut fumi- 9 gentur. Prodesse etiam putat apibus vetustate corruptis examen novem contribuere, quamvis periculosum sit, ne seditione consumantur, verumtamen adiecta multitudine laetaturas.[3] Sed ut concordes maneant, earum apium, quae ex alio domicilio transferuntur, quasi peregrinae plebis [4] submoveri reges [5] debent.[6] Nec tamen dubium, quin frequentissimorum examinum favi, qui iam maturos habent pullos, transferri, et subici paucioribus debeant, ut tanquam 10 novae prolis adoptione domicilia confirmentur. Sed et id [7] cum fiet, animadvertendum est, ut eos favos subiciamus, quorum pulli iam sedes suas adaperiunt,

[1] galla *ac* : galle *SA*.
[2] cuius et *SA* : cibus ex *a* : *om. c*.
[3] laetaturas *scripsi* : laetatura *Aac* : letatur *S*.
[4] plebis *ac* : plebes *SA*.
[5] reges *ac* : regi *SA*.
[6] debent *SA* : debere *ac*.
[7] et id *a* : sed id *c* : sed sit *SA*.

---

[a] See note on p. 260.
[b] Of Soli in Cyprus, who, with Philiscus of Thasos, wrote a book on bees (Pliny, *N.H.* XI. § 9).

happens food is offered them poured into troughs made of reeds, especially boiled honey pounded up with an oak-apple or a dried rose. It is also a good 8 plan to burn *galbanum*,[a] that they may be cured by its odour, and to keep up their strength, when they are exhausted, with raisin-wine and boiled-down must. The root of the starwort, the bushy part of which is yellow and its flower purple, has the best effect of all; it is boiled with old Aminean wine and pressed and then the juice is strained and given as a remedy. Hyginus indeed, in the book which he wrote about bees, says: "Aristomachus [b] is of opinion that help ought to be brought to bees which are sick in the following manner: first, all the diseased combs should be removed and entirely fresh food placed for the bees, and then they should be fumigated." He 9 thinks also that it is beneficial to add a new swarm to the bees who are wasted by old age, although there is a danger that they may be destroyed by sedition, nevertheless they are likely to rejoice because their number is increased. But that they may remain in a state of concord, the kings of those bees which are being transferred from another hive ought to be put out of the way as rulers of an alien people. There is, however, no doubt that the honey-combs of the most populous swarms, which have young bees already matured in them, ought to be transferred and made subject to the less populous swarms that their families may be strengthened by the adoption, as it were, of fresh progeny. But, 10 when this is going to be done, we must remember to put in the care of the old swarm those honey-combs in which the young ones are already opening their cells and putting out their heads and eating away

et velut opercula foraminum obductas ceras erodunt[1] exerentes capita. Nam si favos immaturo[2] fetu transtulerimus, emorientur pulli, cum foveri desi-
11 erint. Saepe etiam vitio quod[3] Graeci φαγέδαιναν[4] vocant, intereunt. Siquidem cum sit haec apium consuetudo, ut prius tantum cerarum confingant, quantum putent explere se[5] posse, non nunquam evenit, consummatis[6] operibus cereis, ut, dum examen conquirendi[7] mellis causa longius evagatur, subitis imbribus, aut turbinibus in silvis opprimatur, et maiorem partem plebis amittat: quod ubi factum est, reliqua[8] paucitas favis complendis non sufficit; tuncque vacuae cerarum partes[9] computrescunt,[10] et vitiis paulatim serpentibus, corrupto[11] melle, ipsae
12 quoque apes intereunt. Id ne fiat, vel duo populi coniungi debent, qui possint adhuc integras ceras explere: vel si non est facultas alterius examinis, ipsos favos, ante quam putrescant, vacuis partibus acutissimo ferro liberare. Nam hoc quoque refert, ne admodum[12] hebes[13] ferramentum (quia non facile penetret) vehementius impressum favos sedibus suis commoveat; quod si factum est, apes domicilium derelinquunt.

13 Est et illa causa interitus, quod interdum continuis annis plurimi flores proveniunt, et apes magis melli-

[1] erodunt *ac*: produnt *S*: produn *A*.
[2] immaturo *ac*: -os *SA*.
[3] quod *ac*: *om. SA*.
[4] φαγέδαιναν A^2: φαγέδεναν *S*: *om. ac*.
[5] se *om. SA*.
[6] consumatis *ac*: cum summas *S*: consummas *A*.
[7] conquirendi *ac*: -is *SA*.
[8] reliqua *ac*: aliqua *SA*.
[9] partes *om. S*.
[10] cum putrescant *c*: partescum *S*: patescunt *A*: patescant *a*.
[11] corrupto *ac*: -a *SA*.
[12] admotum *SAac*.

the wax which was laid upon the top as a kind of covering for their holes. For if we transfer the honey-combs when the brood has not come to maturity, the young bees will die when they cease to be kept warm. For they often die of a distemper which the Greeks call *phagedaina*.[a] For since it is 11 the habit of bees to construct beforehand as many cells as they think they can fill, it sometimes happens that, when their waxen structures are finished, the swarm, while it is roaming too far afield in search of honey, is overwhelmed in the woods by sudden showers and whirlwinds and loses most of the ordinary bees. When this has happened, the few that remain are not enough to fill the combs and then the empty parts of the wax cells become rotten, and since diseases gradually creep in, the honey becomes corrupted and the bees, too, themselves die. To prevent this, either 12 the populations of two hives ought to be united, so that they can fill the waxen cells which are still sound, or, if a second swarm is not available, we must remove the honey-combs from the uninhabited parts, before they go rotten, with a very sharp knife. For it is very important also that a very blunt iron tool, because it does not easily penetrate, should not be pressed with great force and dislodge the honey-combs from their places; for if this has happened, the bees desert their abode.

There is also this cause of mortality among bees 13 that sometimes very many flowers come up during several continuous years and the bees are more eager

[a] Pliny (*N.H.* XXVI. § 11) says that this word has two meanings, either (1) a rodent cancer or (2) voracious hunger. The first is certainly the meaning here.

[13] hebes *Sac* : habes *A*.

ficiis quam fetibus student. Itaque nonnulli, quibus minor est harum rerum scientia, magnis [1] fructibus delectantur, ignorantes exitium apibus imminere, quoniam et nimio fatigatae opere plurimae pereunt, nec ullis iuventutis supplementis con-
14 frequentatae novissime reliquae intereunt. Itaque si tale ver incessit, ut et prata et arva [2] floribus abundent, utilissimum est tertio quoque die exiguis foraminibus relictis per quae non [3] possint exire alvorum exitus praecludi,[4] ut ab opere [5] mellifico avocatae, apes quoniam non sperent se posse ceras omnes liquoribus stipare, fetibus expleant. Atque haec fere sunt examinum vitio laborantium remebia.

XIV. Deinceps illa totius anni cura, ut idem Hyginus commodissime prodidit. Ab aequinoctio primo quod mense Martio circa VIII calendas Aprilis in octava parte Arietis conficitur, ad exortum Vergiliarum dies verni temporis habentur duodequinquaginta. Per hos primum ait apes curandas esse adapertis alveis, ut omnia purgamenta, quae sunt hiberno tempore congesta, eximantur, et araneis, qui favos corrumpunt, detractis fumus immittatur factus incenso bubulo fimo.[6] Hic enim quasi quadam
2 cognatione generis maxime est apibus aptus. Vermiculi quoque, qui tineae vocantur, item papiliones

[1] magis *SAac.*
[2] ut etiam prata parva *a* : et ut prata et arva *c* : et iam parva *SA.*
[3] non *om. SAac.*
[4] praecludit *A.*
[5] ab opere *ac* : alveo fere *SA.*
[6] fimo *Aac* : fimi *S.*

to make honey than to produce offspring. And so some people, whose knowledge of these matters is defective, are delighted at the large production of honey, not being aware of the destruction which is threatening the bees; for, exhausted by too much labour, very many of them are perishing and, as their numbers are not being increased by the addition of young stock, the rest at last die off. And so, if such 14 a spring comes on that both the meadows and the cornfields abound in flowers, it is most expedient every third day to close the exits from the hives (small openings having been left through which the bees cannot pass), so that, called from the activity of making honey, since they have no hope of being able to fill up the waxen cells with liquid honey, they may fill them with offspring. Such then in general are the remedies for swarms suffering from some distemper.

XIV. Next comes the management of bees throughout the year according to the excellent system set forth by the same Hyginus. From the first equinox, which takes place about the twenty-fourth of March in the eighth degree of the Ram, until the rising of the Pleiads, there are reckoned to be the forty-eight days of spring. During these days, he says, the bees ought to receive attention for the first time by opening the hives, so that all filth, which has collected during the winter season, may be removed, and, after the spiders, which rot the honey-combs, have been got rid of, the hives may be fumigated with smoke produced by burning ox-dung; for this smoke is particularly well suited to bees as if some affinity existed between it and them. The little worms also which are called moth-caterpillars and also 2

The management of bees.

enecandi sunt: quae pestes plerumque favis adhaerentes decidunt, si fimo medullam bubulam misceas, et his incensis [1] nidorem admoveas. Hac cura per id tempus quod diximus examina firmabuntur, eaque fortius operibus inservient.

3 Verum maxime custodiendum est curatori, qui apes nutrit, cum alvos tractare debebit, uti pridie castus ab rebus venereis, neve temulentus,[2] nec nisi lotus ad eas accedat, abstineatque omnibus redolentibus esculentis,[3] ut sunt salsamenta, et eorum omnia liquamina; itemque fetentibus acrimoniis alii vel 4 ceparum ceterarumque [4] rerum similium. Duodequinquagesimo [5] die ab aequinoctio verno, cum fit Vergiliarum exortus circa v idus Maias, incipiunt examina viribus et numero augeri. Sed et iisdem diebus intereunt quae paucas et aegras apes habent; eodemque tempore progenerantur in [6] extremis partibus favorum amplioris magnitudinis quam sunt ceterae apes, eosque nonnulli putant esse reges. Verum quidam Graecorum auctores *οἴστρους* [7] appellant ab eo, quod exagitent, neque patiantur examina conquiescere. Itaque praecipiunt eos enecari.

5 Ab exortu Vergiliarum ad solstitium, quod fit ultimo mense Iunio circa octavam partem Cancri, fere examinant alvi: quo tempore vehementius custodiri debent, ne novae soboles diffugiant. Tumque per-

[1] incensis *ac*: impensis *SA*.
[2] temulentus *Ac*: temulentis *a*: temolestus *S*.
[3] estulentis *a*: esculentis *c*: exculentis *S*: excultis *A*.
[4] ceterarumque *ac*: *om. Ac*.
[5] unde quinquagesimo *SAac*.
[6] in *ac*: et *SA*.
[7] *οἴστρους SA*: *om. ac*.

[a] Gadflies or horseflies.

the developed moths must be killed. These pests which generally adhere to the honey-combs fall off, if you mix ox's marrow with dung and, after setting the mixture on fire, bring the smell of burning near them. As a result of this precaution the swarms will be strengthened during the period which we have mentioned and will apply themselves to their work with more vigour.

But very great care must be taken by the man in 3 charge, who feeds the bees, when he must handle the hives, that the day before he has abstained from sexual relations and does not approach them when drunk and only after washing himself, and that he abstain from all edibles which have a strong flavour, such as pickled fish and all the liquids which accompany them, and also from the acrimonious stench of garlic and onions and all other similar things. On the 4 forty-eighth day after the vernal equinox, when the rising of the Pleiads takes place about the 8th of May, the swarms begin to increase in strength and number; but in the same period of days the swarms also which contain few and sickly bees die off, and at the same time in the extremities of the honey-combs bees are born of larger size than the rest, which some people think are king-bees. Some writers among the Greeks, however, call them *oistroi* [a] from the fact that they excite the swarms and do not allow them any rest; therefore they recommend that they should be killed.

From the rising of the Pleiads to the solstice, which 5 takes place at the end of June in about the eighth degree of the Crab, the hives generally swarm. This is a time at which they must be very strictly watched, so that the young brood may not escape. Then,

acto solstitio usque ad ortum Caniculae, qui fere dies triginta sunt, pariter[1] frumenta et favi demetuntur.[2] Sed hi quemadmodum[3] tolli debeant, mox dicetur, cum de confectura mellis praecipiemus.

6 Ceterum hoc eodem tempore progenerari posse apes iuvenco perempto, Democritus et Mago[4] nec minus Vergilius prodiderunt. Mago quidem ventribus etiam bubulis idem fieri affirmat, quam rationem diligentius prosequi supervacuum puto, consentiens Celso, qui prudentissime ait, non tanto interitu 7 pecus istud amitti, ut sic requirendum sit. Verum hoc tempore, et usque in autumni aequinoctium decimo quoque die alvi aperiendae et fumigandae sunt. Quod cum sit molestum examinibus, saluberrimum tamen esse convenit. Suffitas deinde, et aestuantes apes refrigerare oportet, conspersis vacuis partibus alvorum et recentissimi rigoris aqua infusa: deinde si quid ablui non poterit, pinnis aquilae vel etiam cuius libet vastae alitis,[5] quae rigo-8 rem habent, emundari. Praeterea ut tineae[6] everrantur, papilionesque enecentur, qui plerumque intra alvos morantes apibus exitio sunt. Nam et ceras erodunt, et stercore suo vermes progenerant, quos 9 alvorum tineas appellamus. Itaque quo tempore malvae florent, cum est earum[7] maxima multitudo, si vas aeneum simile[8] miliario vespere ponatur inter

[1] pariter *Aac*: pater et *S*.
[2] demetuntur *a*: demetiuntur *c*: demuntur *SA*.
[3] sed hi quem admodum *ac*: sed hiem admodum *SA*.
[4] mago *ac*: magno *SA*.
[5] alitis *ac*: -as *SA*.
[6] *post* tineae *add.* si apparuerint *c*: *om.* *SAa*.
[7] earum *ac*: eorum *AS*.
[8] simile *ac*: -em *SA*.

when the solstice is passed and until the rising of the Dog-star, a period of about thirty days, the harvests of the cornfields and the honey-combs alike are gathered in. How the combs should be removed will be told presently when we give instructions for preparing honey.

Now Democritus, Mago and likewise Vergil have 6 recorded that bees can be generated at this same time of year from a slain bullock. Mago indeed also asserts that the same thing may be done from the bellies of oxen, but I consider it superfluous to deal in more detail with this method, since I am in agreement with Celsus, who very wisely says that there is never such mortality among these creatures, that it is necessary to procure them by this means. But at this time and until the autumn equinox, the 7 hives ought to be opened and fumigated every tenth day. This, though it annoys the swarm, is generally considered to be very wholesome. Then after they have been fumigated and are still heated the bees ought to be cooled by sprinkling the empty parts of the hives and pouring in water which is cold because it is very freshly drawn: then when there is anything which cannot be washed away, it must be cleansed with the feathers of an eagle or of any other large bird which are of a stiff quality. Moreover 8 caterpillars should be swept away and moths killed, which generally linger among the hives and are destructive to the bees; for they both gnaw at the waxen combs and from their dung breed worms which we call "hive-moths." Therefore, at the season 9 when the mallows flower, when the moths are most numerous, if a bronze vessel of the shape of a milestone is placed amongst the hives in the evening

alvos, et in fundum eius lumen aliquod demittatur, undique papiliones concurrunt:[1] dumque circa flammulam volitant,[2] aduruntur, quod[3] nec facile ex angusto sursum evolare,[4] nec rursus longius ab igne possunt recedere, cum lateribus aeneis circumveniantur: ideoque propinquo ardore consumuntur.

10 A Canicula fere post diem quinquagesimum Arcturus oritur, cum irroratis floribus thymi et cunilae thymbraeque apes mella conficiunt: idque[5] optimae notae enitescit[6] autumni aequinoctio, quod est ante calend. Octobris, cum octavam partem Librae sol attigit. Sed inter Caniculae et Arcturi exortum cavendum erit, ne apes intercipiantur violentia crabronum, qui ante alvearia plerumque 11 obsidiantur prodeuntibus. Post Arcturi exortum circa aequinoctium Librae (sicut dixi) favorum secunda est exemptio. Ab aequinoctio deinde quod conficitur circa VIII calend. Octobris ad Vergiliarum occasum diebus XL, ex floribus tamaricis[7] et silvestribus frutectis apes collecta mella cibariis hiemis reponunt. Quibus nihil est omnino detrahendum, ne saepius iniuria contristatae velut desperatione rerum 12 profugiant. Ab occasu Vergiliarum ad brumam, quae fere conficitur[8] circa VIII calend. Ianuarii in octava parte Capricorni, iam recondito melle utuntur examina, eoque usque ad Arcturi exortum sustinen-

[1] concurrant *SA* : -ent *ac*.
[2] volitent *SAac*.
[3] quod *c* : quoniam *a* : quam *SA*.
[4] evolent *SAac*.
[5] idque *ac* : atque *SA*.
[6] enitescit *SAa* : emitescit *c*.
[7] tamaricis *ac* : amaricis *SA*.
[8] conficitur *ac* : confingitur *SA*.

and a light lowered to the bottom of it, the moths rush together from all sides and, flitting round the flame, are scorched because they cannot easily fly upwards from the narrow space or retire to a distance from the fire, since they are hemmed in by the brazen sides of the vessel. They are, therefore, consumed by the burning heat which is near them.

About fifty days from the rising of the Dog-star is the rising of Arcturus, at which time the bees make their honey from the dew-drenched flowers of thyme and marjoram and savory. Honey of the finest quality is at its best at the autumn equinox, which falls before the first of October, when the sun reaches the eighth degree of Libra. But great care will have to be exercised between the rising of the Dog-star and that of Arcturus that the bees are not surprised by violent attacks from hornets, which generally lie in wait in front of the hives for them to come out. After the rising of Arcturus about the time of the equinox, which takes place when the sun is in the Balance (as I have said), the second extraction of honey-combs takes place. Then from the equinox, which occurs about September 24th, until the setting of the Pleiads, a period of forty days, the bees store up the honey which they have collected for winter food from the tamarisk flowers and woodland shrubs. Of this nothing at all must be extracted, lest the bees, disheartened by continual ill-treatment and, as it were, in despair, should take to flight. From the setting of the Pleiads till the winter solstice, which falls about December 23rd in the eighth degree of Capricorn, the bees make use of the honey already stored up and are sustained by it until the rising of Arcturus. I am well acquainted with the reckoning

tur. Nec me fallit Hipparchi ratio, quae docet solstitia et aequinoctia non octavis sed primis partibus signorum confici. Verum in hac ruris disciplina sequor nunc Eudoxi et Metonis[1] antiquorumque fastus astrologorum, qui sunt aptati[2] publicis sacrificiis: quia et notior est ista vetus agricolis concepta opinio; nec tamen Hipparchi subtilitas pinguioribus, 13 ut aiunt, rusticorum literis necessaria est. Ergo Vergiliarum occasu primo statim conveniet aperire alvos, et depurgare quidquid immundi est, diligentiusque curare; quoniam per tempora hiemis non expedit movere aut patefacere vasa. Quam ob causam dum adhuc autumni reliquiae sunt,[3] apricissimo die purgatis domiciliis opercula intus usque ad favos admovenda sunt, omni vacua parte sedis exclusa, quo facilius angustiae cavearum per hiemem concalescant. Idque semper faciendum est etiam in iis alvis, quae paucitate plebis infrequentes sunt.

14 Quidquid deinde rimarum est aut foraminum, luto et fimo bubulo mixtis illinemus extrinsecus, nec nisi aditus, quibus commeent, relinquemus. Et quamvis porticu protecta vasa nihilo minus congestu culmorum et frondium supertegemus, quantumque res patietur, a frigore et tempestatibus municmus. 15 Quidam exemptis interaneis occisas aves intus includunt, quae tempore hiberno plumis suis delitescentibus apibus praebent teporem: tum etiam si

[1] metonis *ac*: mentonis *SA*.
[2] aptati *ac*: aptatis *SA*.
[3] reliquie sunt *c*: relique sunt *a*: requiescunt *SA*.

[a] See note on Book I. 1. 5.
[b] Book I. Preface, § 32.

of Hipparchus,[a] which declares that the solstices and equinoxes occur not in the eighth but in the first degrees of the signs of the Zodiac; however, in these rural instructions I am now following the calendar of Eudoxus and Meto [b] and the old astronomers, which are adapted to the public festivals, because this view, accepted in old times, is more familiar to farmers and, on the other hand, the subtility of Hipparchus is not necessary for rustics of less refined education. On the first rising, then, of the Pleiads it will be advisable immediately to open the hives and clear away any filth that there is and attend to them with particular care, since during the winter time it is not expedient to move or open the hives. For this reason, while there are some remains left of autumn, on a very sunny day, after the bees' habitations have been cleansed, the covers must be put inside close to the honey-combs to prevent there being any empty space within, so that the narrow quarters of the hives may warm up more easily during the winter. This must always be done also in those hives which are sparsely inhabited through lack of bee population. 13

Next any chinks or holes that there are we shall daub outside with a mixture of clay and ox-dung, and we shall only leave entrance by which they may come and go. Also, although the hives are protected by a porch, we shall nevertheless cover them by heaping stalks and leaves on the top of them and fortify them, as far as circumstances allow, against cold and bad weather. Some people kill birds and, after taking out their intestines, shut the birds up in the hives, so that in winter time they may provide a gentle heat for the bees which lurk amongst their 14 15

sunt absumpta cibaria, commode pascuntur esurientes, nec nisi ossa[1] earum relinquunt. Sin autem favi sufficient[2] permanent illibatae, nec quamvis amantissimas[3] munditiarum offendunt odore suo. Melius tamen esse[4] nos existimamus, tempore hiberno fame laborantibus ad ipsos aditus in canaliculis vel contusam et aqua madefactam ficum aridam, vel defrutum aut passum praebere. Quibus liquoribus mundam lanam imbuere oportebit, ut insistentes
16 apes quasi per siphonem succum evocent. Uvas etiam passas cum infregerimus, paulum aqua respersas probe dabimus. Atque his cibariis non solum hieme, sed etiam quibus temporibus, ut iam supra dixi tithymalus, atque etiam ulmi florebunt, sustinendae
17 sunt. Post confectam brumam diebus fere quadraginta quidquid est repositi[5] mellis, nisi liberalius relictum sit[6] consumunt, saepe etiam vacuatis ceris[7] usque in ortum fere Arcturi, qui est ab idib. Februariis, ieiunae favis accubantes torpent more serpentum, et quiete sua spiritum conservant, quem tamen ne amittant,[8] si longior fames incesserit, optimum est per aditum vestibuli siphonibus dulcia liquamina immittere, et ita penuriam temporum sustinere, dum Arcturi ortus et hirundinis adventus commodiores
18 polliceantur futuras tempestates. Itaque, post hoc tempus, cum diei permittit hilaritas, procedere

[1] ossa *ac* : os *S* : oss *A*.
[2] sufficere *SA* : sufficerent *ac*.
[3] amantissimas *ac* : mantissimas *SA*.
[4] esse *om. SA*.
[5] repositi *ac* : -a *S* : -am *A*.
[6] sit *om. SAac*.
[7] ceris *Aac* : cereris *S*.
[8] amittant *ac* : amittam *SA*.

feathers; furthermore, if the stock of food is used up, they can very well feed on these birds, if they are hungry, and leave nothing but the bones. But if the honey-combs supply their needs, the birds remain untouched, nor do they offend the bees with their odour, fond though they are of cleanliness. It is better, however, in our opinion, when they are suffering from hunger in the winter time, to provide them with dried figs pounded and soaked in water or with boiled-down must or raisin-wine placed in little troughs at the very entrance to the hives; and it will be advisable to soak clean wool in these liquids, so that the bees, settling upon it, may draw up the juice as through a small pipe. We shall also 16 do well to give them raisins sprinkled a little with water after we have broken them up. With these foods they must be sustained not only in winter but also at those seasons, when, as we said just now, spurge and also elms are in blossom. When the 17 height of winter is passed, for a period of about forty days, they use up all the honey which is stored, unless an unusually generous allowance is left, and often too, after they have emptied the waxen cells, they lie fasting in the honey-combs in a torpid condition, like snakes, until about the rising of Arcturus, which is on the 13th of February, and by keeping quiet preserve the breath of life; in order, however, that they may not lose it, if too long a fast occurs, it is best to pour sweet liquids through the entrance of the porch by means of small pipes and thus support them during the temporary scarcity until the rising of Arcturus and the coming of the swallow with promise of more favourable weather for the future. And so, 18 after this time, when the more cheerful weather

audent in pascua. Nam ab[1] aequinoctio verno sine cunctatione iam passim vagantur, et idoneos ad fetum decerpunt flores, atque intra tecta comportant.

Haec observanda per anni tempora diligentissime Hyginus praecepit. Ceterum illa Celsus adicit, paucis locis eam felicitatem suppetere,[2] ut apibus alia pabula hiberna atque alia praebeantur aestiva.
19 Itaque quibus locis post veris tempora flores idonei deficiunt, negat oportere immota examina relinqui, sed vernis pastionibus absumptis in ea loca transferri, quae serotinis floribus thymi et origani thymbraeque benignius apes alere possint. Quod fieri ait et Achaiae regionibus, ubi transferuntur in Atticas pastiones, et Euboea, et rursus in insulis Cycladibus, cum ex aliis transferuntur[3] Scyrum, nec minus in Sicilia, cum ex reliquis eius partibus in Hyblam[4]
20 conferuntur. Idemque ait ex floribus ceras fieri, ex matutino rore mella, quae tanto[5] meliorem qualitatem capiunt, quanto iucundiore sit materia cera confecta. Sed ante translationem diligenter alvos inspicere praecipit, veteresque et tineosos, et labantes[6] favos eximere: nec nisi paucos et optimos reservare, ut simul etiam ex meliore flore quam

[1] ab *ac* : *om. SA.*
[2] felicitatem suppetere *ac* : *om. SA.*
[3] transferuntur *c* : transportantur *SAa.*
[4] hyblam *A* : hybleam *Sc* : hibleaem *a.*
[5] ex matutino—tanto *om. A.*
[6] labantis *ac* : labentis *A* : laventis *S.*

allows it, the bees venture to go forth to their pastures; for after the spring equinox they are already roaming about everywhere without hesitation and plucking the produce of flowers which are suitable for the production of their young and carrying it into their dwellings.

These are the principles which Hyginus recommends for the most careful observation throughout the seasons of the year, but Celsus makes the following additions. He says that only in a few places are conditions so favourable as to provide different foods for the bees in winter and summer, and that, therefore, in places where suitable flowers are lacking after the season of spring, the swarms ought not to be left without being moved, but, when the spring foods are consumed, they should be transferred to places which can offer the bees a more liberal diet from the late-flowering blossoms of thyme, marjoram and savory. This, he says, is the practice both in the regions of Achaia, where the bees are transferred to pastures in Attica, and in Euboea, and also in the islands of the Cyclades, when they are transferred from other islands to Scyros, and likewise in Sicily, when they are moved from the other parts of the island to Hybla. The same writer says that the waxen cells are made from flowers and the honey from morning dew, and that, the pleasanter the material from which the wax is made, the better the quality which the honey acquires. He gives instructions to examine the hives carefully before transferring them and to remove honey-combs which are old and wormy and falling to pieces, and to keep only a few and these the best, so that as many as possible may be made at the same time from the better flowers. He

plurimi fiant: eaque vasa, quae[1] quis transferre velit, non nisi noctibus et sine concussione portare.

XV. Mox vere transacto sequitur, ut dixi, mellis vindemia,[2] propter quam totius anni labor exercetur. Eius maturitas intelligitur cum animadvertimus fucos ab apibus expelli ac fugari. Quod est genus amplioris incrementi, simillimum api, sed, ut ait Vergilius, ignavum pecus, et immune,[3] sine industria 2 favis assidens. Nam neque alimenta congerit, et ab aliis invecta consumit. Verumtamen ad procreationem sobolis conferre aliquid hi fuci videntur insidentes seminibus, quibus apes figurantur. Itaque ad fovendam novam prolem familiarius admittuntur. Exclusis deinde pullis, extra tecta proturbantur, et ut idem ait, a praesepibus arcentur. 3 Hos quidam praecipiunt in totum exterminari oportere. Quod ego Magoni consentiens faciendum non censeo, verum saevitiae modum adhibendum. Nam nec ad occidionem gens interimenda est, ne apes inertia laborent, quae, cum fuci aliquam partem cibariorum absumunt, sarciendo damna fiunt agiliores: nec rursus multitudinem praedonum coalescere patiendum est, ne universas opes alienas diri- 4 piant. Ergo cum rixam fucorum et apium saepius committi videris, adapertas alvos inspicies,[4] ut sive

[1] eaque vasa quae *Aac* : eaqueus aquae *S*.
[2] vindemia *ac* : -am *SA*.
[3] et immune *c* : etiam rure *SAa*.
[4] inspicies *a* : aspicies *c* : inspiciens *SA*.

[a] *Georg.* IV. 168. [b] *Ib.*

also says that the hives which anyone wishes to transfer should only be moved at night and without being shaken.

XV. Presently, when spring is over, as I have said, the harvesting of the honey follows, with a view to which the whole year's work is carried out. We conclude that the honey is ripe when we notice that the drones are being expelled and put to flight by the bees. They are insects of a larger growth, very like bees, but as Vergil[a] says " a lazy herd " and idle, sitting near the honey-combs without doing any work; for they do not collect food but consume that which is brought in by others. Nevertheless these drones seem to contribute something to the procreation of the younger generation by sitting on the seeds from which the bees are formed, and so they are admitted on terms of some intimacy in order to sit upon the eggs which produce the new offspring; then, when the young bees are hatched, they are hustled out of the hives and, as the same poet says, " they are kept away from the fold."[b] Some people recommend that they should be entirely exterminated; but I agree with Mago that this should not be done, but that a limit ought to be set to cruelty. For the race ought not to be wholly destroyed, lest the bees suffer from idleness, since, when the drones consume part of their provisions, they become more active in repairing their losses; but, on the other hand, a crowd of robbers ought not to be allowed to form a band, lest they plunder all the wealth of others. Therefore, when you see bees and drones frequently quarrelling with one another, you will open and inspect the hives, so that, if the honey-combs are half-full, they may be let alone for

Of the making of honey.

2

3

4

semipleni favi sint, differantur: sive iam liquore completi, et superpositis ceris tamquam operculis obliti, demetantur.

Dies vero castrandi fere matutinus occupandus est. Neque enim convenit aestu medio exasperatas apes lacessiri.[1] Duobus autem ferramentis ad hunc[2] usum opus est, sesquipedali vel paulo ampliore mensura factis, quorum alterum sit culter oblongus[3] ex utraque parte acie lata, uno capite[4] aduncum habens[5] scalprum; alterum prima fronte planum et acutissimum: quo melius hoc favi succidantur,[6] illo eradantur, et quidquid sordidum deciderit, attra-5 hatur. Sed ubi a posteriore parte, qua nullum est vestibulum, patefactum fuerit alveare, fumum admovebimus factum galbano vel arido fimo. Ea porro vase fictili prunis immixta conduntur: idque vas ansatum simile angustae ollae figuratur, ita ut[7] altera pars sit acutior, per quam modico foramine fumus emanet: altera latior, et ore paulo latiore,[8] 6 per quam possit afflari. Talis olla cum est alveari[9] obiecta, spiritu admoto[10] fumus ad apes promovetur.[11] Quae confestim nidoris impatientes in priorem partem domicilii, et interdum extra vestibulum se conferunt. Atque ubi potestas facta est liberius inspiciendi, fere, si duo sunt examina, duo genera quoque favorum

[1] lacessiri *ac* : lacessi *SA*.
[2] huc *S* : hunc *Aac*.
[3] *post* oblongus *add.* alterum *SA* : *om. ac*.
[4] capite *Sac* : capit *A*.
[5] habens *om. SAac*.
[6] succidantur *ac* : subsecentur *SA*.
[7] ut *ac* : *om. SA*.
[8] latiore *SA* : patentiore *a* : potentiore *c*.
[9] alveario *ac* : albario *SA*.
[10] admoto *ac* : admotu *SA*.
[11] promovetur *ac* : promeourunt (?) *S* : promousit (?) *A*.

a time, but, if they are already full of liquid and sealed up with wax, just as if they had lids over them, the harvest of honey may be gathered in.

The morning should generally be chosen for the removal of the honey; for it is not advisable that the bees should be provoked when they are already exasperated by the midday heat. Two iron instruments are required for this operation, measuring a foot and a half or a little more, one of which should be an oblong knife with a broad edge on both sides and having a curved scraper at one extremity, and the other flat in front and very sharp, so that with the latter the honey-combs may be cut out better, and that with the former they be scraped off and any filth which has fallen upon them may be cleaned away. When the hive has been opened from the back, where there is no porch, we shall apply smoke made from *galbanum* [a] or from dried dung; moreover, these ingredients are mixed with live coals and put into an earthenware vessel. This vessel has handles and is shaped like a narrow pot in such a way that one end of it is sharper through which the smoke may issue through a small aperture, while the other end is broader and has a rather wider mouth, so that the coals can be blown upon through it. When a pot of this kind is applied to a hive, the smoke is conveyed to the bees by the movement set up by the breath. The bees, unable to endure the smell of burning, immediately move to the front part of their abode and sometimes outside the porch. When there is an opportunity of inspecting the hives more freely, usually, if there are two swarms, two kinds of

5

6

[a] See note on Chapter 13. § 7.

7 inveniuntur. Nam etiam in concordia [1] suum quaeque plebs morem figurandi ceras fingendique servant. Sed omnes favi semper cavearum tectis et paululum ab lateribus adhaerentes dependent, ita ne solum contingant: quoniam id praebet examinibus iter. 8 Ceterum figura cerarum talis est, qualis et habitus domicilii. Nam et quadrata et rotunda spatia nec minus longa suam speciem velut formae quaedam favis praebent. Ideoque non semper eiusdem [2] figurae reperiuntur favi. Sed hi qualescunque sint [3] non omnes eximantur. Nam priore messe, dum adhuc rura pastionibus abundant, quinta pars favorum; posteriore, cum iam metuitur hiems, tertia 9 relinquenda est. Atque hic tamen modus non est in omnibus regionibus certus: quoniam pro multitudine florum et ubertate pabuli apibus consulendum est. Ac si cerae dependentes in longitudinem decurrunt, eo ferramento, quod est simile cultro, insecandi sunt favi, deinde subiectis duobus bracchiis excipiendi, atque ita promendi: sin autem transversi tectis cavearum inhaerent, tunc scalprato ferramento est 10 opus, ut adversa fronte impressi desecentur. Eximi autem debent veteres vel vitiosi, et relinqui maxime integri ac melle pleni, et siqui [4] pullos continent, ut examini progenerando reserventur.

[1] in concordia *ac*: in cordia *SA*.
[2] eiusdem *ac*: eius quem *SA*.
[3] sint *Aac*: sunt *S*.
[4] *post* siqui *add.* tamen *SAac*.

honey-combs are also found; for even if they live in 7 harmony together, each community keeps to its own manner of shaping and constructing its waxen cells. All the combs, however, always hang down from the roofs of the hives, adhering very little to the sides and in such a way as not to touch the bottom, thus leaving a passage for the swarms. But 8 the shape of the wax cells depends on the nature of the bee-house; for square and round and also long dimensions impose their own shapes upon the honey-combs as if they were moulds, and that is why the honey-combs are not always found to be of the same shape. But of whatever kind they are, they should not all be removed; for at the first harvesting of honey, when the country still provides plenty of food, one-fifth of the honey-combs must be left; at the later harvesting, when the winter is already causing apprehension, a third part should be left. This, however, is not a fixed rule for all districts, since 9 plans for the bees must be dependent on the abundance of flowers and the richness of the food available. If the hanging waxen cells run into length, the combs must be cut with the iron tool which resembles a knife and must be received by putting your two arms underneath them, and so removed; but if they run horizontally and keep close to the roofs of the hives, then you must use the scraping instrument, so that they may be cut down by the pressure exerted on the side which faces you. But old and defective 10 honey-combs ought to be removed, and those which are soundest and full of honey should be left, as also those which contain young bees, so that they may be preserved for propagating a swarm.

Omnis deinde copia favorum conferenda est in eum locum, in quo mel conficere voles, linendaque sunt diligenter foramina parietum et fenestrarum, nequid sit apibus pervium, quae velut amissas opes suas pertinaciter vestigant, et persecutae consumunt. Itaque ex iisdem rebus fumus [1] etiam in aditu loci faciendus est, qui propulset intrare tentantes.[2]
11 Castratae deinde alvi si quae transversos favos in aditu habebunt, convertendae erunt, ut alterna vice posteriores partes vestibula [3] fiant. Sic enim proxime cum [4] castrabuntur, veteres potius favi quam novi eximentur, ceraeque novabuntur,[5] quae tanto deteriores sunt, quanto vetustiores. Quod si forte alvearia circumstructa et immobilia fuerint, curae erit nobis, ut semper modo a posteriore modo a priore [6] parte castrentur. Idque fieri ante diei quintam horam debebit, deinde repeti vel post
12 nonam, vel postero mane. Sed quotcunque favi sunt demessi, eodem die, dum tepent, conficere mel convenit. Saligneus qualus, vel tenui vimine rarius contextus saccus, inversae metae similis, qualis est quo vinum liquatur, obscuro loco suspenditur: in eum deinde carptim [7] congeruntur favi.[8] Sed adhibenda cura est, ut separentur eae partes cerarum, quae vel pullos habent, vel rubras [9] sordes. Nam sunt mali saporis, et succo suo mella corrumpunt.
13 Deinde ubi liquatum mel in subiectum alveum de-

[1] fumus *ac* : fumis *S* : *om. A.*
[2] itaque—tentantes *om. A.*
[3] vestibula *c* : -o *SAa.*
[4] cum *ac* : *om. SA.*
[5] veteres—renovabuntur *ac* : *om. SA.*
[6] modo a priore *Aac* : *om. S.*
[7] carptim *ac* : -i *SA.*
[8] favi *ac* : favis *SA.*
[9] rubras *ac* : rubas *SA.*

Next the whole store of honey-combs must be collected in the place where you intend to make the honey, and the holes in the walls and windows must be carefully daubed over, so that there may be no passage for the bees which obstinately search as if they were looking for lost wealth, and, if they track down the honey, eat it up. Smoke must, therefore, also be kindled of the same materials as before at the entrance of the place to drive away those that are trying to get in. Then those hives from which the 11 honey has been cut out, if they have combs lying across the entrance, will have to be turned round, so that the hinder parts in their turn become entrances; for in this way, the next time the honey is taken, the old combs rather than the new will be removed, and the waxen cells, which deteriorate as they grow older, will be renewed. But if the hives happen to be surrounded by walls and cannot be moved, we must take care that the combs are cut out, sometimes from the back and sometimes from the front. This process will have to be carried out before the fifth hour of the day and then repeated after the ninth hour or else next morning. But whatever be the 12 number of honey-combs that are harvested, you should make the honey on the same day, while they are still warm. A wickerwork basket or a bag rather loosely woven of fine withies in the shape of an inverted cone, like that through which wine is strained, is hung up in a dark place, and then the honey-combs are heaped in it one by one. But care must be taken that those parts of the waxen cells, which contain either young bees or dirty red matter are separated from them, for they have an ill flavour and corrupt the honey with their juice. Then, when the honey 13

fluxit, transfertur[1] in vasa fictilia, quae paucis diebus aperta sint, dum musteus fructus defervescat, isque saepius ligula purgandus est. Mox deinde fragmina favorum, quae in sacco remanserunt, retractata[2] exprimuntur: atque id[3] secundae notae mel defluit, et ab diligentioribus seorsum reponitur, ne quod est primi saporis hoc adhibito fiat deterius.

XVI. Cerae fructus quamvis aeris exigui non tamen omittendus est, cum sit eius usus ad multa necessarius. Expressae favorum reliquiae, posteaquam diligenter aqua dulci[4] perlutae sunt, in vas aeneum coniciuntur: adiecta[5] deinde aqua liquantur ignibus. Quod ubi factum est, cera per stramenta[6] vel iuncos defusa colatur, atque iterum similiter de integro coquitur, et in quas quis voluit formas aqua prius[7] adiecta defunditur: eamque concretam facile est eximere, quoniam qui subest humor non patitur formis inhaerere.

2 Sed iam consummata disputatione de villatici pecudibus atque pastionibus, quae reliqua nobis rusticarum rerum pars subest, de cultu hortorum, Publi Silvine, deinceps ita, ut et tibi et Gallioni nostro complacuerat, in carmen conferemus.

[1] transfertur *ac* : transferetur *SA*.
[2] remanserunt detracta *ac* : retractata remanserunt *SA*.
[3] id *ac* : in *SA*.
[4] dulci *ac* : dulcis *S* : *om. A*.
[5] adiecta—aqua *om. A*.
[6] stragmenta *a* : stramenta *c* : stramen *SA*.
[7] aqua prius *ac* : aquarius *SA*.

[a] Brother of the younger Seneca and uncle of Lucan the poet. He is mentioned in *Acts of the Apostles* xviii. 12 as proconsular governor of Achaia.

has been strained and has flowed down into the basin put underneath to catch it, it is transferred to earthenware vessels which are left open for a few days until the fresh produce ceases to ferment; and it must be frequently skimmed with a ladle. Next the fragments of the honey-combs, which have remained in the bag, are handled again and the juice squeezed out of them. What flows from them is honey of the second quality and is stored apart by itself by the more careful people, lest any of the honey of the best flavour should deteriorate by having this brought into contact with it.

Of the making of wax.

XVI. The yield of wax, though of little monetary value, must not be overlooked, since its use is necessary for many purposes. The remains of the honey-combs, when they have been well squeezed, after being carefully washed in fresh water, are thrown into a brazen vessel; water is then added to them and they are melted over a fire. When this has been done, the wax is poured out and strained through straw or rushes. It is then boiled over again a second time in the same manner and poured in such moulds as one has thought suitable, water having been first added. When the wax has hardened, it is easy to take it out, since the liquid which remains in the bottom does not allow it to stick to the moulds.

2 Having now finished the discussion of the animals kept at the farmhouse and their feeding, the part of husbandry which still remains to be treated, namely the cultivation of gardens, we will now present in verse in accordance with the desire which both you, Publius Silvinus, and our friend Gallio [a] were pleased to express.

PRINTED IN GREAT BRITAIN BY
RICHARD CLAY AND COMPANY, LTD.
BUNGAY, SUFFOLK.

THE LOEB CLASSICAL LIBRARY

VOLUMES ALREADY PUBLISHED

Latin Authors

AMMIANUS MARCELLINUS. Translated by J. C. Rolfe. 3 Vols. (*2nd Imp. revised.*)
APULEIUS: THE GOLDEN ASS (METAMORPHOSES). W. Adlington (1566). Revised by S. Gaselee. (*7th Imp.*)
ST. AUGUSTINE, CONFESSIONS OF. W. Watts (1631). 2 Vols. (Vol. I. *7th Imp.*, Vol. II. *6th Imp.*)
ST. AUGUSTINE, SELECT LETTERS. J. H. Baxter. (*2nd Imp.*)
AUSONIUS. H. G. Evelyn White. 2 Vols. (*2nd Imp.*)
BEDE. J. E. King. 2 Vols. (*2nd Imp.*)
BOETHIUS: TRACTS and DE CONSOLATIONE PHILOSOPHIAE. Rev. H. F. Stewart and E. K. Rand. (*6th Imp.*)
CAESAR: ALEXANDRINE, AFRICAN and SPANISH WARS. A. S. Way.
CAESAR: CIVIL WARS. A. G. Peskett. (*5th Imp.*)
CAESAR: GALLIC WAR. H. J. Edwards. (*10th Imp.*)
CATO: DE RE RUSTICA; VARRO: DE RE RUSTICA. H. B. Ash and W. D. Hooper. (*3rd Imp.*)
CATULLUS. F. W. Cornish; TIBULLUS. J. B. Postgate; PERVIGILIUM VENERIS. J. W. Mackail. (*12th Imp.*)
CELSUS: DE MEDICINA. W. G. Spencer. 3 Vols. (Vol. I. *3rd Imp. revised*, Vols. II. and III. *2nd Imp.*)
CICERO: BRUTUS, and ORATOR. G. L. Hendrickson and H. M. Hubbell. (*3rd Imp.*)
[CICERO]: AD HERENNIUM. H. Caplan.
CICERO: DE FATO; PARADOXA STOICORUM; DE PARTITIONE ORATORIA. H. Rackham. (With De Oratore, Vol. II.) (*2nd Imp.*)
CICERO: DE FINIBUS. H. Rackham. (*4th Imp. revised.*)
CICERO: DE INVENTIONE, etc. H. M. Hubbell.
CICERO: DE NATURA DEORUM and ACADEMICA. H. Rackham. (*2nd Imp.*)
CICERO: DE OFFICIIS. Walter Miller. (*6th Imp.*)
CICERO: DE ORATORE. 2 Vols. E. W. Sutton and H. Rackham. (*2nd Imp.*)
CICERO: DE REPUBLICA and DE LEGIBUS. Clinton W. Keyes. (*4th Imp.*)
CICERO: DE SENECTUTE, DE AMICITIA, DE DIVINATIONE. W. A. Falconer. (*6th Imp.*)
CICERO: IN CATILINAM, PRO FLACCO, PRO MURENA, PRO SULLA. Louis E. Lord. (*3rd Imp. revised.*)
CICERO: LETTERS TO ATTICUS. E. O. Winstedt. 3 Vols. (Vol. I. *6th Imp.*, Vols. II. and III. *4th Imp.*)

Cicero: Letters to His Friends. W. Glynn Williams. 3 Vols. (Vols. I. and II. *3rd Imp.*, Vol. III. *2nd Imp. revised.*)
Cicero: Philippics. W. C. A. Ker. (*3rd Imp. revised.*)
Cicero: Pro Archia, Post Reditum, De Domo, De Haruspicum Responsis, Pro Plancio. N. H. Watts. (*3rd Imp.*)
Cicero: Pro Caecina, Pro Lege Manilia, Pro Cluentio, Pro Rabirio. H. Grose Hodge. (*3rd Imp.*)
Cicero: Pro Milone, In Pisonem, Pro Scauro, Pro Fonteio, Pro Rabirio Postumo, Pro Marcello, Pro Ligario, Pro Rege Deiotaro. N. H. Watts. (*2nd Imp.*)
Cicero: Pro Quinctio, Pro Roscio Amerino, Pro Roscio Comoedo, Contra Rullum. J. H. Freese. (*2nd Imp.*)
Cicero: Tusculan Disputations. J. E. King. (*4th Imp.*)
Cicero: Verrine Orations. L. H. G. Greenwood. 2 Vols. (Vol. I. *3rd Imp.*, Vol. II. *2nd Imp.*)
Claudian. M. Platnauer. 2 Vols.
Columella: De Re Rustica, De Arboribus. H. B. Ash, E. S. Forster and E. Heffner. 3 Vols. (Vol. I. *2nd Imp.*)
Curtius, Q.: History of Alexander. J. C. Rolfe. 2 Vols.
Florus. E. S. Forster and Cornelius Nepos. J. C. Rolfe. (*2nd Imp.*)
Frontinus: Stratagems and Aqueducts. C. E. Bennett and M. B. McElwain. (Vol. I. *3rd Imp.*, Vol. II. *2nd Imp.*)
Fronto: Correspondence. C. R. Haines. 2 Vols. (Vol. I. *3rd Imp.*, Vol. II. *2nd Imp.*)
Gellius. J. C. Rolfe. 3 Vols. (Vol. I. *3rd Imp.*, Vols. II. and III. *2nd Imp.*)
Horace: Odes and Epodes. C. E. Bennett. (*14th Imp. revised.*)
Horace: Satires, Epistles, Ars Poetica. H. R. Fairclough. (*9th Imp. revised.*)
Jerome: Selected Letters. F. A. Wright. (*2nd Imp.*)
Juvenal and Persius. G. G. Ramsay. (*7th Imp.*)
Livy. B. O. Foster, F. G. Moore, Evan T. Sage, and A. C. Schlesinger. 14 Vols. Vols. I.–XIII. (Vol. I. *4th Imp.*, Vols. II., III., V., and IX. *3rd Imp.*; Vols. IV., VI.–VIII., X.–XII. *2nd Imp. revised.*)
Lucan. J. D. Duff. (*3rd Imp.*)
Lucretius. W. H. D. Rouse. (*7th Imp. revised.*)
Martial. W. C. A. Ker. 2 Vols. (Vol. I. *5th Imp.*, Vol. II. *4th Imp. revised.*)
Minor Latin Poets: from Publilius Syrus to Rutilius Namatianus, including Grattius, Calpurnius Siculus, Nemesianus, Avianus, and others with "Aetna" and the "Phoenix." J. Wight Duff and Arnold M. Duff. (*3rd Imp.*)
Ovid: The Art of Love and Other Poems. J. H. Mozley. (*3rd Imp.*)
Ovid: Fasti. Sir James G. Frazer. (*2nd Imp.*)
Ovid: Heroides and Amores. Grant Showerman. (*5th Imp.*)
Ovid: Metamorphoses. F. J. Miller. 2 Vols. (Vol. I. *10th Imp.*, Vol. II. *8th Imp.*)
Ovid: Tristia and Ex Ponto. A. L. Wheeler. (*3rd Imp.*)

Persius. Cf. Juvenal.
Petronius. M. Heseltine; Seneca Apocolocyntosis. W. H. D. Rouse. (*8th Imp. revised.*)
Plautus. Paul Nixon. 5 Vols. (Vols. I. and II. *5th Imp.*, Vol. III. *3rd Imp.*, Vols. IV. and V. *2nd Imp.*)
Pliny : Letters. Melmoth's Translation revised by W. M. L. Hutchinson. 2 Vols. (*6th Imp.*)
Pliny : Natural History. H. Rackham and W. H. S. Jones. 10 Vols. Vols. I.–V. and IX. H. Rackham. Vol. VI. W. H. S. Jones. (Vols. I. and II. *3rd Imp.*, Vols. III. and IV. *2nd Imp.*)
Propertius. H. E. Butler. (*6th Imp.*)
Prudentius. H. J. Thomson. 2 Vols.
Quintilian. H. E. Butler. 4 Vols. (*3rd Imp.*)
Remains of Old Latin. E. H. Warmington. 4 Vols. Vol. I. (Ennius and Caecilius.) Vol. II. (Livius, Naevius, Pacuvius, Accius.) Vol. III. (Lucilius and Laws of XII Tables.) Vol. IV. (*2nd Imp.*) (Archaic Inscriptions.)
Sallust. J. C. Rolfe. (*3rd Imp. revised.*)
Scriptores Historiae Augustae. D. Magie. 3 Vols. (Vol. I. *3rd Imp. revised*, Vols. II. and III. *2nd Imp.*)
Seneca : Apocolocyntosis. Cf. Petronius.
Seneca : Epistulae Morales. R. M. Gummere. 3 Vols. (Vol. I. *4th Imp.*, Vols. II. and III. *2nd Imp.*)
Seneca : Moral Essays. J. W. Basore. 3 Vols. (Vol. II. *3rd Imp.*, Vols. I. and III. *2nd Imp. revised.*)
Seneca : Tragedies. F. J. Miller. 2 Vols. (Vol. I. *4th Imp.*, Vol. II. *3rd Imp. revised.*)
Sidonius : Poems and Letters. W. B. Anderson. 2 Vols. (Vol. I. *2nd Imp.*)
Silius Italicus. J. D. Duff. 2 Vols. (Vol. I. *2nd Imp.*, Vol. II. *3rd Imp.*)
Statius. J. H. Mozley. 2 Vols. (*2nd Imp.*)
Suetonius. J. C. Rolfe. 2 Vols. (Vol. I. *7th Imp.*, Vol. II. *6th Imp. revised.*)
Tacitus : Dialogus. Sir Wm. Peterson. Agricola and Germania. Maurice Hutton. (*6th Imp.*)
Tacitus : Histories and Annals. C. H. Moore and J. Jackson. 4 Vols. (Vols. I. and II. *3rd Imp.*, Vols. III. and IV. *2nd Imp.*)
Terence. John Sargeaunt. 2 Vols. (*7th Imp.*)
Tertullian : Apologia and De Spectaculis. T. R. Glover. Minucius Felix. G. H. Rendall. (*2nd Imp.*)
Valerius Flaccus. J. H. Mozley. (*2nd Imp. revised.*)
Varro : De Lingua Latina. R. G. Kent. 2 Vols. (*2nd Imp. revised.*)
Velleius Paterculus and Res Gestae Divi Augusti. F. W. Shipley. (*2nd Imp.*)
Virgil. H. R. Fairclough. 2 Vols. (Vol. I. *18th Imp.*, Vol. II. *14th Imp. revised.*)
Vitruvius : De Architectura. F. Granger. 2 Vols. (Vol. I. *2nd Imp.*)

Greek Authors

Achilles Tatius. S. Gaselee. (*2nd Imp.*)
Aeneas Tacticus, Asclepiodotus and Onasander. The Illinois Greek Club. (*2nd Imp.*)
Aeschines. C. D. Adams. (*2nd Imp.*)
Aeschylus. H. Weir Smyth. 2 Vols. (Vol. I. *6th Imp.*, Vol. II. *5th Imp.*)
Alciphron, Aelian, Philostratus Letters. A. R. Benner and F. H. Fobes.
Andocides, Antiphon. Cf. Minor Attic Orators.
Apollodorus. Sir James G. Frazer. 2 Vols. (Vol. I. *3rd Imp.*, Vol. II. *2nd Imp.*)
Apollonius Rhodius. R. C. Seaton. (*4th Imp.*)
The Apostolic Fathers. Kirsopp Lake. 2 Vols. (Vol. I. *8th Imp.*, Vol. II. *6th Imp.*)
Appian: Roman History. Horace White. 4 Vols. (Vol. I. *3rd Imp.*, Vols. II., III., and IV. *2nd Imp.*)
Aratus. Cf. Callimachus.
Aristophanes. Benjamin Bickley Rogers. 3 Vols. Verse trans. (*5th Imp.*)
Aristotle: Art of Rhetoric. J. H. Freese. (*3rd Imp.*)
Aristotle: Athenian Constitution, Eudemian Ethics, Vices and Virtues. H. Rackham. (*3rd Imp.*)
Aristotle: Generation of Animals. A. L. Peck. (*2nd Imp.*)
Aristotle: Metaphysics. H. Tredennick. 2 Vols. (*3rd Imp.*)
Aristotle: Meteorologica. H. D. P. Lee.
Aristotle: Minor Works. W. S. Hett. On Colours, On Things Heard, On Physiognomies, On Plants, On Marvellous Things Heard, Mechanical Problems, On Indivisible Lines, On Situations and Names of Winds, On Melissus, Xenophanes, and Gorgias. (*2nd Imp.*)
Aristotle: Nicomachean Ethics. H. Rackham. (*5th Imp. revised.*)
Aristotle: Oeconomica and Magna Moralia. G. C. Armstrong; (with Metaphysics, Vol. II.). (*3rd Imp.*)
Aristotle: On the Heavens. W. K. C. Guthrie. (*3rd Imp. revised.*)
Aristotle: On the Soul, Parva Naturalia, On Breath. W. S. Hett. (*2nd Imp. revised.*)
Aristotle: Organon. H. P. Cooke and H. Tredennick. 3 Vols. (Vol. I. *2nd Imp.*)
Aristotle: Parts of Animals. A. L. Peck; Motion and Progression of Animals. E. S. Forster. (*3rd Imp. revised.*)
Aristotle: Physics. Rev. P. Wicksteed and F. M. Cornford. 2 Vols. (Vol. I. *2nd Imp.*, Vol. II. *3rd Imp.*)
Aristotle: Poetics and Longinus. W. Hamilton Fyfe; Demetrius on Style. W. Rhys Roberts. (*5th Imp. revised.*)
Aristotle: Politics. H. Rackham. (*4th Imp. revised.*)
Aristotle: Problems. W. S. Hett. 2 Vols. (*2nd Imp. revised.*)

Aristotle: Rhetorica ad Alexandrum (with Problems. Vol. II.). H. Rackham.

Arrian: History of Alexander and Indica. Rev. E. Iliffe Robson. 2 Vols. (Vol. I. *3rd Imp.*, Vol. II. *2nd Imp.*)

Athenaeus: Deipnosophistae. C. B. Gulick. 7 Vols. (Vols. I., V., and VI. *2nd Imp.*)

St. Basil: Letters. R. J. Deferrari. 4 Vols. (*2nd Imp.*)

Callimachus and Lycophron. A. W. Mair; Aratus. G. R. Mair. (*2nd Imp.*)

Clement of Alexandria. Rev. G. W. Butterworth. (*3rd Imp.*)

Colluthus. Cf. Oppian.

Daphnis and Chloe. Thornley's Translation revised by J. M. Edmonds: and Parthenius. S. Gaselee. (*3rd Imp.*)

Demosthenes I: Olynthiacs, Philippics and Minor Orations. I.-XVII. and XX. J. H. Vince. (*2nd Imp.*)

Demosthenes II: De Corona and De Falsa Legatione. C. A. Vince and J. H. Vince. (*3rd Imp. revised.*)

Demosthenes III: Meidias, Androtion, Aristocrates, Timocrates and Aristogeiton, I. and II. J. H. Vince. (*2nd Imp.*)

Demosthenes IV–VI: Private Orations and In Neaeram. A. T. Murray. (Vol. IV. *2nd Imp.*)

Demosthenes VII: Funeral Speech, Erotic Essay, Exordia and Letters. N. W. and N. J. DeWitt.

Dio Cassius: Roman History. E. Cary. 9 Vols. (Vols. I. and II. *3rd Imp.*, Vols. III. and IV. *2nd Imp.*)

Dio Chrysostom. J. W. Cohoon and H. Lamar Crosby. 5 Vols. (Vols. I.–III. *2nd Imp.*)

Diodorus Siculus. 12 Vols. Vols. I.–VI. C. H. Oldfather. Vol. VII. C. L. Sherman. Vols. IX. and X. R. M. Geer. (Vols. I.–III. *2nd Imp.*)

Diogenes Laertius. R. D. Hicks. 2 Vols. (Vol. I. *4th Imp.*, Vol. II. *3rd Imp.*)

Dionysius of Halicarnassus: Roman Antiquities. Spelman's translation revised by E. Cary. 7 Vols. (Vols. I.–IV. *2nd Imp.*)

Epictetus. W. A. Oldfather. 2 Vols. (*2nd Imp.*)

Euripides. A. S. Way. 4 Vols. (Vols. I. and II. *7th Imp.*, III. and IV. *6th Imp.*) Verse trans.

Eusebius: Ecclesiastical History. Kirsopp Lake and J. E. L. Oulton. 2 Vols. (Vol. I. *3rd Imp.*, Vol. II. *4th Imp.*)

Galen: On the Natural Faculties. A. J. Brock. (*4th Imp.*)

The Greek Anthology. W. R. Paton. 5 Vols. (Vols. I. and II. *5th Imp.*, Vol. III. *4th Imp.*, Vols. IV. and V. *3rd Imp.*)

Greek Elegy and Iambus with the Anacreontea. J. M. Edmonds. 2 Vols. (Vol. I. *3rd Imp.*, Vol. II. *2nd Imp.*)

The Greek Bucolic Poets (Theocritus, Bion, Moschus). J. M. Edmonds. (*7th Imp. revised.*)

Greek Mathematical Works. Ivor Thomas. 2 Vols. (*2nd Imp.*)

Herodes. Cf. Theophrastus: Characters.

Herodotus. A. D. Godley. 4 Vols. (Vols. I.–III. *4th Imp.*, Vol. IV. *3rd Imp.*)
Hesiod and The Homeric Hymns. H. G. Evelyn White. (*7th Imp. revised and enlarged.*)
Hippocrates and the Fragments of Heracleitus. W. H. S. Jones and E. T. Withington. 4 Vols. (*3rd Imp.*)
Homer: Iliad. A. T. Murray. 2 Vols. (Vol. I. *7th Imp.*, Vol. II. *6th Imp.*)
Homer: Odyssey. A. T. Murray. 2 Vols. (*8th Imp.*)
Isaeus. E. W. Forster. (*2nd Imp.*)
Isocrates. George Norlin and LaRue Van Hook. 3 Vols.
St. John Damascene: Barlaam and Ioasaph. Rev. G. R. Woodward and Harold Mattingly. (*3rd Imp. revised.*)
Josephus. H. St. J. Thackeray and Ralph Marcus. 9 Vols. Vols. I.–VII. (Vol. V. *3rd Imp.*, Vol. VI. *2nd Imp.*)
Julian. Wilmer Cave Wright. 3 Vols. (Vols. I. and II. *3rd Imp.*, Vol. III. *2nd Imp.*)
Lucian. A. M. Harmon. 8 Vols. Vols. I.–V. (Vols. I. and II. *4th Imp.*, Vol. III. *3rd Imp.*, Vols. IV. and V. *2nd Imp.*)
Lycophron. Cf. Callimachus.
Lyra Graeca. J. M. Edmonds. 3 Vols. (Vol. I. *4th Imp.*, Vol. II. *revised and enlarged*, and III. *3rd Imp.*)
Lysias. W. R. M. Lamb. (*2nd Imp.*)
Manetho. W. G. Waddell: Ptolemy: Tetrabiblos. F. E. Robbins. (*2nd Imp.*)
Marcus Aurelius. C. R. Haines. (*4th Imp. revised.*)
Menander. F. G. Allinson. (*3rd Imp. revised.*)
Minor Attic Orators (Antiphon, Andocides, Lycurgus, Demades, Dinarchus, Hypereides). K. J. Maidment and J. O. Burrt. 2 Vols. (Vol. I. *2nd Imp.*)
Nonnos: Dionysiaca. W. H. D. Rouse. 3 Vols. (Vol. III. *2nd Imp.*)
Oppian, Colluthus, Tryphiodorus. A. W. Mair. (*2nd Imp.*)
Papyri. Non-Literary Selections. A. S. Hunt and C. C. Edgar. 2 Vols. (Vol. I. *2nd Imp.*) Literary Selections. Vol. I. (Poetry). D. L. Page. (*3rd Imp.*)
Parthenius. Cf. Daphnis and Chloe.
Pausanias: Description of Greece. W. H. S. Jones. 5 Vols. and Companion Vol. arranged by R. E. Wycherley. (Vols. I. and III. *3rd Imp.*, Vols. II., IV. and V. *2nd Imp.*)
Philo. 10 Vols. Vols. I.–V.; F. H. Colson and Rev. G. H. Whitaker. Vols. VI.–IX.; F. H. Colson. (Vols. I.–III., V.–IX. *2nd Imp.*, Vol. IV. *3rd Imp.*)
Philo: two supplementary Vols. (*Translation only.*) Ralph Marcus.
Philostratus: The Life of Apollonius of Tyana. F. C. Conybeare. 2 Vols. (Vol. I. *4th Imp.*, Vol. II. *3rd Imp.*)
Philostratus: Imagines; Callistratus: Descriptions. A. Fairbanks.
Philostratus and Eunapius: Lives of the Sophists. Wilmer Cave Wright. (*2nd Imp.*)
Pindar. Sir J. E. Sandys. (*7th Imp. revised.*)

Plato: Charmides, Alcibiades, Hipparchus, The Lovers, Theages, Minos and Epinomis. W. R. M. Lamb. (*2nd Imp.*)

Plato: Cratylus, Parmenides, Greater Hippias, Lesser Hippias. H. N. Fowler. (*4th Imp.*)

Plato: Euthyphro, Apology, Crito, Phaedo, Phaedrus. H. N. Fowler. (*11th Imp.*)

Plato: Laches, Protagoras, Meno, Euthydemus. W. R. M. Lamb. (*3rd Imp. revised.*)

Plato: Laws. Rev. R. G. Bury. 2 Vols. (*3rd Imp.*)

Plato: Lysis, Symposium, Gorgias. W. R. M. Lamb. (*5th Imp. revised.*)

Plato: Republic. Paul Shorey. 2 Vols. (Vol. I. *5th Imp.*, Vol. II. *3rd Imp.*)

Plato: Statesman, Philebus. H. N. Fowler; Ion. W. R. M. Lamb. (*4th Imp.*)

Plato: Theaetetus and Sophist. H. N. Fowler. (*4th Imp.*)

Plato: Timaeus, Critias, Clitopho, Menexenus, Epistulae. Rev. R. G. Bury. (*3rd Imp.*)

Plutarch: Moralia. 14 Vols. Vols. I.–V. F. C. Babbitt; Vol. VI. W. C. Helmbold; Vol. X. H. N. Fowler. (Vols. I., III., and X. *2nd Imp.*)

Plutarch: The Parallel Lives. B. Perrin. 11 Vols. (Vols. I., II., VI., VII., and XI. *3rd Imp.*, Vols. III.–V. and VIII.–X. *2nd Imp.*)

Polybius. W. R. Paton. 6 Vols. (*2nd Imp.*)

Procopius: History of the Wars. H. B. Dewing. 7 Vols. (Vol. I. *3rd Imp.*, Vols. II.–VII. *2nd Imp.*)

Ptolemy: Tetrabiblos. Cf. Manetho.

Quintus Smyrnaeus. A. S. Way. Verse trans. (*2nd Imp.*)

Sextus Empiricus. Rev. R. G. Bury. 4 Vols. (Vol. I. *3rd Imp.*, III. *2nd Imp.*)

Sophocles. F. Storr. 2 Vols. (Vol. I. *10th Imp.*, Vol. II. *6th Imp.*) Verse trans.

Strabo: Geography. Horace L. Jones. 8 Vols. (Vols. I., V., and VIII. *3rd Imp.*, Vols. II., III., IV., VI., and VII. *2nd Imp.*)

Theophrastus: Characters. J. M. Edmonds. Herodes, etc. A. D. Knox. (*3rd Imp.*)

Theophrastus: Enquiry into Plants. Sir Arthur Hort, Bart. 2 Vols. (*2nd Imp.*)

Thucydides. C. F. Smith. 4 Vols. (Vol. I. *4th Imp.*, Vols. II., III., and IV. *3rd Imp. revised.*)

Tryphiodorus. Cf. Oppian.

Xenophon: Cyropaedia. Walter Miller. 2 Vols. (Vol. I. *4th Imp.*, Vol. II. *3rd Imp.*)

Xenophon: Hellenica, Anabasis, Apology, and Symposium. C. L. Brownson and O. J. Todd. 3 Vols. (Vols. I. and III. *3rd Imp.*, Vol. II. *4th Imp.*)

Xenophon: Memorabilia and Oeconomicus. E. C. Marchant. (*3rd Imp.*)

Xenophon: Scripta Minora. E. C. Marchant. (*2nd Imp.*).

IN PREPARATION

Greek Authors

ARISTOTLE: DE MUNDO, ETC. D. Furley and E. M. Forster.
ARISTOTLE: HISTORY OF ANIMALS. A. L. Peck.
PLOTINUS: A. H. Armstrong.

Latin Authors

ST. AUGUSTINE: CITY OF GOD.
CICERO: PRO SESTIO, IN VATINIUM, PRO CAELIO, DE PROVINCIIS CONSULARIBUS, PRO BALBO. J. H. Freese and R. Gardner.
PHAEDRUS. Ben E. Perry.

DESCRIPTIVE PROSPECTUS ON APPLICATION

London — WILLIAM HEINEMANN LTD
Cambridge, Mass. — HARVARD UNIVERSITY PRESS